Agricultural Instability in China
1931–1991

Studies on Contemporary China

The Contemporary China Institute at the School of Oriental and African Studies (University of London) has, since its establishment in 1968, been an international centre for research and publications on twentieth-century China. *Studies on Contemporary China*, which is sponsored by the Institute, seeks to maintain and extend that tradition by making available the best work of scholars and China specialists throughout the world. It embraces a wide variety of subjects relating to Nationalist and Communist China, including social, political, and economic change, intellectual and cultural developments, foreign relations, and national security.

Editorial Advisory Board:

Dr R. F. Ash
Mr B. G. Hook
Professor C. B. Howe
Dr David Shambaugh
Mr David Steeds

Volumes in the Series

Art and Ideology in Revolutionary China, *David Holm*
Demographic Transition in China, *Peng Xizhe*
Economic Trends in Chinese Agriculture, *Y. Y. Kueh and R. F. Ash*
In Praise of Maoist Economic Planning, *Chris Bramall*
Economic Reform and State-Owned Enterprises in China, *D. Hay and D. Morris, with S. Liu and G. Yao*
Rural China in Transition, *Samuel Ho*

Agricultural Instability in China, 1931–1991

Weather, Technology, and Institutions

Y. Y. KUEH

CLARENDON PRESS · OXFORD

1995

Oxford University Press, Walton Street, Oxford ox2 6DP
Oxford New York
Athens Auckland Bangkok Bombay
Calcutta Cape Town Dar es Salaam Delhi
Florence Hong Kong Istanbul Karachi
Kuala Lumpur Madras Madrid Melbourne
Mexico City Nairobi Paris Singapore
Taipei Tokyo Toronto
and associated companies in
Berlin Ibadan

Oxford is a trade mark of Oxford University Press

Published in the United States
by Oxford University Press Inc., New York

British Library Cataloguing in Publication Data
Data available

Library of Congress Cataloging-in-Publication Data
Kueh, Y. Y.
Agricultural instability in China, 1931–1991: weather,
technology, and institutions / by Y. Y. Kueh.
(Studies in contemporary China)
Includes index
1. Agriculture and state—China—History—20th century.
2. Agriculture—Economic aspects—China—History—20th century.
I. Title. II. Series.
HD2098.K83 1994 94-32208
338.1'851—dc20
ISBN 019–828777–1

1 3 5 7 9 10 8 6 4 2

Typeset by Best-set Typesetter Ltd., Hong Kong
Printed in Great Britain
on acid-free paper by
Bookcraft (Bath) Ltd., Midsomer Norton, Avon

To
Cathie
Tan-yuan,
Chien-yuan, and
Hao-yuan

Acknowledgements

My gratitude goes foremost to the late Professor Kenneth R. Walker, who unstintingly afforded me encouragement, advice, and assistance over many years. His sudden death in 1989 and my own wish to pacify his concern in the world beyond lent an added urgency and poignancy to the task of completing this book.

Professor Christopher Howe and Dr Robert Ash have also painstakingly read through the entire manuscript from the first to the final version. I am deeply indebted for the many helpful comments and detailed suggestions that they made, which led to substantial improvements in the book.

My thanks are also due to many other Western specialists who either read one or more chapters of the manuscript at its different stages, or made valuable comments on related articles published earlier, or shared their ideas during a number of seminar presentations. These include Anthony M. Tang, Dwight Perkins, Ishikawa Shigeru, the late Werner Klatt, Joseph C. H. Chai, Bruce Stone, Charles Liu, Fred Surls, Francis Tuan, Frederick Crook, Thomas Wiens, Mark Elvin, Cyril Lin, and an anonymous referee of the British *Journal of Agricultural Economics*.

Part of the preliminary findings were presented at seminars at the School of Oriental and African Studies (University of London), Oxford University, US Department of Agriculture (China Section of its International Economic Services), and the Chinese University of Hong Kong. I am also grateful to the International Economic Association (via Nural Islam) for a similar presentation at the Seventh World Economic Congress held in Madrid 1983. These meetings all helped to expose many of the important hypotheses in my book to expert scrutiny.

I have also benefited a great deal from discussions with many Chinese scholars affiliated to the various institutes of the Chinese Academy of Social Sciences, as well as Chinese government officials in charge of economic policies and agricultural affairs.

My former colleagues at the Chinese University of Hong Kong, Chou Wen-lin and Ho Yin-ping, were always at hand to answer my queries about quantitative methods; and Woo Tun-oy, Tsang Shu-ki and Cheng Yuk-shing of Hong Kong Baptist University carefully checked the major regression estimates against my interpretations and made useful comments. Needless to say, I am solely responsible for my own views and any errors (technical or otherwise) that may remain.

The former Inter-University Council and Association of Commonwealth Universities were instrumental in providing two separate senior fellowships, both held at the School of Oriental and African Studies (in

1977–8 and 1984–5), to enable the major part of the research to be completed in London, where I had ready access to the rich collections of SOAS and the British Library. The Contemporary China Institute (SOAS) made available a travel grant in order to present the second draft of the book to selected members of CCI's Publication Committee in summer 1986. The Chinese University of Hong Kong provided several small faculty grants during the period from 1976 to 1987. In this connection I owe a particular debt to Professor Kuan Hsin-chi for his support and sympathetic understanding of the sustained demands inherent in undertaking such a wide-ranging and long-run study as I have attempted here. Macquarie University granted an extended leave of absence during 1990–1, which allowed me to add the final touches to the book. I am particularly indebted to the Vice-Chancellor, Professor Di Yerbury, for her generous support and kind understanding. Professor Yiu-kwan Ian was also most generous in hosting my visit to Hong Kong Baptist University in that year.

Jean Hung of the Hong Kong-based Universities Service Centre rendered helpful library assistance. Many of my former students assisted in statistical tabulations and regression analyses. They include Herman Y. F. Fung, Vincent W. K. Mok, Yu Ip-Wing, and Ho Ka-kei. Linda Ma, Glennis Yee, and Salina Cheung provided excellent clerical and typing services. A special note is also due to my daughter, Tan-yuan, who, at the age of 13, typed the entire first draft of the book.

None the less, the book could never have been completed without the enduring patience and affection of my wife Cathie and the family. They are all now very relieved and happy that the book has finally been completed.

Hong Kong, 1992   Y. Y. K.

The Editorial Board of the Studies on Contemporary China series would like to express its gratitude to the Chiang Ching-Kuo Foundation for the generous grant to the Contemporary China Institute which supports the publication of the series.

Contents

Part I
An Interpretative Framework

Part II
A History of Chinese Agricultural Instability
since 1931

Part III
Major Factors in Chinese Agricultural Instability since 1931

Part IV
Precipitation and Grain-Yield Variations

Part V
Economic and Policy Implications

Appendices

List of Figures

List of Maps

List of Tables

List of Abbreviations

A. Newspapers in Chinese

AHRB	*Anhui Ribao* (Anhui Daily)
BJRB	*Beijing Ribao* (Beijing Daily)
CJRD	*Changjiang Ribao* (Yangzi River Daily)
CQRB	*Chongqing Ribao* (Chongqing (Sichuan) Daily)
DGBBJ	*Da Gong Bao* (Impartial Daily), Beijing
DGBHK	*Da Gong Bao* (Impartial Daily), Hong Kong
DGBTJ	*Da Gong Bao* (Impartial Daily), Tianjin
DZRB	*Dazhong Ribao* (Mass Daily), Shandong
FJRB	*Fujian Ribao* (Fujian Daily)
GMRB	*Guangming Ribao* (Enlightenment Daily), Beijing
GSRB	*Gongshang Ribao* (Industrial and Commercial Daily), Hong Kong
GSURB	*Gansu Ribao* (Gansu Daily)
GXRB	*Guangxi Ribao* (Guangxi Daily)
GZRB	*Guizhou Ribao* (Guizhou Daily)
HBRB	*Hebei Ribao* (Hebei Daily)
HKSB	*Hong Kong Shi Bao* (Hong Kong Times)
HLJRB	*Heilongjiang Ribao* (Heilongjiang Daily)
HNRB	*Henan Ribao* (Henan Daily)
HUBRB	*Hubei Ribao* (Hubei Daily)
JFRB	*Jiefang Ribao* (Liberation Daily), Shanghai
JLRB	*Jilin Ribao* (Jilin Daily)
JXRB	*Jiangxi Ribao* (Jiangxi Daily)
LNGQTYB	*Liaoning Gongqing Tuanyuan Bao* (Liaoning Communist Youth League Daily)
LNJB	*Liaoning Ribao* (Liaoning Daily)
MBHK	*Ming Bao* (Ming Daily), Hong Kong
NFRB	*Nanfang Ribao* (Southern Daily), Guangdong
NMGRB	*Neimenggu Ribao* (Inner Mongolia Daily)
QHRB	*Qinghai Ribao* (Qinghai Daily)
RMRB	*Renmin Ribao* (People's Daily)
SAXRB	*Shaanxi Ribao* (Shaanxi Daily)
SCRB	*Sichuan Ribao* (Sichuan Daily)
SXRB	*Shanxi Ribao* (Shanxi Daily)
WHBHK	*Wen Hui Bao* (Wenhui Daily), Hong Kong
XHNB	*Xin Hunan Bao* (New Hunan Daily)
XHRBCQ	*Xinhua Ribao* (New China Daily), Chongqing

XHRBNJ	*Xinhua Ribao* (New China Daily), Nanjing
XHS	*Xinhuashe* (New China News Agency Despatch)
XJRB	*Xinjiang Ribao* (Xinjiang Daily)
YNRB	*Yunnan Ribao* (Yunnan Daily)
ZGXWS	*Zhongguo Xinwenshe* (China News Agency Despatch)
ZJRB	*Zhejiang Ribao* (Zhejiang Daily)
ZNRB	*Zhongnan Ribao* (Central South Daily), Hong Kong

B. Journals in Chinese

DLJK	*Dili Jikan* (Collected Geographical Essays), Institute of Geography, Academia Sinica
DLXB	*Dili Xuebao* (Acta Geographica Sinica)
JHJJ	*Jihua Jingji* (Planned Economy)
JHXK	*Jianghai Xuekan* (River and Ocean Academic Journal), Nanjiang
JJYJ	*Jingji Yanjiu* (Economic Research)
JLXB	*Jinling Xuebao* (Nanjing Academic Journal), 1930s
JRDL	*Jinri Dalu* (Mainland Today), Taiwan
LSYK	*Liangshi Yuekan* (Grain Monthly)
NCJJ	*Nongcur Jingji* (Rural Economy)
NYJJWT	*Nongye Jingji Wenti* (Problems of Agricultural Economy)
NYJSJJ	*Nongye Jishu Jingji* (Agricultural-Technical Economics)
NYKX	*Zhongguo Nongye Kexue* (Chinese Agricultural Science), Publication of the Chinese Academy of Agricultural Sciences
NYQX	*Nongye Qixiang* (Agricultural Meteorology)
QSXK	*Qiushe Xuekan* (Seeking Truth Journal), University of Heilongjiang
QXYB	*Qixiang Yuebao* (Meterorological Monthly), 1930s
QZSHKX	*Guizhou Shehui Kexue* (Quizhou Social Science Journal)
SHKXZZ	*Shehui Kexue Zazhi* (Quarterly Review of Social Science), Academia Sinica 1930s
SXDXXB	*Shanxi Daxue Xuebao* (Shanxi University Bulletin), Philosophy and Social Science edn.
TJYB	*Tongji Yuebao* (Statistical Monthly), 1930s
TR	*Turang* (Soil)
TRTB	Turang Tongbao (Soil Bulletin)
TRXB	*Turang Xuebao* (Acta Pedologica Sinica)
TRYK	*Turang Yuebao* (Soil Monthly)
XHBYK	*Xinhua Banyuekan* (New China Bimonthly)
XHYB	*Xinhua Yuebao* (New China Monthly)

XXPL	Xinxiang Pinglun (New Hunan Review)
XXYTS	*Xuexi Yi Tansuo* (Study and Exploration), Heilongjiang Academy of Social Science
ZGZK	*Zuguo Zhoukan* (Motherland Weekly), Union Research Institute, Hong Kong
ZGNB	*Zhongguo Nongbao* (Chinese Agricultural News)
ZGSL	*Zhongguo Shuili* (China's Water Conservancy)

C. Books in Chinese (including Yearbooks, Manuals)

AHNYDL	*Anhui Nongye Dili* (Anhui Agricultural Geography), Dept. of Geography, Anhui Normal University, Hefei: Anhui Kexue Jishu Chubanshe, 1980
AHSQ	*Anhui Shengqing 1949–1983* (Anhui Provincial Situation 1949–1983), compiled by Anhui Government office, Anhui Renmin Chubanshe, 1985
DZYJS	*Dizheng Yanjiusuo* (Research Institute of Land Administration (Taiwan) Series of studies) *Mingguo Ershinian Zhongguo Dalu Tudi Wenti Ziliao*: (Materials on Land Problems in the Chinese Mainland in the 1930s) ed. by Xiao Zheng. The series comprises of separate provincial volumes; Taipei: Cheng Wen Chubanshe, 1960s and 1970s (various years)
FQYJ	*Feiqing Yanjiu* (later renamed *Zhonggong Yanjiu*) (Studies on Chinese Communism), Taiwan
GZJJSC	*Guizhou Jingji Shouce* (Guixhou Economic Handbook), compiled by Guizhou Academy of Social Science, Guizhou Planning Committee, and Guizhou Statistical Bureau, Guizhou Renmin Chubanshe, 1984
HUBNYDL	*Hubei Nongye Dili* (Hubei Agricultural Geography), HUBNYDL Writing Group, Hubei Renmin Chubanshe, 1980
HUNNYDL	*Hunan Nongye Dili* (Hunan Agricultural Geography), Dept. of Geography, Hunan Normal College, Changsha: Hunan Kexue Jishu Chubanshe, 1981
JJNJ	*Zhongguo Jingji Nianjian* (Chinese Economic Yearbook), 1981–1989 issues
KXZTSC	*Kexue Zhongtian Shouce* (Handbook for Scientific Farming), Beijing: Beijing Academy of Agricultural Sciences, 1976

MGNJ	*Zhonghua Mingguo Nianjian* (Yearbook of the Republic of China), Taiwan
MGJJNJ	*Zhonghua Mingguo Jingji Nianjian* (Economic Yearbook of the Republic of China), 1930s
MGTJTY	*Zhonghua Mingguo Tongji Tiyao* (Statistical Survey of the Republic of China), 1930s
MSSWS	*Mao Zedong Sixiang Wansui* (Long Live Mao Zedong's Thought) (China Red Guard Publications), Taiwan Reprint, 1969
NARBCR	*Nongqing Baogao* (Crops Report of the National Agricultural Research Bureau), NARB, 1930s
NCJJTJDQ	*Zhongguo Nongcun Jingji Tongji Daquan 1949–1986* (Compendium of Economic Statistics of Rural China 1949–1986), compiled by Statistical Department, Ministry of Agriculture, Beijing: Nongye Chubanshe, 1989
NCTJNJ	*Zhongguo Nongcun Tongji Nianjian* (Statistical Yearbook of Rural China), 1985–1989 issues
NYKXJSSC	*Nongye Kexue Jishu Shouce* (Agricultural-Scientific-Technical Handbook), compiled by Bureau of Agriculture and Forestry, Shaanxi Provincial Revolutionary Committee, 1975
NYNJ	*Zhongguo Nongye Nianjian* (Chinese Agricultural Yearbook), 1980–1990 issues
NYSCJSSC	*Nongye Shengchan Jishu Shouce* (Handbook of Agricultural Production Techniques), compiled by Shanghai Agricultural Bureau, Shanghai Academy of Agricultural Sciences, and Shanghai Meteorological Bureau, 1979
QXNB	*Qixiang Nianbao* (Meteorological Yearbook), 1930s
RMSC	*Renmin Shouce 1961* (People's Handbook 1961), compiled by DGBBJ publisher
SBNJ	*Shenbao Nianjian* (Shenbao Yearbook), 1930s
TJNJ	*Zhongguo Tongji Nianjian* (Chinese Statistical Year book), 1981–1992 issues
TJZY	*Zhongguo Tongji Zhaiyao* (Chinese Statistical Abstracts)

Introduction

For a country the size of China, which accounts for around a quarter of the world population and some 20 per cent of the world total of cereals output, fluctuations in agricultural production pose serious economic policy problems for the political leadership. Grain output in China per head is still comparatively low and the country still ranks as one of the largest net grain importers in the world. A minor shortfall in production may translate itself into sizeable demands for grain imports.[1] International market repercussions aside, this may result in competitive claims for foreign exchange resources against the planners' programme of machine and equipment imports which are so instrumental in the pursuit of the 'four modernizations'.

Likewise, any policy effort to stabilize and increase agricultural output may call for investment resources to be diverted from the preferred urban industrial sectors into the rural areas. This is precisely the case of the post-Mao policy shifts which led to relatively large-scale intersectoral reallocations in favour of agriculture, when the Maoist approach of rural self-reliance and mass labour mobilizations for Nurkse-type accumulation was shown to be no longer viable.[2]

Moreover, fluctuations in the supplies of agricultural raw materials directly affect industrial planning and production, while the entire urban workforce is dependent upon food supplies from the rural areas. Besides, an overwhelming proportion of the country's foreign exchange earnings is generated from the export of farm or farm-related products. China has been and is still very reliant upon an agricultural surplus in order to finance her ambitious industrialization programme.

Thus, it is of great importance that the planners should seek to minimize the degree of agricultural instability. The problem is, in a way, not entirely dissimilar to that encountered by modern Western countries in their search for steady overall economic growth. However, while adaptations to cyclical business fluctuations in the West, pervasive as they may be, normally proceed in a successive and marginal fashion, interindustrial readjustments caused by a shortfall in agricultural output in China tend to be more drastic, if not larger in scale. This is not surprising, given the heavily agriculture-weighted Chinese industrial linkages. A good case in point is the catastrophic retrenchment in industrial production and investment which occurred in the early 1960s, following the collapse of the Great Leap Forward strategy in rural China.

Macroeconomic instability has also often resulted from harvest failure. Within the Chinese context of austerity, the bulk of urban consumption expenditure is earmarked for wage goods and other farm products, processed or unprocessed. Such supply shortages will inevitably bring about inflationary pressures, open or suppressed, unless the planners are willing to compromise their basic economic strategy by resorting to foreign supplies at the expense of imported producer goods.

The problem of monetary instability may not be as relevant in the rural as in the urban sector. This is because State farm procurements normally vary with harvest conditions. A bad harvest will lead to reduced State expenditure and hence to reduced peasant income. The planners are therefore not seriously confronted with the obligation, as in the urban sector, to soak up the considerable excess purchasing power, which represents a potential source of worker disincentives or, more seriously, social and political instability.[3] Nevertheless, the present Chinese leaders are more aware than anyone else that traditional peasant uprisings were often prompted by widespread harvest failure. Such unrest is potentially much more difficult to contain than its urban counterpart. Any large-scale regional grain shortfall may call for unexpected compensatory inter-provincial transfers, and upset regional production and distribution schedules, as well as impinging upon the overloaded national transportation capacity.

Agricultural instability has remained a persistent phenomenon in China in the past four decades. The relative magnitude of fluctuations in grain production, or in rural output in general, may have been gradually reduced over time, but the absolute size of fluctuations remains substantial. Moreover, the Chinese population has increased greatly, and increased demand for food consumption makes it difficult to divert substantial quantities of resources from agricultural production. Consequently, the established Chinese policy effort for maximizing industrial growth continues to be seriously constrained by agricultural performance. It remains to be seen whether post-Mao agricultural policy reforms will eventually lead China—the leaders and the led—out of this long-run dilemma. There is a final point to be made in this context. That is, the pace and possible outcome of present-day industrial reforms will depend much on the ability of the agricultural sector to ease supply constraints. Agriculture is not only an important source of capital accumulation, but it also supplies the bulk of consumer goods needed to maintain workers' incentives. Besides, any policy attempt to enhance competition among State enterprises necessarily calls for increased supplies of agricultural raw materials to render output markets less monopolistic. In short, a relative abundance of farm output probably represents the single most important material base for any feasible reform programme in the present-day context.

It is noteworthy that in the very midst of the euphoria brought about by spectacular upsurges in agricultural output in the early and mid-1980s Chen Yun reiterated his dictum of 'no stability without agriculture' (*Wu nong bu wen*) to warn against any premature de-emphasis of agriculture (by way of sown-area curtailments, *inter alia*). Chen may be ultra-conservative, but his epigram certainly captures the multifaceted implications of agricultural fluctuations in China. At any rate, it is essential for Chinese policy-makers to identify the sources of agricultural instability, so that appropriate measures may be formulated and adopted in order to avoid consequential losses.

Many attempts have been made in the West to interpret the causes of observed patterns of agricultural fluctuations since 1949. Often the fluctuations have been attributed to the frequent organizational and policy changes which have characterized the Chinese rural scene. Cyclical down-falls and upswings in agricultural output in China are seen to have been caused by unstable rural policy arrangements, alternating between the 'coercive' and 'remunerative' approaches. However, the causal relationships have rarely been explicitly defined except to the effect that the alternative rural collective-distributive approaches either impair or promote peasant incentives.[4]

The weather has also been implicated, but often as a less important factor. It is common for many foreign crop production reports made about China and other countries to begin with a general survey of weather conditions.[5] Hardly any analytical attempts have been made, however, to translate the possible impact of weather disturbance into measured output losses. There are admittedly considerable methodological difficulties involved in quantifying the weather dimension.[6] For a small country, or for agro-climatically well-defined regions, fluctuations can be more easily related to precipitation or temperature changes, for example.[7] But it is extremely difficult, if not meaningless, to construct an aggregate weather index by incorporating similar meteorological variables for a country like China, which embraces a vast territory and great climatic variation.

Yet, it is at the national level that the relative scale of weather influence cannot easily be isolated from the possible effects of changes in economic policies, institutions, and advances in agricultural technology. Thus, even for such an 'exceptional' period as 1959–61, which saw an extraordinary scale of fluctuations in agricultural output in China, it is difficult to quantify the relative effects of bad weather and bad system, which together led to that historic catastrophe.[8] Clearly, for relatively minor annual fluctuations in farm output, the task is all the more tenuous.

This study was originally prompted by intellectual curiosity to solve the 1959–61 puzzle, and determine whether Mao, the mastermind of the Great Leap Forward strategy, did, indeed, on his own bring about disasters which almost proved insurmountable. I began, more than fifteen years ago, to approach the problem from the weather side by examining provincial reports which detailed sown areas affected by natural disasters, as well as crop losses in the three years. No one who is familiar with the statistical fiasco of that time would accept such reports at face value. This, therefore, led me to the collection and tabulation of daily precipitation data for 1959–61 from some 200 major weather stations located all over China. The daily figures, which take the form of telegraphic dispatches to the Royal Hong Kong Observatory from which my collection is generated, are less conducive to falsification. This helps greatly in verifying regional and local weather conditions as a basis for more aggregative analysis.

While weather disturbance in 1959–61 is heavily implied in the precipitation statistics, further attempts to place the three years into proper historical context called for the analysis to be extended to both the 1930s and the whole of the 1950s. This in turn developed into an irreversible pilgrimage, extending beyond the aftermath of the crisis into the 1980s. This prolonged enquiry has proved to be rewarding, for the pre-war period represents a completely different institutional setting, although in terms of agricultural technology it is not entirely dissimilar to the 1950s and early 1960s. Any comparative analysis of the scale of weather disturbances in the 1930s may help to shed some light on the possible extent of human errors in the 1959–61 episode, or for that matter, throughout the period of collectivization.

More important, in the post-war context, a similar analysis may suggest the possible influence of the weather in relation to the long-term impact of advances in agricultural technology from the 1950s to 1980s. Moreover, a comparison of the different policy periods regarding the possible weather implications of agricultural fluctuations may help us evaluate well-established perceptions about the relative influence of policy swings between the 'coercive' and 'remunerative' approaches.

This study, therefore, aims to analyse long-run agricultural instability in China from the 1930s to the 1980s in terms of institutional and policy changes, and the role of agricultural technology. But the main focus is on the weather. Unlike technical farm studies of the relationship between weather and crop yields, however, the analysis is conducted in aggregate terms at both the national and regional or provincial levels. An average Chinese province is sufficiently large, in terms of geographical coverage and population size, to constitute a policy arena for scrutiny.

The thorniest problem is how to measure weather changes in national terms. A full chapter is devoted to this issue and the formulation of a comprehensive 'weather index' which uses the extent (in terms of farm area) of natural disasters as a weather proxy. This index serves as a basis for isolating relative weather influence throughout the study. However, its derivation involves a series of complicated methodological and technical problems, as well as the reconstruction of long-term weather patterns and climatic disturbance in China since the early 1920s. It is in effect a separate study on the weather *per se*. In order not to disturb the integrity and organizational coherency of the main body of analysis, this lengthy chapter is presented as an appendix, along with the many tables of statistical sources underlying the analysis.

The main body of the study is divided into five parts. Part I provides a global perspective by looking at the magnitude of fluctuations in Chinese agriculture in relation to the experience of other major grain-producing countries, notably India, the Soviet Union, and USA. It also furnishes a conceptual framework for interpreting changing agricultural technology, rural institutional arrangements, and peasant behaviour, as non-weather factors which have influenced agricultural stability.

Part II examines long-term trends and patterns of agricultural instability from the 1930s to mid-1980s, by focusing on important measures of grain production, as well as rural output in general. Against the background of climatic changes, the main purpose of the analysis is to examine:

1. how the radical rural reforms of the 1950s might have affected stability, compared with the entirely dissimilar rural environment of the 1930s;

2. whether, within the same socialist setting, the degree of agricultural instability really did alternate between the (presumably stabilizing) 'remunerative' (1952–8 and 1978–84), and (destabilizing) 'coercive' (1970–7) policy approaches, as many Western scholars have suggested; and

3. the influence of long-term changes in agricultural technology from the 1950s to 1980s. In this context we shall also discuss how the increasingly deregulated and monetized rural context blended with improved technology to bear on agricultural instability in China in the early 1980s.

The analysis in Part II is first carried out at the national aggregate level, to be followed by detailed cross-regional and intraregional verifications. It concludes with a summary of the familiar episodes of major instability in modern Chinese history. A comparison of these episodes (notably the Yangzi River basin floods in 1931, 1954, and 1980, the Sichuan basin floods in 1981, the Yangzi and Huai River basin droughts in 1934 and 1959, the Great North China drought in 1972, and not least, the three disastrous years 1959–61), helps to reveal more sharply the comparative long-term impact of agricultural technology and institutional manipulations on the one hand, and possible short-run policy implications on the other hand.

Part III of the study applies the formulated 'weather index' to isolate the relative weather influence. It is basically divided into two sections. The first examines whether, with advances in agricultural technology over time, the scale of weather influence in Chinese agriculture declined successively between the earlier (1931–7 and 1952–8) and later periods (1970–7 and 1978–84), each period being taken as a whole. The second section compares the relative magnitude of year-to-year fluctuations between the different periods. This helps to suggest the impact of any particular weather disturbance in relation to short-run policy changes made under given institutional and technological settings.

Parallel to the analysis in Part II, we also look in this context at the episodes of great instability with a view to examining the relative importance of weather and policy in causing the large-scale fluctuations.

Part IV makes use of the precipitation data collected for the 1930s, and especially for the three years 1959–61 for different regions, in order to supplement the analysis conducted in Part III of the relationship between the 'weather' (defined as changes in the size of farm area covered by natural disasters) and changes in grain yields. The main purpose is, first, to examine whether the relationship can be substantiated by variations in precipitation, which is the single most important weather determinant of the degree of droughts and floods in China; and, second, to make use of the estimated precipitation and yield relationships from the 1930s to investigate the possible scale of losses in 1959–61.

Granted that not all losses during 1959–61 can be substantiated by precipitation shortfalls, an important excursion is made in Part IV to specify the non-weather factors which helped to cause the great agricultural depression and food crisis in the early 1960s.

Finally, Part V is devoted to a study of the post-decollectivization period of 1984–91. Note that the main body of the study traces the Chinese history of agricultural instability only up to the year 1984. This is because both the main body of the statistical analysis was completed in 1985–6 and the year 1984 indeed represents a bench-mark in contemporary Chinese agricultural history, marking the end of the People's Commune and the abolishment *de facto* or *de jure* of the compulsory quota delivery system, which had remained intact since 1953, despite all the rural upheavals experienced in the intervening years.

The years 1984–90 are therefore treated as a separate period in order to examine how recent changes may bear on Chinese agricultural instability against the background of the previous six decades of profound organizational upheavals in the Chinese countryside. We conclude with a brief evaluation of possible future trends, in the wake of apparent uncertainty by the Chinese leadership whether further to push through agricultural reforms with full-fledged privatization of the land tenure system, or reintroduce a degree of centralization in order to address increased instability resulting from increased rural marketization and monetization.

NOTES

1. For a food balance estimate and its possible implications on China's foodgrain import requirements see Y. Y. Kueh, 'China's food balance and the world grain trade: projection, for 1985, 1990, and 2000', *Asian Survey*, 24/12 (Dec. 1984), 1247–74. As a result of the catastrophic droughts in 1982–3, many southern African countries, including notably South Africa were turned from food exporters into importers, see *International Herald Tribune* (14 July 1983), 3.
2. Cf. Kueh, 'China's new agricultural-policy program: major economic consequences 1979–83', *Journal of Comparative Economics*, 8/4 (Dec. 1984), 353–75, for a detailed discussion on this point.
3. The most perceptive analysis of the possible macroeconomic consequences of a harvest failure in China is to be found in Dwight Perkins, *Market Control and Planning in Communist China* (Cambridge: Harvard University Press, 1966), ch. 8.
4. Cf. Anthony M. Tang, *An Analytical and Empirical Investigation of Agriculture in Mainland China, 1952–1980* (Taipei: Chung Hua Institution for Economic Research, 1984), 121–2, and the authors he refers to, including G. W. Skinner and E. A. Winckler, 'Compliance succession in rural communist China: A cyclical theory', in A. Etzion (ed.), *A Sociological Reader on Complex Organization* (New York: Holt, Reinhart & Winston, 1969), 410–38; Alexander Eckstein, *China's Economic Development* (Ann Arbor, Mich.: University of Michigan Press, 1975), 311–22 and 332–8, and Michel Oksenberg, 'Political changes and their causes in China', in *Political Quarterly*, 45 (1974), 95–114.
5. This is the case e.g. with many issues of the annual *China Situation and Outlook Report* of the US Dept. of Agriculture; cf. also the world-wide regular reports of the Food and Agriculture Organization (FAO) of the UN.
6. A truly pioneering work in this respect is Tang, 'Trend, policy cycle, and weather disturbance in Chinese agriculture, 1952–78', *American Journal of Agricultural Economics*, 62 (May 1980), 339–48. See also A. M. Tang and C. J. Huang, 'Changes in input-output relations in the agriculture of the Chinese Mainland, 1952–79', in C. M. Hou and T. S. Yu (eds.), *Agricultural Development in China, Japan and Korea* (Taipei: Academia Sinica, 1982), 319–48. As discussed in Kueh, 'A weather index for analysing grain yield instability in China, 1952–1981', *China Quarterly*, 97 (Mar. 1984), 68–83, the major problem involved here is how to establish a weather index to isolate the weather impact. For an extensive comparative review of Kueh's approach with that of Tang in this respect, see Bruce Stone and Scott Rozelle, 'The composition of changes in foodcrop production variability in China, 1931–1985: A discussion of weather, policy, technology and markets', paper presented at the American Agricultural Economics Association meeting, 2–5 Aug. 1987,

Michigan State University, East Lansing, Mich.; together with the Appendix to this study. Stone and Rozelle apply variance decomposition techniques to provincial crop data to trace the possible sources of output variability, but they do not address the weather problem *per se*. See also B. Stone and Z. Tong, 'Changing patterns of Chinese cereal production variability during the People's Republic period', in J. R. Anderson and P. B. R. Hazell (eds.), *Variability in Grain Yields and Implications for Agricultural Research and Policy* (Baltimore, Md.: Johns Hopkins University Press, 1988).

7. For some useful examples see K. Takahashi and M. Yoshino (eds.), *International Symposium on Recent Climatic Changes and Food Production* (Tokyo: University of Tokyo Press, 1975). This is a collection of papers presented by international climatologists at a Symposium held in Oct. 1976 in Tokyo. Most of the countries and regions dealt with are relatively small in size, such as Sri Lanka, Bangladesh, North-west India, Thailand, and Japan.

8. In the early 1960s official orthodoxy, then still dominated by Mao, was generally to attribute 70% of the losses to the 'three consecutive years of natural disasters', and 30% to 'policy errors and mistakes'. (This is the Maoist *Sanqi Kai* (3 to 7 split) explanation.) This version was strongly re-endorsed during the Cultural Revolution against that, which had presumably been held by the 'Liu-Deng Clique' in the early 1960s, which argued instead that the ratio should be reversed. In the post-Mao years, the 'weather factor' has often been discarded altogether.

I
An Interpretative Framework

1

Measures of and Factors in
Agricultural Instability

An Overview

Agricultural instability can be measured in terms of farm output and farm income. It can also be estimated in terms of the fluctuations in such farm input variables as land, labour, farm machines, diesel oil, chemical fertilizers, etc. Through a comparison of total farm output (income) and inputs, it is possible to estimate fluctuations in agricultural productivity (output/income per hectare or per agricultural worker). Farm instability measures may be extended to cover farm-related non-farm economic activities in order to give a broad measurement of rural economic instability. In China, for example, such a comprehensive measure is the gross value of agricultural output (GVAO). This comprises, besides the farming sector proper, forestry, husbandry, fishery, and cottage-type industries which use agricultural raw materials such as cotton, tobacco, and sugar-cane as inputs. Another measure with a scope even broader than that of the GVAO is the so-called gross value of social output (GVSO) introduced in the mid-1980s. This embraces, in addition to the GVAO, building construction, transportation, and trade sectors. Within the rural context, the volume of undertakings in all these sectors fluctuates with the performance of the agricultural sector.

As an indicator of agricultural instability, the adequacy of all these measures depends critically on their statistical coverage. The greater the degree of statistical aggregation, the less reliable the measures will be as a means of capturing the sources and nature of instability. For example, part of non-farm economic activities, such as handicrafts and farm implement manufacture, included in the rural GVSO or GVAO may not be sensitive to changes in the climatic conditions which influence farm production. Moreover, aggregate measures in value terms are subject to price influences. Farm prices may fluctuate as a result of changes in demand for farm outputs. The changes may involve a general increase or decrease in demand, or a substitution of, say, cotton for grain. In either case the ensuing changes in price aggregates cannot easily be isolated by using any price deflator from fluctuations in farm production and supply.

It seems therefore preferable to use physical rather than value measures to gauge the degree of agricultural instability. Any physical measures adopted must, however, be sufficiently comprehensive in terms of geographic coverage and national quantitative significance in order to be representative of the agricultural sector. For such a large agrarian country as China, total grain output in tonnage represents such a measure, for grain crops are widely planted, and make up the bulk of the agricultural output. This also holds true for many other countries, advanced or less developed, including India, USSR, and USA. In contrast, cotton and other economic crops are not cultivated on the same scale as grain crops. Their relatively small quantitative base may be subject to the random influence of a host of local or regional factors—weather or man-made—which may not be representative of the national scene.

A distinction should also be made between short-run fluctuations and long-run trends in agricultural instability. The short-run measurement concerns year-to-year fluctuations occurring within a certain period in which the farm technological and institutional settings remain basically constant. A comparison of such short-run fluctuations in different Chinese settings, whether related to technology (irrigation and drainage) or rural institutions (socialization or decollectivization) may reveal the long-run stability of farm production relative to the climatic influence. This study examines both short-run and long-run patterns of agricultural instability and analyses their main causes.

There is a myriad of factors which bear on the stability of agricultural production. We may broadly distinguish between those in which the amount of major agricultural inputs tends to vary, and those which affect the total level of output through their impact on farm input–output coefficients. Clearly the weather can be counted in both categories. A flood or drought, for example, may reduce output per hectare sown, and in severe cases farmland (inundated or parched) may be made unsowable altogether. Worse still, farm implements may be destroyed, draught animals starved for want of feed, and labourers forced to emigrate to seek a living elsewhere.

Independent of weather conditions, farm production also fluctuates with changes in peasant incentives and behaviour. Such changes occur by way of work/leisure substitution or controlled variations in land input and such material inputs as chemical fertilizers, pesticides, etc. It is difficult to make a systematic review of the non-weather factors which may underlie such redispositions by peasants. Put simply, the factors concerned normally find their expression in farm demand prices or input costs, either in money terms or in alternative opportunities foregone. A relaxation or dismantling of the compulsory State farm procurement system, for example, may lead to a change in the rural cost–price relationship, thus making farming either more or less lucrative, and prompting the peasant to expand or contract production. Short-run output and productivity variations aside, such cost–price realignments and the changed rural institutional setting may also encourage or discourage peasant savings and farm investment, thereby affecting the pace and pattern of technological progress in the agricultural sector.

The rest of this chapter examines how peasants in China may respond to farm policy and institutional variations in terms of work-effort, factor supplies, and technological investment, as these affect both short-term and long-term agricultural instability. The discussion will be preceded by a review of the possible weather implications in China compared with other major countries of the world. Note that our reference is mainly to the national degree of agricultural instability. For such a large country as China, with its diverse climatic and economic and geographical conditions, variation in total farm output is also subject to the random regional effects of weather disturbance. The last section of the chapter illustrates this spatial factor in order to complete our conceptual framework for interpreting the empirical-analytical findings presented in subsequent chapters on agricultural instability in China.

Weather Disturbance: A Global Perspective

Fig. 1.1 shows the potential magnitude of agricultural instability in China in relation to India, the USA, and the Soviet Union during the three decades 1952–85. As a measure, total grain output is used for China and the comparable, though not identical, statistics of food and cereals production, by FAO definition, for the other countries. Both indicators dominate agricultural output in all these countries. The compiled indices for the three decades reveal substantial output fluctuations from time to time, notwithstanding the fact that in all four countries, grain production showed a long-term steeply increasing trend as a result of continuous technological progress.

Most of the most dramatic yearly fluctuations in grain output are well-known national or international events. The major cause has been identified as large-scale weather anomalies. A good case in point is the world food crisis triggered in the mid-1970s by serious crop failures in the USA, the world's largest grain-exporting country. This caught many US government officials and agronomists by surprise as they 'came very close to believing that modern technological advances had resulted in the development of a grain production system which was almost independent of weather'.[1]

In a wholly monetized market economy like that of the USA, the existence of agricultural instability may be partly attributed to autonomous

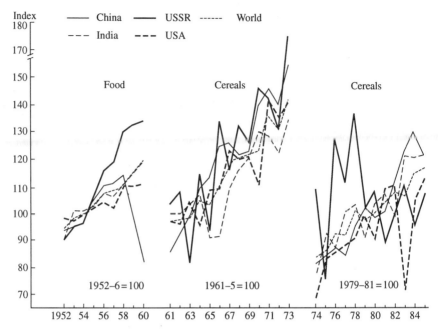

FIG. 1.1. Instability of food and cereals production in China and other major countries of the world, 1952–1985

Source: Table AB.1.

changes in relative prices and farm profitability. However, grain production in the Soviet Union, where central planning prevailed, also fluctuated sharply until the very recent past (Fig. 1.1) in response to weather disturbance.[2] It is well known that Soviet agriculture was long plagued by planning inefficiency and peasant disincentives, but such problems are more of an endemic nature and hardly explain the erratic large upturns or downturns in output.

The example of a less developed country, India, provides another interesting case. In 1965 and 1967 total cereal production fell 20 per cent below the trend values in both years because of serious droughts.[3] A similar dramatic crop failure was recorded in China in the three years 1959–61, although there is less agreement as to whether natural calamities or ill-formulated policy measures associated with the Greap Leap Forward strategy explain the major proportion of the losses.

All these countries are major world grain producers. By their sheer geographical size, one might expect good and bad weather to occur at random over different climatic zones in each country, with the positive and negative effects cancelling each other out, enabling total farm output to remain broadly stable over time. This does not seem to be the case. Indeed, world averages of grain output (as well as yield per sown hectare) also exhibit a significant degree of variability from year to year, as shown in Fig. 1.1. James McQuigg, Director of the US Centre for Climatic and Environmental Assessment, has noted that much of the variability in world output can be associated with large-scale meteorological anomalies.[4] Thus for any single country, small or large, total grain output may at times dramatically deviate from its long-term technological trend.

Weather anomalies which are large in scale (measured by geographic coverage) in duration and intensity are relatively rare. The serious US drought of 1974–5, for example, was preceded by almost fifteen years of remarkably favourable weather in the Corn Belt and in other important agricultural States.[5] Likewise, the severe drought in India of 1965–7 was followed by relatively stable weather conditions until a similar, albeit more localized, drought reoccurred in 1973–5, mainly in Maharashtra, one of the major grain States in India.[6] Similar weather cycles seem to be discernible in China and elsewhere in the world. Thus, a gradual climatic improvement, following the catastrophe of 1959–61, which culminated in the most favourable weather year of 1970, was abruptly interrupted by the great North China drought in 1972–3.[7]

More often, however, bad weather shifts from place to place within a country. The regional scale of output fluctuations can be expected to be larger than the national one. The reason for this is simple: the limited regional base is often entirely covered by adverse weather, while for the country as a whole good harvests elsewhere may wholly or partly compensate for output losses sustained in the stricken area. Extensive farm observations and field experiments conducted in both pre-war and post-war China show that grain crop losses resulting from drought or flood at the local level could easily range from 60 to 100 per cent.[8] In fact, serious drought or flood only needs to occur at critical stages of the growing season in order to bring about disastrous losses. Thus, if adverse weather prevails simultaneously over a relatively large part of the country, especially in the main grain-producing regions, a loss of national grain output by around 20 per cent, such as that experienced by India in 1965–7 and by China in 1959–61, is not impossible.

According to Buck's monumental study of the Chinese farm economy in the pre-war context, famine occurred if around 70 per cent of the cropped area of a *xian* (county) was affected by natural disasters, drought and flood in particular.[9] Buck used the *xian* as the unit of his survey, but bad weather, invisible drought in particular, usually affects one or more provinces, rather than a few isolated *xian*. Thus, with its domain shifting from year to year, one can easily visualize the Chinese land mass to be hopelessly locked in an endless spatial and temporal chain of famines and starvation. This is the agronomic background of the 'land of famine'—the term Mallory applied to imperial and pre-war China.[10] In the same light may be seen the more recent famine tragedy which rotated from the Ethiopian Plateau (about the size of the North China Plain or Sichuan province) to Sudan (not much larger than the Chinese North-east, excluding the Sahara portion), both countries representing the latest link in the prolonged chain of the Sahel African droughts originating in the early 1970s.

If relief-grain transfers are available, a regional crop failure may not develop into a famine. And given the size of China, with possible compensatory harvests from elsewhere, the regional losses may simply show up as a modest fluctuation in the national grain output. Thus, it is important to probe consistently into the possible regional output variations in order to determine the sources of agricultural instability. However, similar marginal output adjustments may also occur when a relatively mild deterioration, or for that matter improvement, in weather conditions across the country, results in a mediocre harvest or a comparatively good one. In such cases, it may be difficult for any large country such as China to isolate exactly the weather impact from possible intervening human factors. Let us now examine, first, how the non-weather factors may bear on agricultural production.

The Institutional Setting and Peasant Behaviour

A distinction should be made between market and Soviet-type economies in terms of the possible institutional implications of agricultural instability. In the USA, for example, rational private economic calculations, in the face of serious drought or flood, may persuade farmers that the real opportunity costs involved do not warrant them making massive efforts to ward off the adverse effects. A good case in point is the record number of farm bankruptcies in the USA during the mid-1980s, caused by persistent pressure of high interest rates for loans taken for expansion purposes a decade or so before, when 'soya beans were like gold' and when soaring land prices prompted the banks to press farmers to take out more and more loans.[11]

When output prices are high, income benefits for US farmers may outweigh marginal costs, for example higher electricity bills for sprinkler irrigation to combat droughts. But when income falls below expenses, the easy answer is not to borrow more, but to take up whatever residual the mercies of nature may leave behind, or to wind up the operation altogether. Likewise, any substantial modification in the massive US system of farm price supports (by cutting down, for example, the 'loan-rate' to below open-market prices in order to reduce the huge federal budget deficit) is likely to affect the input and output decisions of US farmers with or without the impact of bad weather.

The situation in the Soviet Union and China may be quite different. In both countries, agricultural production was dominated by the Stalinist system of compulsory farm deliveries with the agricultural collective serving as its organizational vehicle. This system remained intact from 1928 onward in the Soviet Union and from 1953 in China, at least up to the very recent past.[12] Such physical and administrative controls placed a limit on cost-and-benefit calculation by the peasants, and so helped to minimize market-induced farm instability.

Nevertheless, a distinction is to be made between poor and rich socialist peasants. In the former Soviet Union, per capita grain output stood at nearly 500 kg. in 1928, when Stalin started to collectivize the agricultural sector.[13] This is high compared with 285 kg. in China during 1952, or 378 kg. in 1983. Per capita grain availability on farms, net of forced procurement, was still 415 kg. in 1932 in the Soviet Union, compared with 295 kg. for China in 1957, or 331 kg. in 1977–82.[14] Taking roughly 250 kg. as the per capita subsistence requirement, the farm surplus was obviously much more substantial in the Soviet Union than in China. This extra margin gave Soviet peasants more leeway in their expression of opposition to State extortions whether through higher delivery quotas or lower procurement prices. For the Chinese peasants living on the verge of subsistence, however, output and supply elasticity in this respect were inevitably more restricted.

Moreover, not only were the agricultural constraints confronted by the Soviet planners in the early course of forced industrialization never as serious as in China, but one of the major concerns of Stalin was how to extract a sufficient workforce from the farm sector to meet increasing urban industrial requirements—hence the tractorization of Soviet agriculture.[15] By contrast, enormous population pressures in both the rural and urban sectors made it imperative for the Chinese authorities strictly to control rural labour mobility in order to stave off large-scale rural out-migration. In addition, the drive to stabilize and increase grain output also made it necessary to restrict occupational mobility. Thus, the strong 'subsistence urge' of Chinese peasants was reinforced by a coercive rural institutional environment to hedge against possible output fluctuations arising from bad weather or other circumstances.[16] Any lethargy *vis-à-vis* natural calamities or failure on the part of the peasants to cope with them threatened to be catastrophic.

This stands in sharp contrast to India, as well as pre-war China. In both cases the rural context is dominated by small-scale household holdings and market relations, with hardly any direct government control. The poor Indian or Republican Chinese peasants may not be as sensitive as income-maximizing American farmers to changes in market conditions not caused by weather disturbance. But in the absence of the 'institutional hedge' available in the collectivized context, the same 'subsistence urge' often unfolds promptly under the impact of natural disasters and famines to result in widespread emigration from rural areas. This aggravates the degree of input and output instability.

Within the Chinese framework of collectivized and bureaucratic control, planning and policy arrangements could, however, vary from time to time producing either stabilizing or destabilizing effects. Table 1.1 summarizes, in chronological order, the major features of such changes over more than three decades and shows how peasants' motivation and behaviour may have been affected from the viewpoint of farm stability. Generally speaking, the changes alternated between the familiar policy approaches of centralization (from the early toward the late 1950s, and again from the mid-1960s to the mid-1970s) and decentralization (first half of the 1960s and during the post-Mao era) of rural control. There has thus been an obvious policy cycle in Chinese agriculture, with bureaucratic physical control of input and output targets, restrictive locational and occupational mobility, confiscatory forced siphoning of key farm products, and ideological and political persuasion or coercion increasingly substituted for indirect manipulation through markets, prices, and income incentives, less-harsh State procurement quotas, and a rural environment more congenial to peasants in pursuit of non-farm earnings, and vice versa.[17]

While the centralized (or coercive) policy approach suppresses peasant incentives, it helps, nevertheless, to improve farm output stability through the controlled pro-grain concentration of rural resources and the strengthened 'institutional hedge' against natural disasters. Under the decentralized (or remunerative) approach, on the other hand, it is difficult to mobilize peasants to cope with bad weather, for short-run income benefits cannot be easily defined and provided for such mobilization. Weather aside, manpower and material resources can easily be led away from the collective farming, should any absolute or relative changes in input and output prices occur in favour of private non-farm activities.

In this context, the radical land reforms of the recent past are particularly revealing. The redistribution of collective farmland was certainly a very important incentive contributing to the upsurge in grain output and yield per sown hectare during the 1980s. But amidst a buoyant rural non-farm economy, the abolition of controls over the sown acreage, in favour of control exercised via procurement targets, offered a potential source of

TABLE 1.1.   A chronology of rural institutional and policy changes in relation to
peasant behaviour and farm stability in China, 1949–1990s

Major aspects of institutional and policy changes	Peasant responses and implications for farm stability
1949–52: Land reform Redistribution of confiscated landed properties to self-farming peasant households. Considerable freedom with respect to farm marketing and input and output decision-making in a market environment.	Peasant incentives enhanced by both land redistribution and the introduction of advanced farm purchase contracts furnished by State agencies together with credit advances. Marketing and income risks for peasants reduced. Improved incentives, together with comparatively favourable weather conditions (except in 1949) led to rapid recovery of farm output from the civil war.
1953–7: The collectivization drive Compulsory delivery quotas for cotton and grain introduced in 1953, followed by creeping co-operativization (1953–5) and accelerated collectivization (1955–6) to enforce the procurement scheme. As a planning and accounting unit, the co-operative embraced a *cun* (hamlet with 20 to 30 farm households), and the collective, an entire *xiang* (village with around 250 households). Rural migration and occupational mobility increasingly restricted. Private land titles confiscated under the collective, while owners under the co-operative still allowed to draw 'land dividend' proportionate to land area contributed. The transition also witnessed size of 'private' plots being reduced, thus further limiting the degree of rural economic diversification and occupational choice.	Co-operative sufficiently small in size to subject its leadership to members' pressures for maximizing income benefits. Alternative earning potentials for peasants also comparatively large to place a limit on tightness of compulsory quotas. Leaders of the collective, however, increasingly subjugated to superior command to fulfil key physical delivery targets, with peasant disincentives from loss of land ownership being further compounded by resentment of working for 'an unknown share in a future common pot of an unknown size'. Nevertheless, while farm efficiency tended to vary with peasant incentives and the weather (good in 1955, and bad in 1954 and 1956–7) for the 1953–7 period as a whole *de facto* conscription (near-zero labour mobility) and the 'subsistence urge' (see text) began to work as a powerful hedge against fluctuations in sown area and hence total farm output.

TABLE 1.1.   *cont.*

Major aspects of institutional and policy changes	Peasant responses and implications for farm stability
1958–60: Communization People's communes emerged initially as a merging of agricultural collectives to facilitate labour mobilization for rural capital construction, notably irrigation projects and the 'backyard furnace' campaign. By autumn 1958, it was consolidated not only as a giant planning and accounting unit (with around 5,000 households), but a powerful vehicle of political control for maximizing State farm extractions. System of part-wage (flat rate) and part-supply (food at commune canteen) imposed to minimize differences in income and consumption and, thus, commune expenditure. First phase of disintegration set in by mid-1959, with the production-brigade (size equivalent to the collective) being restored as the economic unit, together with the gradual return of the work-point method, private plot, and rural trade fair. The politico-administrative framework of the commune still intact, however, and indeed vigorously mobilized in late 1959, for the Anti-Rightist Campaign to offset the emerging incentive measures. Second phase of disintegration initiated late 1960 to end up in the new structure since 1961.	Widespread peasant resentment against forced mobilization (for overhead projects) without *quid pro quo* and radical egalitarianism in farm income distribution. Note that commune-wide implementation of the part-wage, part-supply system discriminated not only the able against the less able farmers, but also the rich against the poor villages and hamlets as a whole. Situation aggravated, in winter 1958–9, by excessive grain procurement targets derived from exaggerated output claims made by party cadres. Hence massive peasant effort to conceal the harvests, e.g. by underground storage. Slow restoration of peasant confidence beginning late spring 1959 abruptly interrupted by the Anti-Rightist (an inherently extortionate ideological) Campaign. Fear of harvest and procurement shortfall (widespread bad weather in summer 1959 on top of the unwarranted curtailment of farmland under the three–three system adopted in winter 1958–9) reinforced the campaign to lead to another upsurge in forced siphoning. Incentives aside, amount of grain left too meagre for the peasants to refrain from dipping into seed and foodgrain for survival. Hence, with bad weather continuing through 1960–1, the food crisis became insurmountable in both years.
1961–5: The readjustment The new standard rural structure promulgated under 'the 60 Articles' (May 1961), with the production team (size comparable to the co-operative) now constituting the planning and accounting base under the three-tier	Peasant reorientation toward income benefits, within a decentralized context reminiscent of the co-operative or even the land-reform period (save restoration of private land titles). The grain situation still tense, however,

TABLE 1.1. *cont.*

Major aspects of institutional and policy changes	Peasant responses and implications for farm stability

1961–5 (cont.)
ownership system (comprising, in order of subordination, the commune, brigade, and team). Further decentralization widely experimented with, especially prior to the Tenth Plenum (Sept. 1962), under the Anhui model of the household-responsibility farm (*zerentien* or *baochan daohu*). Tacitly a self-survival solution for peasants to cope with the food crisis, the experiment was supplemented by the expansion of private plots (*ziliu di*), rural trade fairs (*ziyou shichang*), and sideline activities on private account (*zifu yingkui*) to form the well-known 'san (three) *zi yi* (one) *bao*' programme. Both the commune and brigade layers hardly functional, save for conducting, in response to Mao's call for class struggle at the Tenth Plenum, the socialist education campaign (1963–5) to rectify 'capitalist' trends emerging from the drastic rural decentralization.

compelling peasants to tackle more of the subsistence problems, rather than take full advantage of the *sanzi* policy to maximize cash income. Moreover, reduced State extractions (due to accelerated increases in grain imports to balance urban needs), coupled with bulk of State chemical fertilizer supply, made farming comparatively favourable. The weather also supportive of the steady recovery. Also, Mao's ideological campaign thwarted by tactical Liu-Deng manipulation to minimize its incentive-impairing effort, although the potential conflicts finally led to the Cultural Revolution.

1966–76: The Cultural Revolution
The standard rural structure established in 1961 remained formally intact and indeed was codified in the New Constitution of 1974. However, Mao's twin call for regional grain self-sufficiency and to rid the 'capitalist remnants' virtually reduced the *sanzi* provisions to nil (let alone family farming), thus compelling the peasants to concentrate on physical grain targets. The same campaign, coded *Dazhai* after the familiar brigade in Shanxi province, prompted large-scale labour mobilization and coerced egalitarian farm distribution in terms no less traumatic than the early

Peasant disincentive and farm inefficiency aside, massive concentration of rural resources at the expense of the non-grain sectors helped, nevertheless, to stabilize grain production, and thus the agricultural sector at large. The most notable exception being 1972, the year of the great North China drought, with national grain output reduced remarkably midway through the Cultural Revolution when the tumult was already over. The pro-grain strategy itself also a possible source of instability in that grain-sown area was unduly expanded into ecologically

TABLE 1.1.   *cont.*

Major aspects of institutional and policy changes	Peasant responses and implications for farm stability

1966–76 (cont.)

communization drive. Commune and brigade now also reactivated as mobilization vehicles for the 'five-small industries' (viz. chemical fertilizers, farm machinery, metal-making, cement, and energy) launched locally in support of agriculture, i.e. grain production.

vulnerable regions—in disregard of potential frost or drought threats in the North-east or North-west, for example.

1977–1990s: The post-Mao reforms

Conventional incentive measures, notably the work-point system restored (1977–8), to be followed by drastic increases in State farm procurement prices (1979–80), and subsequently by various forms of household-responsibility systems (1981–3), so that the price incentives could filter more directly to hardworking families (least lazy families might receive, through the egalitarian commune distributive structure, the same price subvention). The restrictive commune system discarded in the New Constitution (Dec. 1982) to ease further reforms to culminate in the leasehold right (of landed properties) being extended to over 15 years, and made transferable as well (1984); whereby the single obligation relegated to the individual family concerned (*daohu*) being the *baogan* targets, i.e. the mutually agreed and predetermined deliveries to the State (compulsory procurement) and the collective (for common welfare and overhead expenditure). Beginning 1985, forced siphoning replaced by advanced purchase contracts, virtually bringing the rural situation back to the land-reform context of 1949–52. Parallel liberal provisions also made to

Peasant motivations for income maximization within the household context now resemble their counterparts in market-type economies. Note particularly the *baogan* targets virtually identical to the lump-sum tax system familiar to non-socialist peasants. Income manœuvrability of peasants also enhanced by improved per capita grain availability on farm, following relative reductions in State procurements and drastic increases in grain imports to meet urban demands. However, accelerated economic diversification and monetization of the rural economy, coupled with increased price deregulation (generally in favour of non-farm products), made farm production increasingly subject to volatile market conditions and price fluctuations. Moreover, the pro-grain strategy of the Cultural Revolution reversed and resulted in cotton (and in other cash crops) substituting for grain cultivation in wide areas. This, together with the weather alternating its good (1979, 1984) and bad (1980, 1981) years, tended to exacerbate the degree of agricultural instability for the post-Mao era as a whole.

TABLE 1.1. *cont.*

Major aspects of institutional and policy changes	Peasant responses and implications for farm stability
1977 (cont.) promote rural non-farm branches, viz. industries, trade, construction, and transportation in order to (1) absorb labourers rendered redundant by the increasingly efficient farm sectors, and (2) provide consumables to meet rural demands generated by increased cash income. Thus, non-farm businesses renamed *xiangzhen* (a neutral geographical term meaning village and market town) enterprises, to be freed from the *shedui* (commune and brigade) administrative tutelage, and pave the way for rural residents to set up any individual or co-operative ventures. Consistently bureaucratic control of population movement also relaxed to facilitate inter-regional marketing, transportation, and co-operation.	

output uncertainty. Thus, it was necessary for the State to readjust frequently its terms of procurement, by changing output and input prices or by manipulating peasants' obligatory sales quotas to allow for a greater volume of higher-priced, voluntary sales. Such measures ultimately (1985) underlined the need to substitute a new system of procurement contracts for the conventional method of forced collection, in place since 1953.[18]

In addition to the problems of short-term fluctuations, both the coercive and renumerative policy approaches also have far-reaching implications for long-term agricultural stability in China. This is discussed in the following section.

Agricultural Technology and Investment

For a traditional agriculture like that of China to be modernized, thereby enabling farming to enter a stage of sustained and steady growth, it is imperative to raise farm accumulation for investment in irrigation and drainage capacity. Improved irrigation and drainage facilities not only help to stabilize farm output in a drought- and flood-prone country like China, but they are also the prerequisite for increased application of chemical fertilizers necessary for raising yield per hectare. Moreover, many high-yielding seed strains are developed to be responsive to chemical fertilizers.[19]

Irrigation and drainage technology also helps to expand multiple-cropping and so indirectly contributes to improving farm stability. This is because harvest failure in one crop may be offset by success in another, since very often drought or flood lasts for only one cropping season or even shorter. This 'risk-aversion function' can be made operative only when rainfall in the wet seasons can be stored for redistribution in the dry seasons, and flooded or inundated farmland can be effectively drained to prepare for the next sowing. Thus, clear correlations have been observed between increases in the multiple-cropping index (MCI) and in irrigated areas in the Chinese mainland, in Taiwan, and in other countries.[20] The 'risk-aversion function' also explains why, according to Buck's pre-war survey, famines were more frequent in China's wheat region (with a lower MCI) than in the rice region (higher MCI), averaging about once in every seven years and once in a decade respectively, with the Yangzi rice and wheat area showing 'a definite transition to the better conditions of the Rice Region in that the effect of famine was not so severe as in the Wheat Region, but more severe than for the other areas of the Rice Region'.[21]

Farm mechanization of tilling, sowing, transplanting, and harvesting also works to raise the multiple-cropping index and farm stability. For a country like China, which seems to have an abundance of labour, the need for farm mechanization may not appear to be urgent. However, rapid mechanized operations certainly help to accelerate the process from the first to the second and third crops and thus to squeeze successive harvests into the limited frost-free growing period.[22] This is a process which cannot easily be handled by any amount of labour mobilization. Of course, the possibility of multiple-cropping very often depends on whether early-maturing seed strains are available.

Farm mechanization hardly existed in the 1930s. Irrigation and multiple-cropping are, however, long-standing and widespread practices in Chinese history. A distinction should be made between the traditional method of manual water-pumping or gravity flow, and the use of modern, power-driven irrigation and drainage equipment.[23] Water resources can more easily be tapped with electrical or mechanical pumping from wells and siphoning from river streams or irrigation canals. In the face of flooding and inundation, draining machines simply cannot be replaced by any amount of labour mobilization in order to rescue the submerged crop.[24]

Moreover, large-scale water reservoirs, and more reliable flood containment dykes and flood diversion structures, cannot easily be built without essential modern supplies such as cement, steel bars, and hydraulic engineering know-how.

The transition from traditional to more modernized irrigation and drainage technology has certainly been an important factor in the observed increases in the multiple-cropping index for China as a whole, and the warmer rice region in particular, in the post-war period.[25] The same technological disparity may also explain the peculiar pre-war phenomenon, also observed by Buck, whereby famines occurred more often in the wheat region than in the rice region, while it still took an average of over ten months for natural disasters to enter the famine stage in the drought-prone North, compared with only 3.6 months in the water-rich South.[26] Three and a half months may not seem a long period, but flooding in the South normally occurs in the summer months, frequently in early summer or midsummer, at the time of the busy transition period from the first to the second rice crops. In such circumstances, lack of electrical or mechanical drainage facilities easily results in prolonged inundation, thereby unexpectedly reducing the multi-cropped area to render the 'risk-aversion function' inoperative.

, Advances in modern irrigation and drainage technology in post-war years have considerably reduced the impact of natural disasters in this respect. This is reflected in the steady decline since 1952 in sown areas statistically classified as so severely affected as to suffer a crop loss of more than 30 per cent.[27] The decline has been more substantial in flood than in drought areas. This implies accelerated improvement in farm stability, because floods are normally more damaging than drought in terms of output losses, and occur more frequently in the fertile southern provinces, which weigh much more heavily than the North in national agricultural output.

The background to changing technological standards and long-term agricultural stability is again to be found in rural institutional arrangements. In a free and marketized rural context, the high income elasticity of consumption of peasants makes it difficult to generate substantial farm savings. Thus, investments in irrigation and drainage capacity are bound to be limited and slow in assuming sufficient magnitude effectively to withstand severe droughts and floods. This was the case in pre-war China and India.

By contrast, centralized control by way of compulsory State deliveries and collective retentions helped to institutionalize forced savings in Maoist China. Moreover, bureaucratic restrictions on population movements—locational and occupational—not only coerced peasants into fighting against floods and drought to minimize short-run output fluctuations, but facilitated mass labour mobilization to construct large-scale reservoirs, dams, and irrigation canals, all of which constitute the infrastructure for installing powered water-siphoning stations, hydropower plants, sprinkle irrigation, etc. In this respect, Mao was another example of Wittfogel's 'oriental despotism'—the linchpin of the 'hydraulic society' of ancient China.[28]

Maoist-style 'labour accumulation' can of course hardly be imposed upon peasants operating independently in a market economy. Thus, there is clear demographic evidence of large-scale rural emigration in pre-war China. In many provinces, notably Hebei and Jiangsu, the large urban centres were flooded with young peasants, while the age-structure of the rural population became increasingly gourd-shaped, i.e. dominated by the elderly and children.[29] A nation-wide survey, conducted in 1935, reveals that around two-thirds of rural emigrants were urban-bound. And as shown in Table 1.2 from the same source, three-quarters of all the emigrants were moving under the impact of socio-economic pressures (39.3 per cent) or bad weather (33.5 per cent), especially floods (9.8 per cent) and droughts (13.2 per cent). Notice that the weather in 1935 was not at all bad for China as a whole. By contrast, in 1931, the disastrous Yangzi and Huai River floods alone forced 40 per cent of the rural population from the two basins to abandon their farmsteads.[30] This inevitably affected farmwork in ensuing crop seasons, and wide tracts of arable land could not be sown; nor could mass mobilization for construction take place.

However, the scale of accumulation—whether in terms of sheer labour mobilization à la Nurkse, or in the more usual form of capital savings—may vary within the same socialist collective framework. Thus, following the communization drive and the Cultural Revolution—both periods being characteristic of the coercive approach—policy emphasis shifted to small-scale irrigation and drainage projects appropriate to the limited potential of labour mobilization under the remunerative approach (see Table 1.1). Seen in this light, it is precisely the coercive, rather than the remunerative approach, which has contributed more to long-term agricultural stability. The remunerative approach would have involved potentially prohibitive labour costs to sustain the scale of investment feasible under the coercive approach.

Yet a distinction should be made between the two periods of decentralized rural control. The same remuneration approach may fare differently depending upon the degree of rural decentralization and the standard of peasants' incomes. During the readjustment period 1961–5 and the gradual collectivization of the 1950s, peasants' income incentives were still capable of being compromised by increased collective mobilization, to a greater or lesser extent. Under the post-Mao reforms, however, improved income and a higher savings capacity made the

TABLE 1.2. The frequency of floods and droughts in relation to the socio-economic factors causing peasant migration from their home villages in China, 1935 (%)

	Basic natural factors					Socio-economic factors							Others (including local insurgencies in parenthesis)
	Floods	Droughts	Other calamities	Bad harvests	Subtotal	Land and population pressures	Financial difficulties and bankruptcies	Poverty and livelihood difficulties	High land rentals & taxes	Low farm product prices	Decling sideline production	Subtotal	
	(1)	(2)	(3)	(4)	(5)	(6)	(7)	(8)	(9)	(10)	(11)	(12)	(13)
North-west													
Chahar	2.6	7.7	18.0	2.6	30.9		5.1	15.4	5.1	5.1		30.7	38.4 (33.3)
Suiyuan	20.6	14.7	2.9		38.2		5.9	17.7	8.8			32.4	29.4 (17.7)
Shaanxi	9.5	21.8	8.4	3.3	43.0	2.8	2.2	6.7	5.6	1.1		18.4	38.6 (28.5)
Gansu	7.6	18.5	5.9	10.1	42.1		1.7	11.7	19.3	1.7		34.4	23.5 (15.1)
Qinghai	6.3	12.5	28.1	3.1	50.0		3.1	21.9	9.4			34.4	15.6 (3.10)
Ningxia			20.0		20.0			10.0	60.0			70.0	10.0 (–)
North													
Hebei	8.8	10.7	6.5	4.3	30.3	10.8	11.2	16.3	2.2	0.8		41.5	28.2 (12.3)
Shanxi	4.9	12.9	4.4	4.9	27.1	15.8	5.0	24.2	9.4	4.0	0.2	58.4	14.5 (4.0)
Shandong	12.9	9.0	2.3	2.3	26.5	11.8	4.5	31.8	0.9		0.3	49.3	24.2 (7.6)
Henan	11.4	11.2	10.5	2.7	35.8	8.8	2.5	13.3	3.6	0.7		28.9	35.3 (22.7)
Central													
Jiangxi	14.9	22.0	6.4	4.3	47.6	2.8	4.2	12.1	5.0	0.7	0.7	25.5	26.9 (17.0)
Hubei	16.9	17.6	7.1	4.9	46.5	0.7	2.1	19.7	5.6	0.7		28.8	24.7 (18.3)
Hunan	14.8	18.1	6.6	3.8	43.3	7.6	5.5	15.4	2.3		0.6	31.4	25.3 (15.4)
East													
Jiangsu	10.6	11.9	2.8	7.3	32.6	6.7	6.8	24.6	2.3	1.8	2.6	44.8	21.8 (17.0)
Zhejiang	6.6	13.7	4.6	4.1	29.0	8.1	5.1	19.8	0.5	5.1		38.6	32.4 (11.2)
Anhui	17.6	28.7	4.6	1.4	52.3	1.8	7.5	13.3	2.9	2.1	0.4	28.0	19.7 (14.7)

TABLE 1.2. *cont.*

	Basic natural factors					Socio-economic factors							Others (including local insurgencies in parenthesis)
	Floods	Droughts	Other calamities	Bad harvests	Subtotal	Land and population pressures	Financial difficulties and bankruptcies	Poverty and livelihood difficulties	High land rentals & taxes	Low farm product prices	Declining sideline production	Subtotal	
	(1)	(2)	(3)	(4)	(5)	(6)	(7)	(8)	(9)	(10)	(11)	(12)	(13)
South-west													
Sichuan	2.0	11.4	11.0	1.6	26.0	5.0	10.2	14.9	18.0	2.0	0.4	50.5	23.5 (15.7)
Guizhou	10.3	10.3	8.8	1.5	30.9		4.4	7.3	23.5		1.5	36.7	32.4 (26.5)
Yunnan	5.6	9.9	8.5		24.0	8.4	4.2	11.3	15.5	2.8		42.2	33.8 (11.3)
South													
Fujian	5.2	6.2	12.4		23.8	2.0	5.2	18.5	6.2	2.1		34.0	42.2 (33.0)
Guangdong	4.9	4.2	2.8	5.6	17.5	7.0	10.5	23.7	9.8	3.5	4.9	59.4	23.1 (9.1)
Guangxi	4.2	6.3	9.0	2.8	22.3	6.3	6.9	32.6	3.5	0.7		50.0	27.7 (5.5)
All China	9.8	13.2	6.8	3.7	33.5	7.3	6.4	18.2	5.4	1.5	0.5	39.3	27.2 (14.3)

Notes: Column 13 covers besides 'local insurgencies', migration for further study, occupational changes, and other unaccountable factors.

Sources: *NARBCR* 4/7 (July 1936). 179, as reproduced in Zhang Yuyi, *Zhongguo Jindai Nongyeshe Ziliao, 1927–1937*, iii (Beijing: Sanlian Shadian, 1957). 892.

peasants less dependent on the collectives, and helped to reinforce the effect of decollectivization in generating a powerful income-maximizing drive that was detrimental to collective manipulation of farm investment.

The consequence was twofold: first, chemical fertilizers, diesel oil for farm tractors, and similar essential modern farm inputs now tended to fluctuate negatively with alternative non-farm earning potentials. Second, peasants' apathy towards collectivism manifested itself in terms of wide-spread disinvestment in irrigation and drainage facilities.[31] The first aspect lends credence to some observations made about the Indian ex-perience that modern technologies could well lead to increased farm instability.[32] The second aspect raises the fundamental question of whether, after all, it would have been strategically better not to substitute 'the road of serfdom' for the more moderate land-reform programme, but to resort to indirect manipulation through market and price incentives in order gradually to achieve agricultural modernization.

The Spatial Factor

In any analysis which attempts to relate the human element to the potential scale of weather influence, one should consider how different regions or agricultural areas in China may respond to the random geographic incidence of weather disturbance. Specifically, the same drought or flood may have a substantially different impact on aggregate output stability, depending on whether it occurs in North or South China.

Table 1.3 shows the basic hydrological, economic and geographical parameters of the seven major river basins, which embrace virtually the whole of China proper. Taking the Yangzi River as the dividing line, the contrast between the North and the South is most striking. The South enjoys far greater water resources relative to both population and cultivated area. If other non-basin areas are considered as well, South China (including Tibet) makes up 80 per cent of the annual volume of surface run-off, compared with 18 per cent for the North. But this must be contrasted with the distribution of the nation's arable land of roughly 35 per cent and 65 per cent respectively, as well as with the fact that 55 per cent of the total population live in the South and 45 per cent in North China.[33]

Moreover, rainfall variability is considerable in the North China plains, averaging, apart from the coastal areas, 25 per cent. In the area farther north and west, it generally exceeds 30 per cent. It was observed in pre-war years that 'the crop yield in any region is unstable where the rainfall variability exceeds 20 per cent, and farming becomes a total failure where variability exceeds 40 per cent.'[34] Thus, agriculture in the North is in an

TABLE 1.3. Hydrological, economic and geographical indicators of the seven major river basins in China, 1982

	Basin area		Annual surface run-off		Cultivated area		Population		Per capita water availability		Average water availability per ha. of cultivated area	
	Million sq. km.	As % of national total	100 million m³	As % of national total	Million ha.	As % of national total	100 million persons	As % of national total	m³/year	As a ratio of national average	m³/100 ha./year	As a ratio of national average
	(1)	(2)	(3)	(4)	(5)	(6)	(7)	(8)	(9)	(10)	(11)	(12)
National total	9.60	100	26,300	100	99.33	100	9.7	100	2,700	1	263.0	1
(1) Yangzi River	1.80	18.8	9,790	37.2	24.67	24.8	3.45	35.6	2,840	1.05	397.5	1.51
(2) Pearl River	0.45	4.7	3,410	13.0	5.2	5.2	0.76	7.8	4,487	1.66	655.8	2.49
(3) Yellow, Huai Hai, and Liao Rivers	1.57	16.3	1,509	5.7	41.53	41.8	3.34	34.4	452	0.17	36.3	0.14
Yellow River	0.75	7.8	560	2.1	13.07	13.2	0.82	8.5	683	0.25	42.9	0.16
Huai River	0.27	2.8	500	1.9	12.53	12.6	1.25	12.9	400	0.15	39.5	0.15
Hai-Luan Rivers	0.32	3.3	292	1.1	11.33	11.4	0.98	10.1	298	0.11	25.8	0.10
Liao River	0.23	2.4	157	0.6	4.60	4.6	0.29	8.0	541	0.20	34.2	0.13
(4) Sunghuajiang River	0.55	5.7	760	2.9	11.67	11.7	0.47	4.8	1,617	0.60	65.1	0.25

Sources: Lei XiLu, *Woguo de Shuili Jianshe* (Water Conservancy Construction in Our Country) (Beijing: Nongye Chubanshe, 1984), 4.

inferior position as far as the amount and variability of rainfall are concerned compared with the southern provinces. According to Zhu Kezhen, the most distinguished Chinese meteorologist, if the isohyetal line of 500 mm could be shifted northward by 10 or even 5 degrees latitude, the north-western part of China would be able to support a considerably larger population, and the frequency and seriousness of natural disasters in that area would be greatly reduced.[35]

This study is not the place for a detailed examination of the agro-climatic conditions of the Chinese land mass. But the basic regional contours, as briefly sketched, should be sufficient to show why the multiple-cropped areas have been expanded much more rapidly in the South in the past three decades, why yield per hectare has been consistently higher in the South, and why the South as a whole contributes substantially more than the North to national grain output. By virtue of the 'risk-aversion function' of multiple-cropping, it should also be clear that its expansion in the South has served as an important stabilizing factor in the Chinese rural economy.

Nevertheless, to the extent that farm output is still subject to weather influence, a flood or drought occurring in the South will result in greater absolute losses and will thus show up in national aggregate output statistics more substantially than any comparable random disturbance in the North. Of course, the amplitude of such weather-caused output fluctuations may be smaller, compared with that occurring in circumstances in which multiple-cropping is less widespread, or agricultural technologies remain at a primitive stage. The points made here can be similarly applied to intraregional differences. Within the North in the broader sense, the North China plain is certainly very different from the North-west, for example, in terms of agro-climatic parameters, and geomorphological and edaphological structure.

Thus, in any attempt to interpret the changing national average scale of weather disturbances in terms of farm output, it is always important to check consistently against disaggregated regional studies of the weather–output relationships. Otherwise, the conclusions drawn may give a quite misleading impression of the relative impact of weather and human factors on agricultural stability in China.

In this context, the greater post-Mao policy emphasis on regional specialization in Chinese agriculture is noteworthy. Wide stretches of arable land devoted to grain cultivation, notably in North China, have been replaced by cotton and other cash crops, or afforestation and soil conservation in the drive to redress the previous pro-grain bias of the Maoist strategy.[36] Short-term fluctuations aside, such cropping shifts along the lines of regional comparative advantage may imply long-term instability as well. At least two points may be made in this respect. First, the reduced area may make grain production more vulnerable to random weather disturbance—a point that is by no means unimportant, considering the emerging significance of the North China plain as a grain region. Second, if the inverse 'subsistence urge' function holds true, then a higher per capita grain output in the areas concerned may result in increased peasant lethargy *vis-à-vis* the impact of natural disasters.

It may also be noted that during the initial post-Mao reforms in 1978–9 a number of large-scale schemes for regional specialization were formulated, involving, among other things, the abandonment of large stretches of the loess plateau considered unsuitable for grain cultivation. All of them apparently foundered, in part because of practical difficulties associated with procurement and transportation in order to bridge the grain requirements of these vast deficit areas, but also probably because of fears that any potential surplus from the grain regions might be easily wiped out by adverse weather.[37] This may explain why, no sooner had the twelve major regional commercial grain bases been proposed by the central authorities than they were modified in 1982–3 into a much more modest programme, comprising fifty narrowly defined *xian* grain bases.[38]

NOTES

1. James D. McQuigg, 'Climatic constraints in food grain production', in Frank Schaller (ed.), *Proceedings of the World Food Conference 1976, 27 June–1 July* (Ames, Ia.: Iowa State University Press, 1977), 390. Cf. also L. M. Thompson, 'Weather variability, climatic change, and grain production', in P. H. Abelson (ed.), *Food: Politics, Economics, Nutrition and Research* (Washington, DC: American Association for the Advancement of Science, 1975), 43.
2. Russell A. Ambroziak and David W. Carey, 'Climate and grain production in the Soviet Union', in *Soviet Economy in the 1980s: Problems and Prospects* (Pt. 2), Joint Economic Committee (JEC) of US Congress (Washington, DC: US Government Printing Office, 1983). See also David W. Carey, 'Soviet agriculture: recent performance and future plans', in *Soviet Economy in a New Perspective*, JEC US Congress (Washington, DC: US Government Printing Office, 1976), 575–99. The catastrophic loss in grain output in 1975 (total output down to 140 m. tonnes from an average of 190 m. tonnes for 1971–4), was also seen as a consequence of the severe droughts occurring in that year.
3. Peter B. R. Hazel, *Instability in Indian Foodgrain Production* (Washington, DC: International Food Policy Research Institute, 1982), 44.
4. McQuigg, 'Effective use of weather information in projections of global grain production', in Douglas Ensminger (ed.), *Food Enough or Starvation for Millions* (New Delhi: McGraw-Hill, 1977), 312.
5. McQuigg, 'Climatic constraints in food grain production', 390.

6. Elisabeth Oughton, 'The Maharashtra droughts of 1970–73: An analysis of scarcity', *Oxford Bulletin of Economics and Statistics*, 44/3 (Aug. 1982), 169–97.

7. The droughts are well documented not only by the Chinese themselves (see Appendix), but also by Western and Japanese sources as a world-scale climatic phenomenon, similar to the great droughts in the Soviet Union in 1972 and 1975, in the USA in 1974, and in West Africa which lasted 6–7 years starting in the late 1960s. See Yasoji Tsuboi and Jyunkichi Nemota (eds.), *Qixiang Yichang Yi Nongye* (Climatic Abnormality and Agriculture), in Chinese trans. from the Japanese by Lin Zhenyao, Sun Shenqing, Ji Yuanzhong, and Chen Enjiu (Beijing: Kexue Chubanshe, 1983), 156, 166, 168, and 179.

8. Y. Y. Kueh, 'A weather index for analysing grain yield instability in China, 1952–1981', *China Quarterly*, 97 (Mar. 1984), 75.

9. John L. Buck, *Land Utilization in China* (Nanjing: Nanjing University Press, 1937), 125.

10. Walter H. Mallory, *China: Land of Famine* (New York: American Geographical Society, 1926).

11. *Sunday Times* (30 Dec. 1984), 9.

12. For an elaboration of the relationship between the entire Stalinist development strategy as adopted in China and the organizational accommodations in Chinese agriculture, see Robert F. Dernberger, 'Agriculture in Communist development strategy', in Randolph Barker and Radha Sinha, *The Chinese Agricultural Economy* (Boulder, Colo.: Westview Press, 1982), 68–76. See also Benjamin Ward, 'The Chinese approach to economic development', in Robert F. Dernberger (ed.), *The Chinese Experience in Comparative Perspective* (Cambridge, Mass.: Harvard University Press, 1980), for the adoption of the Soviet model in general and adaptions made in China. A broader description of the institutional changes in rural China since 1952 is available in Carl Riskin, *China's Political Economy: The Quest for Development since 1949* (New York: Oxford University Press, 1987), 123–5, 170–4, 220–1, and 286–9.

13. The figure is cited by Anthony M. Tang, 'Policy and performance in agriculture', in Alexander Eckstein, Walter Galenson, and Ta-chung Liu (eds.), *Economic Trends in Communist China* (Chicago: Aldine Press, 1968), 466 from Oleg Hoeffding, 'State planning and forced industrialization', in *Problems of Communism*, 8 (Nov./Dec. 1959), 38–46.

14. Kueh, 'China's food balance and the world grain trade', *Asian Survey*, 24/12 (Dec. 1984), 1251.

15. Dwight Perkins, *Market Control and Planning in Communist China* (Cambridge, Mass.: Harvard University Press, 1966), 60.

16. For the development of the idea of a 'subsistence urge' and 'institutional hedge' see Kueh, 'Weather, technology, and peasants' organization as factors in China's foodgrain production', *Economic Bulletin for Asia and the Pacific*, 34/1 (June 1983), 15–26.

17. See Nicholas Lardy, *Agriculture in China's Modern Economic Development* (Cambridge: Cambridge University Press, 1983), ch. 2 for a good account of

34   *An Interpretative Framework*

the interchanging roles of planning and relative prices control in Chinese agriculture from the 1950s through the early 1980s.

18. See Robert Ash, 'The evolution of agricultural policy', and esp. Terry Sicular, 'Agricultural planning and pricing', both in Kueh and Ash (eds.), *Economic Trends in Chinese Agriculture: The Impact of Post-Mao Reforms* (Oxford: Oxford University Press, 1993) for a more detailed discussion of the changing role of agricultural prices in relation to the decollectivization drive during the post-Mao era.
19. Bruce Stone, 'Basic agricultural technology under reform', in Kueh and Ash, *Economic Trends in Chinese Agriculture*.
20. For the case of Taiwan see T. H. Lee, 'Food supply and population growth in developing countries: A case study of Taiwan', in Nurul Islam (ed.), *Agricultural Policies in Developing Countries* (London: Macmillan Press, for the International Economic Association, 1974), 184–5.
21. Buck, *Land Utilization in China*, 127.
22. Kenneth R. Walker, 'Organization of agricultural production', in A. Eckstein, W. Galenson, and Ta-Chung Liu, *Economic Trends in Communist China* (Edinburgh: Edinburgh University Press, 1968), 406–9. Cf. also Thomas Rawski, 'Agricultural employment and technology', in Barker and Sinha (eds.), *The Chinese Agricultural Economy*, 127–8.
23. Cf. Mark Elvin, 'The technology of farming in late-traditional China', in Barker and Sinha, *The Chinese Agricultural Economy*, 21–4.
24. This is illustrated by one of the most disastrous floods in the Yangzi River basin in summer 1980, when a massive amount of drainage equipment and electricity supplies, totalling 1.5 m. kW and 1.8 m. hp, were rapidly mobilized to drain a total of 30 bn. m^3 of water from 2.7 m. ha. of farmland; see *NYNJ 1981*, 126.
25. The national average multiple-cropping index increased from 131 in 1952 to 145 in 1983, with the 13 southern provinces taken together showing an increase from 152 in 1952 to 203 in 1979; see Kueh, 'A weather index for analysing grain yield instability', 77, and 'Technology and agricultural development in China: regional spread and inequality', *Development and Change*, 16/14 (1985), 549. Cf. also Walker 'Trends in agricultural production', in Kueh and Ash (eds.), *Economic Trends in Chinese Agriculture* for more recent developments in this respect.
26. Buck, *Land Utilization in China*, 125.
27. As a proportion of total sown area, the size of area covered by floods and droughts and sustaining a crop loss of more than 30% declined from an average of 63% and 40% in 1952–66 to 45% and 33% in 1970–83 respectively; see *TJNJ 1984*, 190.
28. Karl A. Wittfogel *Oriental Despotism* (New Haven, Conn.: Yale University Press, 1957); see also his 'Class structure and total power in oriental despotism', in E. Stuart Kirby (ed.), *Contemporary China*, iii (1958–9) (Hong Kong: Hong Kong University Press and Oxford University Press, 1960), 1–10.
29. Cf. Kueh, 'Population growth and economic development in China', in Hermann Schubnell (ed.), *Population Policies in Asian Countries: Con-*

temporary Targets, Measures, and Effects (Draeger Foundation, Germany and Centre of Asian Studies, Hong Kong University, 1984), 445–7.

30. Nanjing (Jinling) University (Dept. of Agricultural Economics, School of Agriculture), 'Zhonghua Minguo Ershinian Shuizai qiyi zi jingji tiaozha' (The 1931 flood in China: An economic survey), *JLXB* 2/1 (1932), 216.

31. This was clearly brought to light by Li Bening, Vice-Minister for Irrigation and Power Generation as early as late 1983. At the same National Rural Work Conference held in Nov.–Dec. 1983, after Wan Li (a Vice-Premier responsible for agriculture) delivered the keynote speech to prepare for the effective reparcellization of collective farmland (viz. the extension of peasants' leasehold right from three to fifteen years starting Jan. 1984), Li Bening explicitly spoke of the serious difficulty not only in increasing the irrigated areas, but also in maintaining existing irrigation facilities. See *XHYB* 1 (1984), 117–18.

32. Hazel, *Instability in Indian Foodgrain Production*, 9–10.

33. Jan Erik Gustafsson, *Water Resources Development in the Peoples' Republic of China* (Stockholm: Royal Institute of Technology, 1984), 41; see also Perkins, 'Constraints influencing China's agricultural performance', in *China: A Reassessment of the Economy*, JEC US Congress (Washington, DC: US Government Printing Office, 1975), 350–65.

34. T. H. Shen, *Agricultural Resources of China* (Ithaca, NY: Cornell University Press, 1951), 17.

35. Kezhen Zhu, *Zhu Kezhen Wenji (Collected Essays of Zhu Kezhen)* (Beijing: Kexue Chubanshe, 1979), 176. He made this point in a widely read article, 'Huabei zhi ganhan jiqi qianyin houguo' (Drought in North China and its causes and consequences) which was originally published in *DLXB* 1/2 (1934).

36. The national grain-sown area declined by 5.8% from 121 m. ha. in 1978 to 114 m. ha. in 1983; whereas the areas sown to cotton and oilseeds increased respectively by 3.5% (i.e. from 4.5 to 6.1 m. ha.) and by 19% (i.e. 7.1 to 8.4 m. ha.) in the same period. Note also that from 1979 to 1982, the combined grain-sown area for the three important cotton-producing provinces, Hebei, Shandong, and Henan, declined by 7.9% (compared with the national average of 5.8%), in favour of an impressive 69% increase in the cotton-sown area (compared with the national average of 29%); see *NYNJ 1980*, 100 and 107, and *NYNJ 1983*, 37–42.

37. Cf. *RMRB*, 26 Nov. 1978 and 13 Feb. and 22 Mar. 1979, for a lengthy discussion of the problems involved. The issue dominated academic circles for a few years in the early 1980s. See Zhang Weibang, 'Huangtu gaoyuan guotu zhengzhi fangan chutan' (A trial exploration into the problem for ameliorating loess plateau), in *SXDXXB* 3 (July 1984), 15–26 for a more recent survey and proposals for coping with the problems.

38. See Jiang Yishan and Liu Jianxing, 'Dui fazhan liangshi shangpin shengchan wentide tantao' (An exploratory study on the problems of developing commercial grain production), in *QSXK* 4 (Aug. 1983), 73–7 for a classification and discussion of the commercial grain bases; cf. also Huan Xiyi, 'Shilun lao shangpin liang jidi fazhan liangshi shengchan yi baohu nongye shengtai de

guanxi' (A trial discussion on the relationships between developing grain production and protecting agricultural ecology in the old commercial grain bases), in *JHXK* 2 (1984), 26–30, for concerns raised about the declining rates of grain exports from the old bases.

II

A History of Chinese Agricultural Instability since 1931

2

Methods and Problems of Measurement

Part II of our study examines long-term trends of agricultural instability in China from 1931 to 1984. We focus on the three major physical measures, i.e. total grain output, grain-sown area, and grain yield per sown hectare, in addition to the 'gross value of agricultural output' (GVAO) series, to show how the stability of farm output and input has changed over time. In other words, a particular time-period, say 1952–8, is compared with another period, say 1931–7 or 1979–84, in order to determine whether farm production has become more stable or unstable. Such a comparison between different periods promises to show historical trends and patterns of farm stability.

Three methodological problems are involved. The first concerns the definition of the periods. The second relates to the adequacy and relationships of the four measurement standards used. The third addresses the methodology of estimating the degree of farm instability for the periods chosen. All three problems are fundamental to our entire study of agricultural instability in China. They will therefore be discussed at some length here.

Periods Chosen for Comparison

The policy periods as defined in Table 1.1, i.e. 1949–52, 1952–7, 1958–60, 1961–5, 1966–76, and 1977 to the 1990s, plus the pre-war period 1931–7, provide a good basis for tracing the possible weather influence on farm stability relative to technological and institutional policy factors. Relevant factors are as follows.

First, the periods 1931–7 and 1952–7 are quite different in terms of rural organization. The former was dominated by private farming and market relations and the latter by a process of accelerated collectivization. The technological setting was, however, very similar. A comparison of the two periods may therefore help to show how, in the absence of major technological change, the radical institutional transition from the 1930s to the 1950s may have affected Chinese agricultural stability in relation to the weather influence.

Second, the different post-war periods differed in terms of the intensity of collective and bureaucratic control. Put simply, the remunerative approach alternated with the coercive approach of control within the same overall framework of rural socialization. Any observed variations in the degree of farm instability may reflect changing peasant incentives and behaviour, and so help to verify the comparative effectiveness of the 'institutional hedge' against weather disturbance.

Third, technological progress in agriculture did not gain momentum until the mid-1960s. However, both before and after this watershed, basic rural institutional and policy parameters were not entirely dissimilar for some of the periods shown in Table 1.1. This may facilitate a comparison of possible technological influence on farm stability by allowing some abstraction from institutional changes.

It is, however, not possible to adhere completely to the proposed periodization for the study of long-run farm instability. We have to omit entirely some of the periods and slightly modify others to suit the availability of either national or provincial grain statistics or weather data. This will become apparent in due course.

Standards used for Measurement

A distinction should first be made between the broad value measure of GVAO and the physical measure of national grain output. As shown in Figs. 2.1 and 2.2, the GVAO for which statistics are available in post-war periods, tends to vary closely with the grain output series. This suggests that the more comprehensive GVAO is largely dominated by the physical grain output measure, and may as such serve as an alternative for a comparative evaluation of farm instability in China.

Nevertheless, the relative growth rates of the GVAO and grain output series do differ from time to time. The gaps between the two measures widened at first in the 1950s in favour of grain output (Fig. 2.1). This reflects accelerated collectivization and increased concentration on grain production at the expense of non-farm activities, which are also included in GVAO. By the late 1950s, however, the same gap had markedly narrowed, for GVAO was then dominated by grain output. This trend was reinforced by the massive diversion of rural resources to solve the food crisis through 1962. The subsequent rural redecentralization resulted in increased economic diversification, gradually widening the gap from 1963, but in favour of GVAO rather than the grain output series. The Cultural Revolution with its pro-grain strategy nearly closed the gap once again, until it was forcefully reopened following the death of Mao, to bring about the most spectacular growth in rural non-farm industries from the late 1970s (Fig. 2.2).[1]

The changing quantitative relationship between GVAO and grain

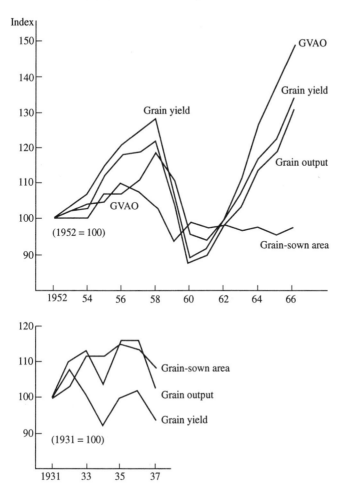

Fig. 2.1. Index of grain yield, grain output, grain-sown area, and gross value of agricultural output (GVAO) in China, 1931–1937, 1952–1966

Sources: Tables AB.2 and AB.12.

output consistently substantiates the points made in Table 1.1 about the periodic policy shifts which occurred after the early 1950s. For the purposes of this study, the inference is that amidst such changes, GVAO may fare differently as an instability measure. It may become less suitable, compared with the grain output series, for measuring the weather disturbance if an accelerated process of rural economic diversification results in increased size of the less weather—(or non-weather-) sensitive components of GVAO. The larger the size of the non-grain GVAO com-

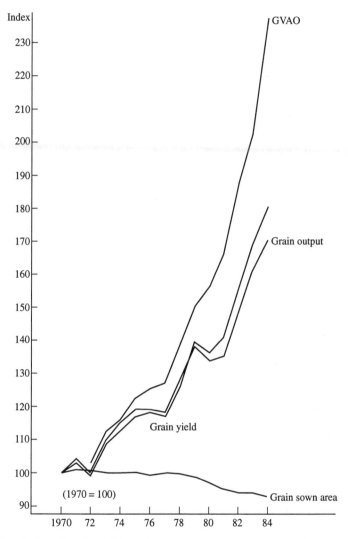

FIG. 2.2.   Index of grain yield, grain output, grain-sown area, and gross value of agricultural output (GVAO) in China, 1970–1984

Sources: Tables AB.2 and AB.12.

ponents, the greater tends to be the discrepancy in the degree of instability between GVAO and grain output measures.

Turning to the three physical measures, it should be noted that grain output is the product of sown area multiplied by yield. Thus output instability can be traced to the two latter variables. Both of them are of course subject to a host of factors—policy, institutional, and weather.

Nevertheless, apart from the pre-war years 1931–7 and the 1950s, which witnessed some expansion (Fig. 2.1), the grain-sown area remained quite stable for most of the post-war period, until marked cropping shifts took place in the early 1980s in favour of cash crops (Fig. 2.2).[2] Variation in yield per sown hectare therefore seems to be the key to understanding instability in grain output and the agricultural sector at large.

The instability of grain yield per sown hectare may reflect changing peasant incentives in terms of labour input and such essential current inputs as chemical fertilizers, combined with the long-run effect of technological change in agriculture. However, two other important factors must be mentioned in the case of post-war China. One is the marked increase in multiple-cropping, which by virtue of its 'risk-aversion function' may have helped to stabilize yield per hectare. The other is the changing qualitative composition of the national grain-sown area. Thus, less fertile land (expansion of grain area in the North-west, for example) may be substituted, in absolute or relative terms, for more fertile areas set aside for industrial use (perhaps in East and South China). This inevitably affects the national average yield per sown hectare.

The 'Instability Index'

How can the degree of agricultural instability be measured for different periods, given the four standards of measurement, i.e. GVAO, grain output, grain-sown area, and yield per sown hectare? Note, first of all, that grain-sown area apart, the other three measures exhibit a consistently upward trend for most of the post-war periods (Figs. 2.1 and 2.2). This is common to most economic time-series. It reflects steady progress in agricultural technology. Of course, both grain output and GVAO may also be expanded through increases in farm labour and sown area. This seems to be the case in the 1950s, when collectivization combined with rural migration control helped to raise the labour participation rate.

Whatever the factors sustaining the upsurges in output and yield, agricultural instability is initially best understood as the year-to-year fluctuations around the rising trend. Once determined, the yearly amplitudes of fluctuations can then be averaged to form a summary statistical measure of instability for any particular period. A good method of measuring the degree of instability in this way is by way of Coppock's index of instability which was originally designed to analyse export fluctuations and was extensively applied by Walker to study grain output and consumption instability in China.[3] Coppock's index measures instability in relation to the trend value. It has the advantage that it can be consistently applied to all the other measurement series, whether displaying (in Figs. 2.1 and

2.2) a falling trend or a basically horizontal one (e.g. the 1931–7 output and yield series).

Statistically, the index can simply be defined as the annual average of the yearly percentage deviations of output, yield, or sown area, from the respective trend values, ignoring the plus and minus signs. In the extreme case, it may assume the minimum of zero, implying perfect stability. Thus, the degree of instability varies positively with the computed index. Based on the Chinese experience of grain output fluctuations, Walker has defined four different categories of instability index: 'very unstable' (instability index above 6), 'unstable' (index 4–6), 'fairly stable' (index 1.5–4), and 'very stable' (index below 1.5).[4] Subject to certain qualifications, we shall make use of this classification in our subsequent discussions of agricultural instability in China.

It is worth recording that Walker's classifications were primarily designed to identify provincial instability for grain output only. It may not be appropriate to apply them to sown-area instability. More importantly, the amplitude of output and yield fluctuations are necessarily larger at the provincial level than at the national. Thus the four descriptive bands adopted by Walker may understate the computed national degree of instability. One could argue, for example, that an average annual fluctuation of 4 per cent, rather than 6 per cent (net of technological influence), during 1952–8, or 1970–7, should be considered as 'very unstable' from the national perspective of grain production.

Clearly, no simple theoretical principles can be used to designate what is 'fairly stable' or 'very unstable'. This is because the implications of agricultural output fluctuations are multi-faceted, affecting peasant consumption levels, the performance of industry and agricultural trade, the collection of State revenue, and its allocation for investment purposes. All variations should also be understood in relation to the existing stage of economic development and income level, as well as the chosen development strategy.

National and Regional Analysis

Our analysis of agricultural instability in China will be carried out at both national and regional levels, with a view to verifying the regional sources of national disturbances. Three regional levels are distinguished. The highest level is defined by the dichotomy between North and South China. This can also be interpreted in terms of a distinction between essentially northern crops (viz. wheat, maize, potato, soya bean, etc.), and the rice crops which dominate the southern provinces. This represents a cropwise regional differentiation.

The lowest level is represented by individual provinces, while the entire country is divided into seven familiar regions (the North-east, North-west, North, Centre, East, South-west, and South), to form the mid-layer. Each region comprises a number of provinces as referred to in the course of our analysis. No cropwise differentiation is made for provinces or regions. Instead, we use grain as the basis of analysis. This is necessary to keep the disaggregations at a manageable level.

Against this background, two major aspects of the differences between the regional or provincial approach and the broad regional cropwise analysis should be noted for ease of interpretation in the subsequent discussion.

The first concerns the fact that the grain-sown area or output of a province (and hence a region) is likely to embrace a variety of crops, as for example, both wheat and rice, in the Yangzi basin. These crops are planted in different seasons and may therefore be differently affected by the weather. For example, a shortfall in one crop may be compensated for by a good harvest in another. Thus, for reasons similar to the 'risk-aversion function' of multiple-cropping, annual regional or provincial all-grain output statistics may not reflect the full scale of any random temporal incidence of weather disturbance.

The second aspect of the problem is that the regional or provincial borders represent historically inherited administrative demarcations which may not coincide with any agro-climatic entity. This implies that the total impact of adverse weather, say floods, may be spread over a number of provinces (regions), without affecting any one in its entirety. In such cases, good harvests achieved in one part of the same province or region may balance out the losses incurred elsewhere. Because of this 'spatial risk dispersal effect', it may not always be possible to detect the random weather disturbance in our regional analysis.

The cropwise study of agricultural instability is the subject of Chapter 3, together with the national aggregate analysis, and is the first step towards tracing the regional evidence of disturbances. The problems of regional instability and their intraregional differences are discussed in Chapter 4. This also helps to verify provincial sources of regional fluctuations in farm output. In Chapter 5, we show how large-scale regional disturbances can generate episodes of great instability on a national scale.

NOTES

1. For the economic background to the diversification drive and a study of the early post-Mao period in this respect, see Kueh, 'China's new agricultural policy program', 358.

2. See Ch. 1 n. 36 for detail.
3. Kenneth R. Walker, *Food Grain Procurement and Consumption in China* (Cambridge: Cambridge University Press), 1984, and Joseph D. Coppock, *International Trade Instability* (New York and Farnborough: Saxon House, 1977), 4–10. The formula for the index is:

$$\sum_{t=1}^{N} = (Y_t - Y'_t)/Y'_t \cdot 100/N$$

where Y_t = the observed value of the variable in year t, Y'_t = the logarithmic least square estimate of the trend value for year t, and N = the number of years.
4. Walker, *Food Grain Procurement and Consumption in China*, 35–6.

3

National Trends and Patterns of Instability, 1931–1984

The computed instability indices for the national grain-sown area, yield, and output, are presented in Fig. 3.1 for 1931–7 and the different postwar policy periods. Breakdowns for the major grain crops are also given. The indices are plotted against the four categories of instability indices defined by Walker in order to suggest the relative magnitude of long-run changes in agricultural instability. Note that the periods are slightly different from those set out in Table 1.1. The modifications are necessary in order to align the national series with available weather and provincial data. This facilitates consistent verification of the national trends of instability against the regional sources of disturbance.

The discussion of long-run trends in national agricultural instability is divided into five sections. The first examines the possible impact of rural collectivization in the 1950s in relation to the pre-war period of 1931–7. The second compares 1970–7 with 1952–8 in order to show how enhanced rural control under the pro-grain strategy of the Cultural Revolution may have influenced agricultural stability in China. The third discusses emerging trends and patterns of instability under the accelerated process of agricultural decollectivization. The fourth gives a long-run review of the instability trends between the 1930s and 1980s in order to indicate institutional and technological influences. Finally, the fifth section uses the broad measure of GVAO to examine the changing degrees of instability of the broader rural economy in relation to the agricultural sector proper.

Radical Rural Reform between 1931–1937 and 1952–1958

Several important findings emerge from Fig. 3.1. First and most strikingly, for virtually all the adopted physical measures, i.e. sown area, yield, and output of wheat, rice, and all-grain, the computed instability index declined considerably from 1931–7 to 1952–8. In the absence of any significant technological progress in the 1950s, such consistent improvements lend support to our earlier contention about the role of

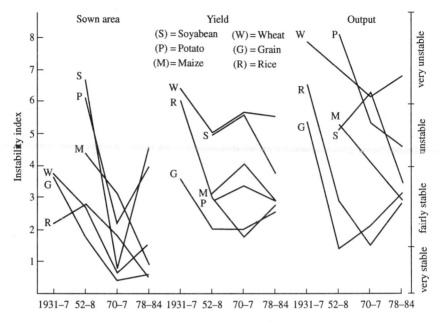

FIG. 3.1. Long-run trends in agricultural instability in China by major grain crops: sown area, yield, and output instability indices, 1931–1984

Sources: Tables AB.3 and AB.4.

collectivization as an 'institutional hedge' against agricultural fluctuations. This is bound up with the entire mechanism of bureaucratic control in rural areas, comprising the imposition of physical delivery quotas, migration restrictions, and compulsory labour days. Theoretically, improved stability could have coincided with smaller weather disturbance, but this was not the case in the 1950s.

Second, the only exception to the consistent decline in agricultural instability from the 1930s to the 1950s was the rice-sown area. Its instability index rose from 2.22 in 1931–7 to 2.85 in 1952–8. The difference is marginal, but the exception warrants an explanation. It is to be found in the severe Yangzi River floods of 1954, which rendered wide stretches of area unsowable.[1] In Hubei province alone the rice-sown area was reduced by 13 per cent in 1954.[2] The floods seem to have halted the growth of the total rice-sown area in both 1954 and 1955 when, in contrast to the abortive expansion, by over 14 per cent in 1956, there was hardly any increase. The pre-war period was also hostage to similar Yangzi River floods in 1931, and to disastrous Yangzi and Huai River basin droughts in 1934. But compared with 1952–8, rice-sown areas in the 1930s, covered by our statistics, represent the richer and more stable areas with good water and transport conditions, and are more accessible to statistical surveys.[3]

Third, the striking improvements in output stability from the 1930s to the 1950s represent, in all three cases—rice, wheat and all-grain—the combined effect of both sown area and yield stabilization (disregarding again the minor exception of the notably small rice-sown area instability index for 1931–7). Yields were, however, more susceptible to stabilization than sown area. In the case of rice, the degree of yield stabilization easily offset the slight increase in sown-area instability and generated the impressive decline in rice output instability. These relative trends in yield and sown-area stability imply that bureaucratic collective control in the 1950s fared comparatively better as a hedge against fluctuations in yield than in sown area. This is understandable: in the absence of modern flood prevention and drainage technology, mass mobilization in the 1950s could do little to stave off severe floods or drain inundated fields in time for sowing, although migration restrictions did help to stop the remaining arable area from being abandoned altogether. By contrast, once the area was sown, physical grain target control within the collective framework eliminated a host of fluctuating market and price factors which could have affected peasants' input incentives and hence yield per hectare in the 1950s.[4]

Fourth, the degree of both output and yield instability for rice declined to a much greater extent than in the case of wheat, from 1931–7 to 1952–8. This seems to suggest that the 'institutional hedge' present in the 1950s, worked less effectively in the wheat areas, because the geoclimatic conditions in northern regions (rainfall shortage and extremely large precipitation variability) placed greater limits on human manipulation. This rice–wheat contrast is all the more striking when we remember that the 1950s were comparatively wet, with flooding constituting the principal source of weather disturbance and occurring mainly in the southern rice areas, rather than in the northern wheat areas.

Fifth, for the all-grain measure, the instability indices of output, yield, and sown area are all consistently smaller than those for wheat and rice in both 1931–7 and 1952–8 (disregarding again the comparison with the 1931–7 instability index for rice-sown area). This contrast underlines the earlier point that the less aggregative the statistical measures are, the greater will be the degree of instability, and vice versa. Thus, we may expect the regional instability indices to be smaller than those for provinces, but greater than those for the country as a whole. This conclusion is borne out by the subsequent regional or provincial analysis.

These observations suggest that agricultural collectivization in the 1950s helped significantly to stabilize grain production. The degree of stabilization tends to differ, however, from region to region, depending on the geoclimatic limits to human intervention. Nevertheless, despite the evident stabilizing trends from 1931–7 to 1952–8, remaining instability can, in most cases, still be traced to weather disturbance.

Socialization and the Pro-Grain Strategy: 1952–1958 to 1970–1977

We now examine how within the same collective framework, rural policy variations and technological change may influence agricultural stability in relation to the weather influence. In addition to Fig. 3.1 which shows how the instability indices changed over the long run, i.e. from period to period, Fig. 3.2 shows short-run, i.e. year-to-year, fluctuations in grain-sown area, yield, and output within each period. The yearly percentage fluctuations given in Fig. 3.2 are calculated by taking the respective trend values as the base. Since the instability indices shown in Fig. 3.1 represent only the average of such 'detrended' yearly percentage fluctuations for each relevant period, Fig. 3.2 helps to identify the sources of inter-period variations in agricultural stability. Such a verification procedure is neces-sary, because after the dramatic decline between 1931–7 and 1952–8, instability indices between different post-war periods varied within a relatively small range, making it increasingly difficult to distinguish exactly the factors underlying these marginal variations.

It is against this same background that in addition to rice and wheat, the instability index is computed separately for maize, soya beans, and potatoes (Fig. 3.1), together with their year-to-year percentage fluctua-tions in sown area, yield, and output (Fig. 3.2). Note that the combined share of rice and wheat in total grain output was only 54 per cent in 1952–8 and 60 per cent in 1970–7, but that with the other three major grain crops, the share increased to around 80 per cent and 95 per cent respectively.[5]

Several interesting points emerge from Figs. 3.1 and 3.2. Some initial general observations are, however, in order. As shown in Fig. 3.1, the sown-area instability index for the five major grain crops declined con-sistently from 1952–8 to 1970–7. Their yield instability index moved, however, in the opposite direction. The only exception was that of rice which saw its yield instability index markedly reduced from 1952–8 (index 2.85) to 1970–7 (1.64), reversing the trend of increasing instability after 1931–7 (Fig. 3.1).

The consistent decline in sown-area instability from 1952–8 to 1970–7 outweighed the comparatively small increases in yield instability to result in reduced output instability for all crops. The only notable exception is

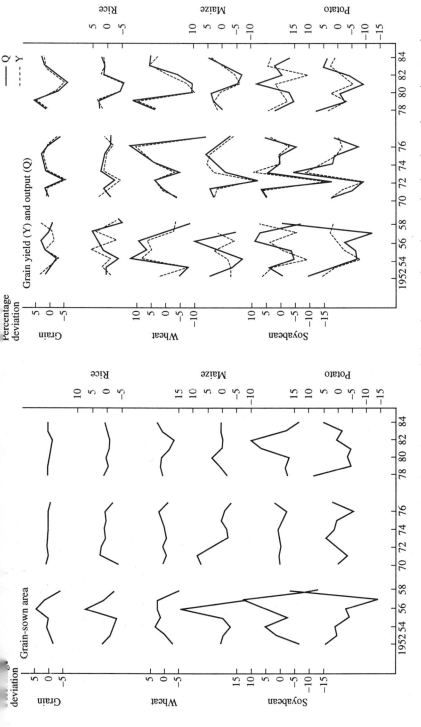

FIG. 3.2.   Yearly percentage fluctuations in grain-sown area, yield, and output from the trend value, by major grain crops in China, 1952–1984

Source: As Fig. 3.1.

that of soya beans. As with wheat, the yield instability index for soya beans increased only marginally from 1952–8 (5.05) to 1970–7 (5.71). Yet despite the fact that the sown-area instability index for soya beans fell dramatically (from 6.73 to 0.83), compared with that of wheat, its output instability index increased (from 5.18 to 6.29), while that of wheat declined (Fig. 3.1). A similar observation can be made in respect to the all-grain measure. Its sown-area instability index declined significantly from 1952–8 (1.81) to 1970–7 (0.40), but its output instability index increased, albeit in a less pronounced fashion (from 1.45 to 2.16), despite the fact that its yield instability index remained constant (at 2.02). An explanation of these trends is offered below.

First, the pervasive decline in sown-area instability from 1952–8 to 1970–7 undoubtedly reflects the continuous process of harnessing the major river streams and improving irrigation and drainage facilities.[6] Note that, with the minor exception of maize, the sown-area instability indices for all the northern crops (wheat, soya bean, and potato), became comparable with, or even smaller than that of rice (Fig. 3.1). This reflects the celebrated drive towards grain self-sufficiency in the North,[7] and indeed the entire pro-grain strategy of the Cultural Revolution, whereby rural resources were forcibly concentrated on grain production at the expense of cash crops and related non-farm undertakings. The stabilization trends are, in fact, very consistent, in that for all the grain crops the deviations of the sown area from the trend values are much smaller for virtually all of the years 1970–7 compared with 1952–8 (Fig. 3.2).

Second, the increases in yield instability for all the northern grain crops, except perhaps for wheat, from 1951–8 to 1970–7 (Fig. 3.1) can be attributed to the severe North China drought of 1972. This is clearly shown in the year-to-year yield fluctuations in 1970–7 (Fig. 3.2). Maize, soya bean, and potato all recorded very substantial negative yield deviations in 1972. Rice yields were also affected, as the drought affected part of the Yangzi River basin. The impact was, however, not as serious as for maize, soya bean, and potato, and the instability index for rice yields continued to decline from 1952–8 (index 3.01) to 1970–7 (1.81). Wheat yields were affected not in 1972, but in spring 1973 (Fig. 3.2), under the continuing impact of the same drought which had begun in summer 1972 after the wheat was harvested. Had it not been for the unusual incidence of drought in 1972–3, the 1970–7 yield instability indices for all these crops would certainly have further declined from the 1952–8 level. Note that the year-to-year yield fluctuations for virtually all the grain crops shown in Fig. 3.2 were, 1972–3 aside, much narrower in 1970–7 than in 1952–8.

The catastrophic 1972–3 drought also rendered wide areas of North

China unsowable for the autumn crops, predominantly maize. Where it could be sown, high-yield crops were substituted for lower-yield ones in order to minimize output losses.[8] This explains the sharply downward and upward sown-area deviations for maize and potato in 1972 and 1973 (Fig. 3.2). As a result, the decline in the sown-area instability index from 1952–8 to 1970–7 was not as pronounced as in the case of the other grain crops, as indicated above.

Third, turning to output trends, notice first that the output of virtually all grain crops fluctuated closely with yield per sown hectare throughout the years 1970–7 (see Fig. 3.2). With sown area remaining basically constant, grain output in 1970–7 followed yields, which were strongly influenced by the North China drought of 1972. By contrast, in 1952–8, yield reductions/increases tended to be accompanied by sown area expansions/contractions for many crops (Fig. 3.2). Such an 'area-for-yield'/'yield-for-area' substitution naturally helped to stabilize yearly fluctuations in output. This process, rather than any lack of weather disturbance, probably explains the lower degree of grain output instability in 1952–8 (index 1.45) compared with 1970–7 (2.16), when saturation in sown-area expansion left no room for any area manipulation. As shown in Fig. 3.2, the 'area-for-yield'/'yield-for-area' substitution was particularly consistent in the case of soya beans during 1952–8. As a result, despite significant yearly fluctuations in both its sown area and yield, compared with 1970–7 (hardly any changes in sown area, and only one major yield loss in 1972), soya bean output for 1952–8 was remarkably more stable (Fig. 3.1).

This is not the place to examine, in detail, the policy background to the sown-area fluctuations of 1952–8. Briefly, the greatest fluctuations occurred within the three years of 1956–8. The abortive expansion of the rice-sown area towards the North in 1956 is a well-known episode (cf. Fig. 3.2), as is the contraction in 1957 (partly a correction of the excessive sown-area expansion of 1956) and especially in 1958 (mainly as a result of the diversion of farm labour to support the rural non-farm crash programmes).[9] These shifts were all accompanied by grain yield movements in the opposite direction (Fig. 3.2). The contraction in 1958 was also coupled with massive cropping shifts from low-yielding crops, such as gaoliang and millet, to high-yielding potatoes.[10]

The observed substitution of area-for-yield/yield-for-area also helped to stabilize the yearly fluctuations in grain yield per sown hectare in 1952–8, with or without substituting potatoes for gaoling or millet. The reason is twofold. First, average yield per sown hectare is likely to be higher on newly-sown land than in areas already affected by natural disasters, especially if the expanded area has been planted with high-yield crops, as was the case in 1956. Second, the contraction in sown area normally took place at the expense of relatively infertile marginal land. This helped to raise the average yield of area sown.

In the context of 1952–8, there is yet another important (mainly statistical) factor which helped to stabilize the annual average yield in the wake of natural disasters. Grain yield per sown hectare is based on area sown net of farmland rendered unsowable by serious floods or droughts. To the extent that it is unlikely that crops planted within the same season should be similarly affected by bad weather, the realized yield is bound to be higher. The effect is similar to the policy-induced contraction of sown area. Such compensatory elements were clearly no longer present in 1970–7, especially against the background of massive collective, coercive mobilization of peasants during the Cultural Revolution, when, for example, the completion of planting in the wake of serious drought was mandatory for the maximization of grain output.[11] This explains why the degree of instability of grain yield—as well as output—in 1952–8 is relatively lower than in 1970–7. It also implies, however, that had it not been for the stabilizing 'area-for-yield' and 'yield-for-area' substitutions, the degree of yield instability would have been higher in 1952–8 than in 1970–7.

Nevertheless, the observed yield fluctuations during 1952–8, which happened to be associated with sown-area movements in the opposite direction, can in most cases still be identified with known weather events, in much the same way as the dramatic yield reductions during 1972–3. The sharp decline in rice, as well as all-grain yields in both 1954 and 1956 (Fig. 3.2) was clearly caused by the disastrous Yangzi River floods of 1954, and a similar disaster in 1956 which occurred south of the Yangzi in May, and in its lower reaches in the autumn. In many areas, notably in the middle reaches of the basin, the 1956 floods were followed by serious drought.[12] Likewise, the disastrous floods in the North-east in July and August 1957 probably accounted for most of the serious soya bean yield loss in that year, which again saw a very substantial compensatory expansion in soya bean-sown area (Fig. 3.2). Without the 'area-for-yield' substitution, the resulting yield instability in 1952–8 could have turned out to be even greater than in 1970–7.

A related point will serve to conclude this discussion of the comparative instability of Chinese agriculture between 1952–8 and 1970–7. One wonders how the highly coercive approach adopted in 1970–7 might have affected peasant incentives *vis-à-vis* farm stability. Intensive bureaucratic physical control helped to ensure adherence to the imposed sown-area targets—hence the remarkable area stability—but once sown, peasant apathy could make yield per hectare more susceptible to weather influence in 1970–7 than in the less harsh conditions of 1952–8.

This question touches on the relative intensity and effectiveness of collective cohesion during the Cultural Revolution. As shown earlier, the yield instability index for rice did decline from 1952–8 to 1970–7, and, for almost all the northern crops, the amplitude of year-to-year yield deviations was also reduced in the latter period, apart from abrupt fluctuations in 1972–3. It is worth asking whether the losses would not have been much more substantial had a drought on a disastrous scale similar to that of 1972–3 occurred in the 1950s. On balance, it seems more likely that such 'hard' factors as weather and technology rather than speculative assumptions about peasant incentives and behaviour explain the short-run yield fluctuations and the long-run stability trends between 1952–8 and 1970–7. The coercive resource mobilization and its concentration on grain production during 1970–7 may obviously count as another 'hard' factor in this respect. The richer rice area, which traditionally enjoyed a greater degree of economic diversification compared with the North, could certainly rally greater rural resources to comply with the requirements of the pro-grain strategy for stabilizing and maximizing grain yield and output. This may also help to explain why the yield instability index for the northern crops was still consistently higher than that for rice in 1970–7 (Fig. 3.1).

The Transition to Decollectivization: 1970–1977 to 1978–1984

How may the radical post-Mao reforms have affected agricultural stability? The divergent trends of the instability indices for the various grain crops from 1970–7 to 1978–84 (Fig. 3.1) suggest a mixed picture. Thus, declining output instability indices for maize, soya bean, and potato contrast with increased instability indices for wheat, rice, and the all-grain measure. But how are the divergent trends related to the variations in sown-area and yield instabilities? And how important was the weather as an instability factor in the wake of years of technological advances in Chinese agriculture? Several points may be made.

First, except for maize, the sown area for the other three northern grain crops, i.e. wheat, soya bean, and potato, became less stable in 1978–84 compared with 1970–7 (Fig. 3.1). This clearly reflects the cropping shifts which took place with particular force in 1981 and 1982 (Fig. 3.2), as a result of the deliberate policy to correct the excesses of the pro-grain strategy of the Maoist period.[13] Table 3.1 shows the year-to-year changes in sown areas and the process of intercrop area substitution in 1978–84. Laxity in crop acreage control as a result of the decentraliza-

TABLE 3.1. Changes in sown area and intercrop area substitutions in China, 1978–1984: reduction or increase since 1978 (1,000 ha.)

	Grain	Economic crops	Ratio (2)/(1)	Rice	Wheat	Maize	Soya bean
	(1)	(2)	(3)	(4)	(5)	(6)	(7)
1978	120,587	1,444		33,421	29,183	19,961	7,144
1979	−1,325	+327	0.25	−548	+174	+171	+103
1980	−2,029	+1,154	0.57	+6	−129	+262	−20
1981	−2,276	+1,639	0.72	−584	−921	−970	+797
1982	−1,495	+1,233	0.82	−233	−351	−881	−59
1983	+585	−1,033	1.77	+65	+1,095	+281	−851
1984	−1,163	+1,527	1.31	+42	+527	−287	−281

	Potato	Cotton	Peanut and oilseeds	Combination of crops		
				(6) + (8)	(7) + (10)	(4) + (5)
	(8)	(9)	(10)	(11)	(12)	(13)
1978	11,796	4,867	4,367			
1979	−844	−355	+468	−374	−673	+571
1980	−799	+408	+335	−123	−537	+315
1981	−533	+265	+1,103	−1,505	−1,503	+1,900
1982	−251	+643	+265	−584	−1,132	+206
1983	+32	+249	−668	+1,160	+313	−1,519
1984	−414	+846	−35	+569	−701	−316

Notes: The correlation between the variations in sown area for cash crops and grain crops may be estimated by the following simple regression equations. Three major points may be made: (1) while the increases in sown area for all economic crops combined are closely related to the decreases in grain-sown area, the substitution of oil-bearing crops (rather than cotton), for grain crops seems to be particularly pronounced, indicating the intensive demand for improving the dietary structure in the post-Mao era; (2) the demand for such an intercrop substitution seems to be stronger, as might have been expected, in the northern wheat area than in the southern rice area; (3) the positive correlation between the variation in sown area for soya bean and peanut and oilseed suggests that soya bean was treated as oil-bearing, rather than a grain crop.

Sources: Tables AB.3 and AB.4, and *TJNJ 1990*, 357.

Cash (X) for grain (Y) crops substitution

(Y)/(X)	Estimated regression equations	t-value	r^2	t-significance
(1)/(2)	$Y = -574.07073 - 0.878600X$	3.737	0.7773	0.0202
(6)/(9)	$Y = -50.85518 - 0.544197X$	0.852	0.1535	0.4423
(8)/(9)	$Y = 0.58567 + 0.342924X$	0.937	0.1799	0.4019
(12)/(9)	$Y = 386.59026 - 0.565440X$	0.426	0.043	0.6921
(4)/(10)	$Y = -108.45224 - 0.409601X$	2.726	0.6500	0.0527
(5)/(10)	$Y = 340.73245 - 1.123566X$	5.230	0.8724	0.0064
(6)/(10)	$Y = -98.99049 - 0.565434X$	1.413	0.3330	0.2305

TABLE 3.1. *cont.*

(7)/(10)	$Y = -275.42140 + 0.913848X$	22.045	0.9918	0.0081
(8)/(10)	$Y = -376.87509 - 0.373126X$	1.730	0.4279	0.1587
(12)/(10)	$Y = -475.86558 - 0.938560X$	3.969	0.7975	0.0165
(12)/(13)	$Y = -610.65105 - 0.491870X$	4.048	0.8038	0.0155
(11)/(13)	$Y = 11.40993 - 0.799879X$	7.288	0.9300	0.0019

tion drive in the early 1980s may also have allowed income-maximizing peasants to expand area sown to cash crops such as cotton and oilseeds, at the expense of grain, notably wheat, potato, and, not least, maize.[14]

Second, the maize-sown area also fluctuated with rural institutional and policy changes during 1978–84. The fluctuations were, however, not as pronounced as those caused by the extraordinary drought of 1972–3. This explains why the sown-area instability index for maize could still be reduced from 1970–7 (index 3.27) to 1978–84 (0.97), so maintaining its long-term declining trend since 1952–8 (Fig. 3.1). Note, however, that despite the reversal towards a greater degree of instability, the comparable indices for wheat, soya bean, and potato for 1978–84 were still consistently lower than the respective figures for 1952–8. This suggests that sown-area fluctuations resulting from deliberate cropping shifts in 1978–84 were more limited than in the 1950s, when in the absence of technological progress, extensive areas of farmland could be rendered unsowable by serious floods and droughts.

Third, in contrast to the northern crops, maize aside, the sown-area instability index for rice continued to decline from 1970–7 to 1978–84 (Fig. 3.1). This indicates that the rice sown area was less subject to manipulated cropping shifts than in the North, where the post-Mao economic strategy dictated the reinstatement of traditional crops, namely cotton, soya bean, peanuts, etc. (cf. Fig. 3.2 and Table 3.1). More importantly, the reduced instability in rice-sown area in 1978–84 was achieved in the face of the catastrophic floods in the Yangzi River basin, especially in the middle reaches in 1980, and again in the Sichuan basin in 1981.[15] The floods were comparable in scale to those of 1954 and 1931, but potential sown-area losses were avoided by large-scale mobilization, not in the main of labour for dyke protection, but rather of powered water-siphoning and drainage equipment.

Fourth, turning to yield instability, the most striking finding is that the index for rice increased markedly from 1970–7 (index 1.81) to 1978–84 (2.73), so reversing its previous long-run decline since the 1930s (Fig. 3.1). By contrast, virtually all the northern crops saw their yield instability indices reduced in 1978–84 from the relatively high levels of 1970–7 (caused by the North China drought of 1972). The only minor exception was that of wheat, for which the comparable index hardly dropped at all (Fig. 3.1). These divergent yield instability trends between rice (increasing) and the northern dry crops (decreasing) from 1970–7 to 1978–84 are all the more noteworthy, if we compare them with the comparable transition between 1952–8 and 1970–7, when declining rice yield instability was exactly offset by the increased instability of all the northern crops (Fig. 3.1). The main reason for increased rice yield instability is readily apparent from Fig. 3.2, which shows abrupt negative fluctuations in rice yield in both 1980 and 1981 as a result of the floods in the Yangzi and Sichuan basins. The northern crops, except for wheat, were, however, relatively free of weather disturbance in 1978–84, compared with 1970–7, and so displayed declining yield instability. Admittedly, wheat yields were adversely affected in 1980 (Fig. 3.2) by widespread spring drought, coupled with two serious late spring frosts in mid-April and mid-May of the same year: hence the relatively high yield instability index for 1978–84 (Fig. 3.1).[16]

Fifth, with the wheat yield instability index in 1978–84 remaining nearly as high as in 1970–7, the increase in the wheat output instability index between 1970–7 and 1978–84, shown in Fig. 3.1, was mainly a result of the sharp sown-area adjustments made in the early 1980s, which as already mentioned, led to a greater degree of sown-area instability for wheat, among northern crops, in 1978–84. For maize, soya bean, and potato, however, the observed decline in yield instability in 1978–84 outweighed, or reinforced in the case of maize, the increase/decline in sown-area instability and so generated output instability from 1970–7 to 1978–84 (Fig. 3.1). Rice presents a very different story in that improved sown-area stability was greatly offset by increased yield instability, thereby contributing to a considerable increase in rice output instability.

Sixth, in terms of the all-grain measure, the marked increase in output instability from 1970–7 (index 2.16) to 1978–84 (3.18) (Fig. 3.1) clearly represents the combined effects of deteriorating output instability of rice and wheat, given the overwhelming preponderance of these two crops in total grain production. That is, overall grain output became less stable in 1978–84, partly as a result of drastic post-Mao cropping policy shifts (wheat area fluctuations), and partly because of the Yangzi floods in 1980 and 1981 (serious rice-yield losses), the latter factor carrying a greater weight than the former, in the light of the predominant share of rice output. Had it not been for the contribution of modern flood-fighting technology in stabilizing the rice-sown area during the 1980–1 Yangzi floods, the fluctuations in rice output, and hence the instability of overall grain output for 1978–84 as a whole, would certainly have been much more pronounced than shown in Figs. 3.1 and 3.2.

Finally, what influence might the post-Mao rural organizational reforms have had on agricultural stability in China in terms of sown area, yield, and output? All the reform measures, especially those which followed the dismantling of the commune system, point to a reparcellization or quasi-reprivatization of collective farmland in support of a greater degree of rural marketization.[17] Should sown area and yield have become less stable in the 1980s, compared with the highly centralized rural conditions of the 1970s, as a result of greater pursuit of financial and income benefits by peasants? It is difficult to estimate such influences. Nevertheless, the evidence of weather conditions is too clearly visible to be nullified by changes in peasant behaviour.

Thus, against the background of a uniform decollectivization drive throughout China, it is difficult to explain, without resorting to the weather factor, why, during 1978–84 there was improved stability for all the northern crops, with the minor exception of wheat, but increased rice yield (and output) instability. Undoubtedly, the Yangzi and Sichuan basin floods in 1980 and 1981, affecting rice crops, and the spring droughts in 1980, and to a lesser extent in 1981, affecting wheat crops only, explain most, if not all the observed discrepancies of yield and output instabilities in 1978–84. The situation is in fact similar to the highly coercive rural context of 1970–7, except that in the later period the 1980–1 floods played the role of the 1972–3 North China drought to reverse the divergent instability trends between rice and the northern crops.

A Long-Term View: 1931–1937 to 1978–1984

We have shown that grain output instability in different periods can be consistently traced to the intercrop variations in sown area and yield, and that the weather and farm technology are implicated in the changing magnitude of such variations. From a current vantage point, a summary evaluation of long-run trends and patterns of instability in grain production can be made, which may also serve as a proxy for agriculture as a whole.

First, in terms of overall grain-sown area instability, the two later periods, 1970–7 (index 0.40) and 1978–84 (index 0.61), are quite similar in that both saw the sown areas for virtually all individual grain crops, and so for grain as a whole, became significantly more stable than in 1952–8 (index 1.81). This underlines the fact that China has gradually strengthened its irrigation and drainage infrastructure as a hedge against possible area losses from serious droughts and floods. Nevertheless, grain-sown area instability in 1952–8 still compares very favourably with that of 1931–7 (index 3.53). The implication is that in the absence of technological progress, the collective institutional hedge was already at work in the 1950s to mitigate sown-area fluctuations.

With sown area remaining remarkably stable, grain output in 1970–7 and 1978–84 varied mainly with yields per sown hectare. The North China droughts of 1972–3 and the Yangzi and Sichuan basin floods in 1980 and 1981 nevertheless caused large-scale yield losses. As a result, the degree of instability of grain output became even higher in both 1970–7 (index 2.16) and 1978–84 (index 3.18) than in 1952–8 (index 1.45). By Walker's standard, it had moved from the category of 'very stable' (index below 1.5) during 1952–8 to that of 'fairly stable' (index 1.5 to 4) in both the later periods. The seriousness of the situation in 1978–84 was compounded by massive cropping shifts in the North, hence the greater degree of grain output instability compared with 1970–7. However, the grain output instability indices for all three post-war periods remain far below the 1931–7 record (index 5.4).

As for the degree of grain yield instability, the period 1952–8 (with an index of 2.02) is, even in the absence of technological advances, no less stable than 1970–7 (index 2.02), and is even somewhat more stable than 1978–84 (index 2.56). But all three post-war periods are substantially more stable than the pre-war period 1931–7 (index 3.57). The marketized rural context of pre-war China clearly lacked not only the benefit of collective coercion, but also the necessary technological means to help stabilize yield fluctuations.

The comparatively low instability of grain yield and grain output in 1952–8 should, however, be understood against the background of stabilization of yield or output through 'area-for-yield'/'yield-for-area' substitution, rather than attributed to any lack of weather disturbance. In fact, if we look at rice alone, the degree of yield instability was greater for 1952–8 (index 3.01) than 1978–84 (index 2.73), both periods being hostage to the Yangzi River floods in 1954 and 1980–1 respectively. Even so, these figures compare very favourably with the extremely high rice yield instability for 1931–7 (index 6.06), during which a similar Yangzi River flood (1931) and widespread summer and autumn droughts in both the Yangzi and Huai River basins (1934) occurred.

The long-run significance of the institutional and technological hedge in the post-war period, relative to possible weather influence, is also highlighted by the changing comparative yield instabilities of rice and wheat. The instability index for wheat was consistently higher than that of rice throughout the pre-war and post-war periods (Fig. 3.1). This reflects the fact that agro-climatic conditions, especially rainfall variability, are much less favourable in the North than in the South. Yet while in 1931–7 the yield instability of rice was brought, by the bad weather occurring in 1931 and 1934, to a level (index 6.06) nearly as high as that of wheat (index

6.45), the comparable discrepancy between the two crops widened substantially in favour of rice in all the post-war periods (Fig. 3.1). This happened despite disastrous floods in 1954 and 1980–1 and droughts in 1972, which were no less severe than those of 1931 and 1934. The post-war institutional and technological edge was quite effective in areas susceptible to man-made policy manipulation.

Finally, let us return to the question of the relationship between peasant aspirations and behaviour, and agricultural instability. It has been shown that, from the point of view of long-run trends or short-term fluctuations, weather, technology and institutional change are *all* reflected in agricultural production in China since 1931. This does not of course negate the potential significance of peasant behaviour, under the changing influence of a sometimes coercive, sometimes cohesive collective framework, as a source of instability. It merely highlights the importance of determining the scale, and impact, of changing weather conditions during the various periods under investigation. This task defines the focus of Part III. Meanwhile, our interim judgement is that against the background of prevailing technological and institutional conditions, the weather—not behavioural factors—provides the best (if not a total) explanation of observed agricultural fluctuations.

Instability of the Rural Economy at Large: 1952–1958 to 1978–1984

We now use the gross value of agricultural output (GVAO) measure in order to examine how fluctuations in the overall rural economy compare with those in the agricultural sector proper, as represented by grain production. Note again that GVAO embraces, besides grain and other cropping activities, forestry, animal husbandry, fisheries, and sideline production. The latter has come to be increasingly dominated by small-scale rural non-farm industries, at the expense of the cropping share in the GVAO, as shown in Fig. 3.3. However, in view of the overwhelming share of the cropping branch which was in turn dominated by grain production throughout the periods under study, GVAO can be expected to vary closely with grain output. This is borne out by the observed year-to-year fluctuations, shown in Fig. 3.4, of both grain output and the GVAO in 1952–8, 1970–7, and 1978–84 (GVAO data for 1931–7 are not available). Another factor which enhances this close correlation derives from sideline production's heavy reliance on raw material supplies from the cropping sectors. Moreover, fluctuations in grain production may also affect feed supplies, and hence the output of animal husbandry.

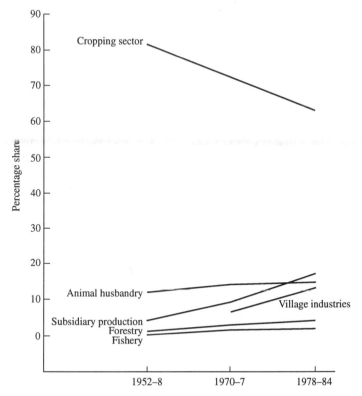

Fɪɢ. 3.3. Percentage shares of the different agricultural branches in the gross value of agricultural output in China, 1952–1984

Sources: *TJNJ 1980*, 151; *1985*, 241, and *1990*, 335.

Thus, the instability indices of GVAO and grain output moved consistently in the same direction as shown in Fig. 3.4. One or two further points may be mentioned briefly.

First, grain output has been consistently less stable than GVAO. This is understandable, for the latter is more comprehensive in its coverage and contains a number of less weather-sensitive components, such as forestry and non-farm industries, compared with grain production. With few exceptions, GVAO declined or increased from year to year more slowly than grain output in all three periods, 1952–8, 1970–7, and 1978–84 (Fig. 3.4).

Second, the discrepancy between grain output and GVAO instability widened in favour of GVAO in 1970–7 (index 2.16 vs. 1.31), compared with 1952–8 (index 1.45 vs. 1.14), and 1978–84 (index 3.18 vs. 3.08).

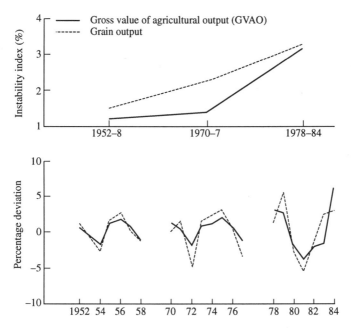

FIG. 3.4. Comparative instability indices of the gross value of agricultural output and grain output, and their yearly percentage fluctuations from the trend values, 1952–1984

Sources: Tables AB.2 and AB.12.

This underlines the fact that under the pro-grain strategy and the associated ideology of the Cultural Revolution, non-farm industries, among other non-grain branches, were increasingly reduced to production activities other than the processing of agricultural raw materials for consumption purposes. The proliferation of the 'five small industries' (i.e. farm machinery, cement, chemical fertilizers, metal-manufacture, and energy), clearly symbolized this trend.[18] These are basically weather-neutral industries, which helped to stabilize GVAO relative to the fluctuations in grain output. By contrast, the emergence of the 'new five small industries' (i.e. cotton-spinning, crocheting, cigarette-manufacture, sugar-refining, and wine-brewing), designed to reap the price and income benefits offered by the post-Mao policy reforms,[19] made GVAO more sensitive to output changes in the cropping sectors, and so generated a closer relationship between GVAO and grain output instability in 1978–84. The same applies of course to the 1950s, when the cropping sectors constituted the largest share of GVAO in 1952–8, compared with 1970–7 and 1978–84 (Fig. 3.3).

Finally and most importantly, the rural economy as covered by GVAO, was still vulnerable to weather disturbance in the 1980s, as in the earlier periods. Note, for example, the abrupt fluctuations in GVAO in 1980 and 1981 in response to the Yangzi and Sichuan basin floods, and the comparable phenomenon in 1972 (the North China drought) and 1954 (the Yangzi flood).

In short, disturbance in the agricultural sector proper tends naturally to spill over to other parts of the rural economy, excluding basically weather independent economic branches, such as building construction, transportation, and non-farm-associated rural industries which are, in addition to GVAO, also covered by the new broader Chinese measure of rural social value output (RSVO). It is to be expected that accelerated economic diversification since the early 1980s will increasingly help to convert the non-cropping or non-grain branches into a system *sui generis*, making GVAO less sensitive to weather changes, compared with grain production.

NOTES

1. The 1954 flood in China is well documented. It was a year of world-wide climatic abnormality. For example, Chicago city was also seriously flooded, as was Wuhan, capital of Hubei province. For this and other records of major floods and droughts referred to in this study, interested readers are referred to the detailed examination given in the Appendix A. For an overview of the temporal and spatial distribution of floods and droughts see Figs. AA.2 and AA.3. Descriptive verbal accounts specific to their occurrence at provincial and local levels are available in most cases in relevant provincial newspapers, national press, and yearbooks as cited in connection with the disaster-covered area statistics shown in Tables AB.14 and AB.16. For the 1954 flood, good summary accounts are available from the former Hong Kong-based Union Research Institute as published in *ZGZK* 7/5 (2 Aug. 1954), 9–12; 13 (27 Sept. 1954), 10–13; 8/10 (6 Dec. 1954), 12–14; and 9/2 (10 Jan. 1955), 16–18 and 40–2. In the standard texts of provincial geographical surveys published in the 1980s, the 1954 floods which had their most serious impact in Hubei, Anhui and Hunan (cf. Table 5.1) were also singled out as a classic case, together with the 1931 Yangji-Huai basin floods; see e.g. *HBNYDL* 17, *AHNYDL* 15; and *HNNYDL* 18.
2. See K. R. Walker, *Foodgrain Procurement and Consumption in China*, 222.
3. Cf. Thomas B. Wiens, 'Agricultural statistics in the People's Republic of China', in Alexander Eckstein (ed.), *Quantitative Measures of China's Economic Growth* (Ann Arbor, Mich.: University of Michigan Press, 1980), 290 and 301.
4. For an elaboration of how the physical output targets worked to control farm production in the 1950s see Dwight Derkins, *Market Control and Planning in Communist China*, 68–9.

5. Same sources as Tables AB.3 and AB.4.
6. For an earlier study of the Chinese effort to tame the major rivers see Leslie T. C. Kuo, *The Technical Transformation of Agriculture in Communist China* (New York: Praeger, 1972), 72–9. For a brief special report (in Chinese), see *MBHK* 13 Oct. 1982, 5.
7. Cf. Walker, 'Grain self-sufficiency in North China, 1963–1975', in *China Quarterly*, 71 (1978), 555–90.
8. Y. Y. Kueh, 'A weather index for analysing grain yield instability in China, 1952–81', 78, and the original sources given therein. Such forced intercrop area substitutions have been a widespread practice throughout China in both the post-war and pre-war periods. Thus it was observed in Sichuan, e.g. in the 1930s, that the ratio of coarse grain to paddy fields was negatively correlated with the amount of precipitation, rather than being determined by any changes in their relative prices; for a detailed dicussion see *DJYJS* series 45 (*Sichuan* vol.), Taipbei: Dec. 1967, 25157–8.
9. Walker, *Foodgrain Procurement and Consumption in China*, 132–3.
10. Ibid.
11. The Chinese press at the time was full of reports about how peasant masses were mobilized (or coerced) into a phenomenal fight-drought operation. Apart from Chinese sources about the familiar Dazhai Brigade, information about the various extremely labour-intensive techniques applied (e.g. to press out the subsoil moisture by massive stepping-*zhenya*) are available in such standard manuals as *KXZTSC*, *NYKXJSSC*, and *NYSCJSSC*.
12. The 1954 and 1956 disasters are also noted by Walker, *Foodgrain Procurement and Consumption in China*, 31 and 34, with statistics on the farm area affected, extent of output losses sustained, and population suffering.
13. The readjustment was of course initiated prior to 1981; see Yu Guoyao, 'Some problems on adjusting the structure of Chinese agriculture', in *NYJJWT* 1 (Jan. 1981), 24–8.
14. Cf. Terry Sicular, 'Ten years of reform: progress and setbacks in agricultural planning and pricing', in Kueh and Ash, *Economic Trends in Chinese Agriculture*, 47–96.
15. The 1980 Yangzi River floods affected virtually all the provinces within its basin, except perhaps Sichuan; see *NYNJ 1980*, 26. The events together with the disastrous spring droughts in North China were at the time widely noted outside China; see *South China Morning Post* (11 Mar. 1981), 9. The number of people affected in Hebei (by the drought) was given as 23 million (15 million most seriously), and in Hubei (by the floods) 20 million (6 million). The Chinese government sought relief aid from the UN for the first time; see *SCMP*, ibid. The 1981 Sichuan basin floods were equally destructive and also received sensational press coverage in and outside China; see *SCMP* (18 July 1981), 4. The floods also led to extensive discussion of the long-term problems of environmental disruption in terms of deforestation, soil erosion, and over-reclamation (from lakes) in Hubei and Huhan provinces in particular, see e.g. Luo Ruda 'Jingti da zirande baofu—cong qunian woguo de teda shuizai xiqu jinyuan' (Beware of the revenge of great Nature—learn a lesson from the extraordinary floods in China of the last year), in *XXPL* 1 (Jan.

1982), 52–3; Wu Zhonglun, 'Further discussion on the relation of forest and floods', in *NYJJWT* 3 (Mar. 1982), 34–5; and He Siwei, 'Relationship between forests and water conservancy—an afterthought on the ravages of flood in Sichuan provinces', *NYJJWT* 12 (Dec. 1981), 3–8. However, given the weather patterns, the floods could have occurred without the creeping ecological disturbance which took place over the years.
16. *NYNJ 1981*, 71; *JJNJ 1981*, IV–11; and *South China Morning Post*, 11 Mar. 1981.
17. For an elaboration on this point, see Kueh, 'The economics of the "second land reform" in China', in *China Quarterly*, 101 (1985), 122–31.
18. See Kueh, 'China's new agricultural-policy program: major economic consequences, 1979–1983', 365–6. For the broader background of the development of rural non-farm industries prior to the recent reforms, see Kueh, *Economic Planning and Local Mobilization in China* (London: Contemporary China Institute, School of Oriental and African Studies, 1985).
19. Kueh, 'China's new agricultural-policy program', 365.

4

Regional Variations

This chapter examines regional trends in agricultural instability and its intraregional differentiation in terms of changes in provincial instability indices from 1931–7 to 1979–84. We focus exclusively on the three physical grain measures, i.e. sown area, yield per sown hectare, and total output, in order to show, first, how interregional differences in the degree of instability changed over time and whether there was an equalizing trend as a result of technological progress and enhanced collective control in the post-war years. In the second section, we examine the various provincial trends in each of our seven regions, with a view to verifying the sources of change in regional instability trends.

A Cross-Regional View

The computed regional instability indices for grain-sown area, yield, and output for the different periods, 1931–7, 1952–7, 1974–9, and 1979–84, are plotted in Fig. 4.1. Note that two of the three post-war periods, namely 1952–8 and 1978–84, which have already been used for purposes of national analysis, have been slightly modified, each being reduced by one year (1958 and 1978). The modification is dictated by the availability of the regional data. The missing standard period 1970–7, which was used to represent the pro-grain strategy of the Cultural Revolution, has in turn been replaced by a new period, 1974–9. The new period is in fact a poor substitute, for it omits the North China drought of 1972 which was so important in our national analysis. Moreover, in order to make it comparable in length to the other periods, it has been extended to the year 1979. As such, it embraces the initial post-Mao reforms, designed to eliminate the excesses of the pro-grain strategy. For these reasons we shall merely use this questionable period as a side reference, where necessary.

To adhere consistently to the procedure adopted in the previous analysis, we examine the long-run agricultural instability trends in successive order of grain-sown area, yield, and output stabilities.

Grain-sown area instability

It is clear that fluctuations in grain-sown area in nearly all regions were substantially stabilized from 1931–7 to 1952–7, as shown in Fig. 4.1. This

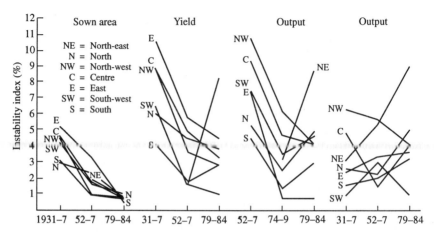

Fig. 4.1. Regional trends of agricultural instability in China by grain-sown area, yield, and output, 1931–1984

Source: Table AB.11.

is closely in line with observations made at the national level. It under-lines, once again, the particular edge of the emerging collective coercion in rural China in the 1950s. Yet amidst such forceful stabilizing trends, the effects of large-scale weather disturbance, notably the 1954 Yangzi floods, are still visible. Thus, for both the Centre and the East, which are dominated by more stable rice cultivation, sown-area instability declined much less impressively from 1931–7 to 1952–7, compared with the North and North-west, whose agro-climatical conditions are much less favourable.

The year-to-year changes in sown area in 1952–7, shown in Fig. 4.2, clearly reveal that the sharp fluctuations in 1954 were the single most important factor contributing to the comparatively high degree of sown-area instability for the Centre, and perhaps also the East, during the period as a whole. Note, however, that in 1931–7, grain-sown areas in these two regions were already as unstable as in the North-west, and even somewhat more unstable than in the North, judging by the comparative instability indices in Fig. 4.1. Clearly the disastrous 1931 Yangzi floods and the great 1934 summer droughts, which prevailed mainly in the Centre and East regions, are implicated in this respect (see Fig. 4.2).

From 1952–7 to 1979–84, grain-sown area in all the regions continued to be stabilized to a very low level of fluctuations. More importantly, there were hardly any interregional differences in sown-area instability in 1979–84 (index all below 1), compared with 1952–7, or for that matter,

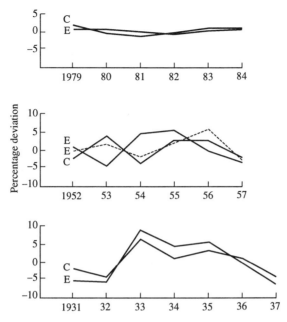

FIG. 4.2. Yearly percentage fluctuations in grain-sown area from the trend value in the Central and East regions in China, 1931–1984

Notes: The dotted line for East Region for 1952–7 is based on Chao's figures. It shows a substantial difference from the solid line for 1954. The difference is essentially accountable by Jiangsu province, for which the sown area size per Walker for 1954 (8.809 million ha.) is by 35% higher than that cited by Chao. In terms of grain output, the loss incurred in Jiangsu from the 1954 floods as measured against either the trend output or 1953 record was much smaller than Hubei (in the Centre Region) and Anhui (East) which both stood at the centre of the floods (see Table 5.1).

Source: As Fig. 4.1.

1931–7 (Fig. 4.1). These pervasive equalizing trends undoubtedly re-sulted from widespread improvements in irrigation and drainage capacity, and especially in flood prevention technology. The performance is all the more remarkable, when viewed in conjunction with the Centre region during the Yangzi floods in 1980. Unlike what happened during similar flood disasters in 1931 and 1954, there were virtually no sown-area losses in this region during the 1980 floods (see Fig. 4.2).

Instability in yield per hectare

Turning to grain yield instability, the transition from 1931–7 to 1952–7 resembles almost exactly the sown-area stabilizing trends, including the pattern of interregional discrepancies. The same applies basically to the transition from 1952–7 to 1979–84, except for two notable departures from the general trend. The first is the comparatively high yield instability for the Centre and East regions in 1979–84, viewed again *vis-à-vis* the North and North-west. The second is, more curiously, the dramatically deteriorating yield instability which characterized the North-east from 1952–7 to 1979–84 (see Fig. 4.1). This is in sharp contrast to the mutually equalizing regional trends in sown-area instability.

The sources of these divergent instability trends can again be identified through an examination of the yearly yield fluctuations in 1979–84, shown in Fig. 4.3. For both the Centre and East, the Yangzi basin floods of 1980 accounted for the large negative yield deviations in that year, and hence their relatively high instability for 1979–84 taken as a whole. As for the North-east, the sharp grain yield losses in 1981 and 1982, by 8.7 and 14.6 per cent respectively from the trend values, can also be explained by severe weather anomalies. The region, especially Heilongjiang, was affected in both years by disastrous floods (in 1981) and droughts (1982) on an apparently unprecedented scale, affecting 38 and 85 per cent of the provincial sown area.[1] The extremely poor harvest in the North-east in 1982 was very different from the positive yield gains in nearly all other regions, except for the neighbouring North (see Fig. 4.3). There is, however, no reason to suggest any deliberate policy discrimination against the North-east in 1982 which was one of the best harvests in recent years.

Nevertheless, cold and frost damage was also cited as a reason for yield losses in the North-east in 1981. This highlights another dimension of the problem which seems to have been unique to this region. According to one source, the drive to increase grain production prompted local authorities in the North-east to ignore the natural agro-meteorological demarcations for sowing wheat, maize, soya bean, and other grain crops.[2] The planting of certain varieties of high-yielding grain crops without adequate attention to temperature constraints often resulted in enormous losses from frost. This resembles the situation in the North-west, where the ploughing up of pasture apparently rendered average grain yields increasingly unstable. Such complications are, however, of a long-term nature, and could not have pre-empted the weather as an explanatory factor for the observed short-run fluctuations. Note that shortly before and after the large-scale yield losses in 1981 and 1982, there were powerful upsurges in grain yield in the North-east as well.

Virtually all significant yearly yield variations in the other regions in 1979–84 can be associated with known weather events. The South-west, which comprises mainly Sichuan province, is a good example. The 1981 Sichuan basin flood is easily detected by the sharp negative yield fluctuation shown in Fig. 4.3 in that year. In short, while grain yield

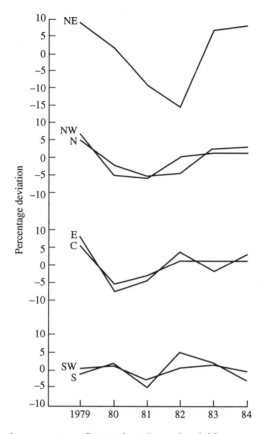

FIG. 4.3. Yearly percentage fluctuations in grain yield per sown hectare from the trend value in the different regions in China, 1979–1984

Source: As Fig. 4.1.

instability was consistently reduced across the regions in 1979–84, compared with 1952–7 or 1931–7, grain production in the broad regional context in China was still by no means weather-proof in the early 1980s.

Grain output instability

With the fluctuations in sown area becoming more stable between 1931–7 and 1952–7, and on to 1979–84, regional trends in grain output may be expected to vary, as at the national level, with yield instability. In other words, improving or deteriorating output stability should be consistently associated with declining or increasing yield instability. This is indeed generally borne out by the yield and output instability trends shown earlier in Fig. 4.1. But there are one or two regions which display an output instability trend contrary to the observed positive correlation between regional output and yield instabilities. These exceptional cases deserve detailed consideration in order to understand fully the possible sources of interregional variations in output stability.

The exceptions concern the North and the East. In both regions grain output instability increased from 1952–7 (index respectively 2.5 and 3.36) to 1979–84 (index 4.87 and 4.55). The increases occurred despite declining yield instability (index down from 4.44 to 3.81, for the North, and from 5.68 to 4.35 for the East) and improved sown-area stability (index respectively down from 1.04 to 0.54, and 5.36 to 0.52). How can these startling discrepancies be explained?

Notice first that in both the North and East, the output instability indices for relevant individual provinces did tend to converge strongly from 1952–7 to 1979–84. This is common to virtually all the regions and is discussed in greater detail in the next section on intraregional variations. Viewed against the background of 1952–7, such convergence clearly signals equalizing interprovincial technological advances. The really important question yet to be posed is why large interprovincial differences in output instability in the two regions could result in a comparatively low average regional output instability for each region for 1952–7 *vis-à-vis* 1979–84. The background to this is one of simple arithmetic.

Theoretically, output losses in some provinces due to bad weather can be compensated for by good harvests elsewhere within the same region. If this happens more or less consistently from year to year, the fluctuations for the regions as a whole may be dampened within any particular period. This is exactly the case with the North, and to a lesser extent the East in 1952–7, as revealed in Table 4.1. Consider for example the two years 1953 and 1957, which saw negative output growth for the entire North region. In 1953, grain output increases in Henan alone helped to offset one-third of the losses incurred elsewhere, notably Hebei (mainly due to flood) and Shandong (floods as well as drought).[3] As in 1957, the gains in Hebei in turn offset 60 per cent of the total regional losses occurring mainly in Shandong and Henan (both due to disastrous floods in the Yellow River and Huai River basins).[4] Likewise, while the 1954 Yangzi floods reduced the provincial grain output of Anhui in the East by some 14 per cent from the 1953 level, relatively good harvests in neighbouring Jiangsu (comparable loss of around 2 per cent only), and Zhejiang (hardly any loss), helped to check the fluctuations at the regional level.

This interprovincial gains/losses compensatory process somewhat resembles the 'area-for-yield' substitution which as illustrated earlier, already accounted for the comparatively low national instability for rice, as well as all grains, during 1952–7 *vis-à-vis* 1979–84. The fact that bad

TABLE 4.1. Provincial and regional changes in grain output and interprovincial gains/losses compensation in the North and East regions, 1952–1957 (million tonnes)

	Total grain output	Amount of yearly increases (+) or decreases (−) in grain output				
	1952	1953	1954	1955	1956	1957
North						
Beijing	0.113	+0.010	−0.040	+0.004	+0.033	+0.041
Tianjin	0.158	−0.024	+0.011	+0.049	+0.024	+0.032
Hebei	9.440	−1.315	−0.135	+1.135	−0.865	+1.840
Shanxi	3.845	+0.477	−0.207	−0.390	+0.610	−0.770
Shandong	12.584	−1.507	+2.110	+0.322	+1.166	−1.707
Henan	10.285	+0.955	+0.560	+1.050	+0.025	−0.575
Net regional change		−1.404	+2.299	+2.206	+0.993	−1.139
Total regional gains (+)		+1.442	+2.681	+2.596	+1.858	+1.913
Total regional losses (−)		−2.846	−0.382	−0.390	−0.865	−3.052
East						
Shanghai	0.041	+0.003	−0.006	+0.004	+0.006	−0.015
Jiangsu	10.885	+0.865	−0.250	+1.370	−0.870	+0.229
Zhejiang	7.000	+0.155	−0.070	+0.537	+0.042	+0.240
Anhui	8.810	+0.275	−1.315	+3.759	−0.620	+1.441
Net regional change		+1.295	−1.641	+5.670	−1.442	+1.895
Total regional gains (+)		+1.295	0	+5.670	+0.048	+1.910
Total regional losses (−)		0	−1.641	0	−0.707	−0.015

Source: As Table AB.11.

harvests in one province tended to be wholly or partly offset by good harvests elsewhere also applies to other regions. There are, nevertheless, two factors—one historical, the other geographical—which assumed particular importance in the North and East regions and contributed to their low regional average levels of grain output instability in 1952–7. The historical factor is that the prevalent weather hazards in these two regions in 1952–7 were floods rather than droughts. While the impact of floods may have been more disastrous, their geographical extent was normally much more limited than that of droughts, thus allowing more room for compensatory gains in non-stricken areas. This historical pattern of weather incidence occurred against the geographical background of the North and East which constitute the most heterogeneous regions in terms of agro-climatic criteria, at least according to the classical demarcations made by Buck (see Map 10.1).

Perhaps even more important, the output-stabilizing 'area-for-yield' substitution process took place especially markedly in the North and East in the 1950s. The rice acreage, for example, was expanded most rapidly in northern Jiangsu and Anhui (both part of East China), as well as in Henan and Hebei (in the North).[5] Inappropriate area expansion may impair average yields per hectare for crops concerned. But if this involves higher-yielding crops, as was the case with rice, or indeed maize in the 1950s, then the multiplicative gains on area and yield might help compensate for any losses incurred as a result of bad weather in any particular year. Thus overall grain output for any province or region taken as a whole could become more stable than the scale of weather disturbance would suggest. This helps explain why the degree of grain output instability could increase in 1979–84 compared with 1952–7, despite improved agricultural technology.

With regional grain output in the North and East remaining comparatively stable in 1952–7, it follows that under the 'area-for-yield' substitution process, both possible output quotients, i.e. sown area and yield per sown hectare, were bound to be less stable, with or without weather disturbance. In relation to 1979–84, sown-area and yield instability in the North and East in 1952–7 would then naturally seem relatively high, quite apart from the fact that both regions, as elsewhere in China in the 1950s, lacked the necessary technological hedge against sown-area and yield fluctuations.

It should be recalled, however, that while sown-area fluctuations were significantly stabilized, large-scale weather disturbances could still cause considerable yield deviations in 1979–84. This was true in the East, for example, much of which was affected by the 1980 Yangzi floods. As a result, regional yield instability during 1979–84 was not much lower than in 1952–7 (see Fig. 4.1). Yet the decisive difference is that in the 1980s the East no longer benefited from compensatory 'area-for-yield' substitution, hence its higher regional output instability in 1979–84 than in 1952–7, despite somewhat reduced yield instability. The same rationale applies to the North and Centre regions. The latter, with Hubei province, was the centre of the 1980 floods and it recorded hardly any decline in regional output instability from 1952–7 to 1979–84 (see Fig. 4.1).

Nevertheless, the observed reversals in regional output instability trends from 1952–7 to 1979–84 are, apart from that of the North-east, rather limited in scale. Since 1931–7, all regions had exhibited a persistent long-run trend of declining instability. This no doubt reflects the institutional impact of the post-war period and the secular technological influence. If the period 1974–9 is included—inadequately representative as it may be of the pro-grain strategy of the Cultural Revolution and its radical policy approach towards rural control—the general picture of long-run improvement in regional agricultural instability remains unaltered. This is closely in line with the conclusions drawn from the national-level analysis.

Intraregional Differences

Consistent with the regional trends, it is to be expected that interprovincial differences in sown area, yield, and output instabilities within the different regions would be successively reduced from 1931–7 to 1952–7, and to 1979–84. Yet Fig. 4.4 shows that there were clear reversals, some very radical indeed, in provincial instability trends. In what follows, we can only attempt to highlight some of the more typical cases in order to suggest the relative importance of weather *vis-à-vis* long-run institutional and technological influence.

The North-east and North-west

Data for 1931–7 are not available. The degree of sown-area instability for the three North-eastern provinces, Liaoning (index 4.26), Jilin (index 1.33), and Heilongjiang (index 2.94) for 1952–7, converged strongly in 1979–84 to around unity (Fig. 4.4). However, the respective yield and output instability trends diverged equally strongly from 1952–7 to 1979–84. Both Jilin and Heilongjiang were especially unstable in 1979–84, compared with Liaoning. This was clearly associated with the disastrous floods (in 1981), and droughts (1982), and serious frost damage (1981), as a result of the excessive sown-area expansion into the ecologically vulnerable far north. Liaoning, which lies south of Jilin and Heilongjiang, was much less seriously affected, however, hence the considerably smaller increase in its yield and output instability.

As for the North-west, the most notable departure from general regional trends was the increased yield instability for Shaanxi from 1952–7 to 1979–84. This left the province with an extremely high output instability (index 7.60) for 1979–84, compared with the regional average (index 4.00) and all the other provinces within the North-west (Fig. 4.4). Examination of the yearly data for Shaanxi reveals that the average grain yield for the province was reduced dramatically by 17 per cent in 1980 from the 1979 record.[6] Widespread natural disasters were indeed

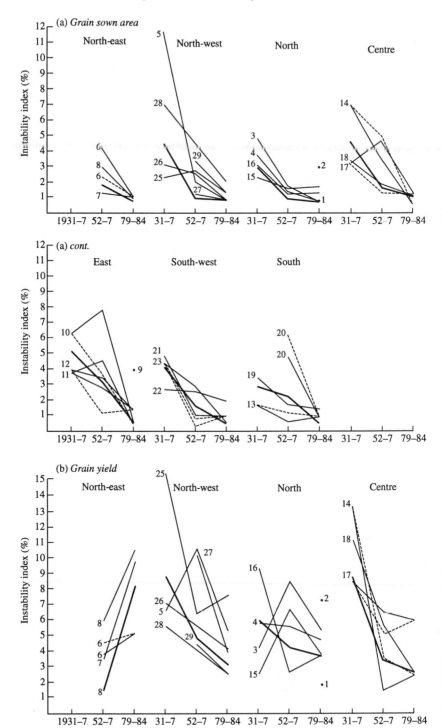

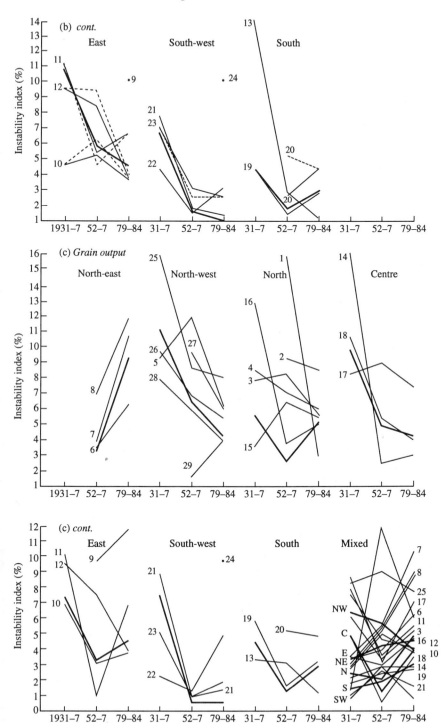

monitored in 1980, with droughts and floods affecting nearly half of the province's sown area. In 1981, floods of similar severity occurred over roughly the same fertile southern areas of Shaanxi. It was not until 1982 that the losses were recovered following a startling annual yield increase of 22 per cent.[7] In contrast to Shaanxi, most other provinces in the North-west were relatively free of weather disturbance in 1979–84, hence their lower degree of yield and output instability.

Another peculiar case in the North-west is that of Xinjiang which saw a considerable increase in output instability from 1952–7 to 1979–84 (Fig. 4.4). This occurred despite improved sown-area and yield stability. The province is too small a grain producer to warrant lengthy investigation of the reasons for the observed discrepancy. Nevertheless, even a casual glance at the yearly pattern of fluctuations suggests that in a similar manner to that of the North and East regions, there was a clear trend of 'area-for-yield' (or vice versa) substitutions between 1952 and 1957. This helped relatively to stabilize the province's grain output for 1952–7 in comparison with 1979–84.

The North China plain

Shandong and Henan are the only two provinces, not only within the North region, but virtually throughout the entire country, which saw

Fig. 4.4.   Intraregional trends in agricultural instability in China, 1931–1984

Notes: bold line: regional mean; thin line: provincial indices

The dotted lines are drawn by joining the alternative 1952–7 instability indices derived from Chao Kang's figures to the given 1931–7 and 1979–84 indices.

Provincial codes

North-east		North		Jiangsu	10
Liaoning	6	Beijing	1	Zhejiang	11
Jilin	7	Tianjin	2	Anhui	12
Heilongjiang	8	Hebei	3		
		Shanxi	4	*South-west*	
North-west		Shandong	15	Sichuan	21
Suiyuan	5	Henan	16	Guizhou	22
Neimenggu	5			Yunnan	23
Shaanxi	25	*Centre*		Xizang	24
Gansu	26	Jiangxi	14		
Qinghai	27	Hubei	17	*South*	
Ningxia	28	Hunan	18	Fujian	13
Xinjiang	29			Guangdong	19
		East		Guangxi	20
		Shanghai	9		

Source: Table AB.11.

increased provincial grain-sown area instability, albeit marginally, from 1952–7 to 1979–84 (Fig. 4.4). This is not surprising, given that post-Mao cropping shifts, as illustrated earlier, occurred mostly in these two provinces in an attempt to restore the traditional cotton production at the expense of the grain crops.

Perhaps more noteworthy are the mutually opposing trends in yield instability between Henan (declining) and Hebei and Shandong (both increasing) between 1931–7 and 1952–7 (see Fig. 4.4). In Hebei, the increased yield instability could still be offset by marked improvements in sown-area stability, or indeed, by 'area-for-yield' substitution in 1952–7, so that output instability hardly experienced any increase in the latter period. Shandong, however, seems to have lacked such compensatory effects and its grain output became noticeably more unstable in 1952–7, compared with 1931–7. In this respect, Shandong, together with Hubei in the Centre region, constitute virtually the only two provinces in the whole country in which the provincial degree of output instability increased, rather than declined, from 1931–7 to 1952–7 (Fig. 4.4).

The observed interprovincial variations in the North, affecting sown-area, yield, and output instability, are again clearly associated with agro-climatic dynamics unique to the region. This deserves detailed consideration, not least against the historical background of frequent disasters in this region—the heartland of China.

The annual percentage fluctuations in sown area, yield, and output in the three major northern provinces, Hebei, Shandong, and Henan, are plotted in Fig. 4.5 in relation to the respective trends, for the three periods 1931–7, 1952–7, and 1979–84. They invite the following comments.

First, there is a striking difference between 1931–7 and the two post-war periods in that in the former, grain output in all three provinces varied more closely with sown area, than with yield per sown hectare. Exactly the reverse is true, however, for the post-war years. There are only two notable exceptions. One is that of 1936 when a reduction in sown area in all three provinces was strongly offset by sharp increases in yields to result in a substantial increase in grain output. The other exception is 1982 and applies mainly to Shandong, where deliberate downward readjustments in grain-sown area in favour of cash crops caused grain output to decline, despite a marked increase in yield per sown hectare. In a similar fashion, parallel sown-area readjustments in Henan, also in 1982, were strongly reinforced by yield losses to cause the most abrupt negative output deviation in a single year during 1979–84.

Secondly, while sown-area fluctuations were substantially stabilized in the post-war years, yield per sown hectare in the North as represented by the three major provinces under study, was still capable of demonstrating year to year changes that were as dramatic as those of the 1930s (see Fig.

4.5). This singles out the North as the one region which witnessed a
consistent long-term improvement in yield stability from the 1930s to
the 1980s. More importantly, there were considerable differences in the
timing of the yearly yield and output deviations, negative or positive,
amongst the three provinces, in both pre-war and post-war periods. For
example, the abrupt downturn in yield and output trends in 1953 in both
Shandong and Hebei contrasted with an upturn in Henan in the same
year. Likewise, while Shandong was by the same measures more stable in

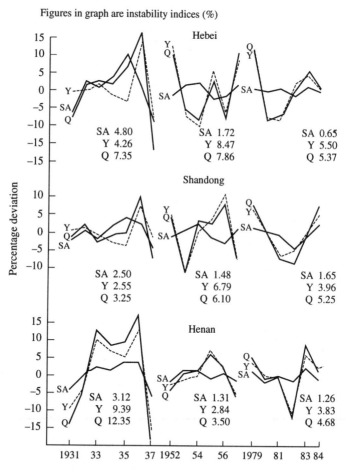

FIG. 4.5. Yearly percentage fluctuations in grain-sown area, yield, and output
from the trend values in Hebei, Shandong, and Henan provinces in China,
1931–1984. Y = grain yield, SA = grain-sown area, Q = grain output

Source: As Table AB.11.

1931–7, compared with both 1952–7 and 1979–84, neighbouring Hebei and Henan displayed an exactly opposite intertemporal relationship.

All these interprovincial variations suggest that the North, more than anywhere else, was highly susceptible to the random geographical incidence of adverse weather. This is the agro-climatic background to the recurrent phenomenon of large-scale rural emigrations in the past, which, under the impact of poor harvests, occurred within the relatively limited context of the North China plain.

Thirdly, all the major yield and output fluctuations in Hebei, Shandong, and Henan, observed in Fig. 4.5, in both pre-war and post-war periods, are attributable to large-scale weather anomalies. In Shandong, for example, the sharp negative yield and output deviations by 11 per cent in 1953, and again by around 8 per cent in 1957, largely reflect the combined impact of catastrophic floods and droughts which together covered respectively 50 and 68 per cent of the provincial grain-sown area in these two years.[8] Floods were more widespread in 1953 than in 1957. This probably explains the greater extent of losses in the earlier year.

Most strikingly of all, in 1956, in sharp contrast to 1953 and 1957, Shandong witnessed the greatest yield and output gains ever recorded in 1952–7. Yet 1956 was the year of the 'high tide of rural socialization', which swept through China—including Shandong. Whatever the disincentive implications of this institutional upheaval, there was a rare absence of reports in 1956 of natural calamities in the province. Interestingly, in the same year Hebei experienced a sharp decline in both yield and output, mainly as a result of floods during the high crop seasons of June and July 1956 (see Fig. 4.5).[9] In 1957, however, Hebei exchanged its lot with Shandong. In the aftermath of the 1957 disasters, peasants in some parts of Shandong were reduced to eating grasses and bark, conditions reminiscent of the distant past, except for the absence of spontaneous rural exodus.[10] By contrast Hebei reaped a bumper harvest comparable in scale with that of 1952.

Similar interprovincial harvest disparities within the North are also clearly discernible in 1979–84. Thus, amidst the much vaunted nationwide bumper harvests of 1982, Henan province suffered a harvest failure, with average grain yield and output being reduced by over 10 per cent from anticipated trend levels (Fig. 4.5). The background of these losses was 'a province-wide drought occurring in the summer, and prolonged drizzles plus widespread floods and water-logging in the autumn, which altogether covered more than 90 *xian* and some 1,330 thousand farm hectares' (i.e. equal to around 15 per cent of the provincial grain-sown area).[11]

Finally, turning to the pre-war period 1931–7, the climatic background of observed year-to-year provincial fluctuations in the North, in yield, output, but also in sown area, can, in most cases, be fairly readily identified. Thus, in 1931 Henan province fared in terms of yield and output instability less well than Shandong and Hebei (see Fig. 4.5). This is because the southern half of Henan lies more in the fertile Huai and Yangzi River basins than in the North China plain proper. In other words, Henan was greatly affected by the catastrophic 1931 floods, whereas Hebei and Shandong, which both border the province to its north and north-east were not.

The impact of the 1935 Yellow River floods is also clearly visible. All three provinces, Hebei, Shandong, and Henan were affected. As shown in Fig. 4.5, grain yields were consistently reduced. Sown areas were spared, however, for the floods occurred mainly after the spring and summer sowing. Similar flood havoc was recorded for 1933. Shandong was described as 'most calamitously affected, with flood waters from Henan, Hebei, and Jiangsu converging simultaneously into the province'.[12] This verbal summary is borne out by the divergent yield trends, also shown in Fig. 4.5, between Shandong (declining) and the other two provinces, Hebei and Henan (both still increasing), for 1933.

Between the two major Yellow River floods of 1933 and 1935 was the great 1934 summer drought. This mainly affected the Yangzi and Huai River basins, but its periphery also extended well into the North. As a result, grain yields for all three provinces declined markedly in 1934 to almost the same low as in 1935 (see Fig. 4.5).

But by far the most spectacular fluctuation in the North China plain during 1931–7 occurred from 1936 to 1937. As is shown in Fig. 4.5, a surge in grain yield and output in Hebei, Shandong, and Henan in 1936 was dramatically followed by an equally strong downturn in 1937, not only in yield and output, but in grain-sown area as well. The fluctuations were especially pronounced in Hebei and Henan, although in Shandong too its amplitude was unprecedented during the period 1931–7. The upsurge in 1936 is readily understood, this reportedly being a year of good weather not only in the North, but throughout most of China.[13] The truly catastrophic downturn in 1937 may seem to have reflected the Japanese invasion, although the Sino-Japanese war was not formally declared until the Marco Polo Bridge incident in suburban Beijing, then the provincial capital of Hebei on 7 July 1937.

A careful scrutiny of the initial war chronology,[14] and especially of the provincial sown-area and yield data presented in Table 4.2, reveals, in fact, that the war had little to do with the illustrated losses. For Henan, for example, virtually all the losses can be easily accounted for by reductions in both sown area and yield for the cross-winter (1936/7) crops, especially wheat. The winter crops were all harvested in early summer 1937, well before the Japanese invasion. The National Agricultural

TABLE 4.2. Changes in provincial grain-sown area, yield, and output in the North China plain from 1936 to 1937

	All grain			Winter grain crops						Summer grain crops		
				Wheat			Other crops					
	Area (A) (1,000 ha.)	Yield (Y) (kg./ha.)	Output (Q) (1,000 tonnes)	Area (1,000 ha.)	Yield (kg./ha.)	Output (1,000 tonnes)	Area (1,000 ha.)	Yield (kg./ha.)	Output (1,000 tonnes)	Area (1,000 ha.)	Yield (kg./ha.)	Output (1,000 tonnes)
Hebei												
1936	5,914	1,178	6,967	1,982	773	1,533	364	948	345	3,568	1,426	5,089
1937	5,153	926	4,774	1,420	661	939	285	884	252	3,448	1,039	3,583
Percentage changes	-12.9	-21.4	-31.5	-28.4	-14.6	-38.7	-21.7	-6.8	-27.0	-3.4	-27.1	-29.6
A and Y shares in Q change	37.6	62.4	100	66.0	34.0	100	76.1	23.9	100	11.1	88.9	100
Absolute change	-761	-252	-2,193	-562	-112	-594	-79	-64	-93	-120	-387	-1,506
Crop shares in all-grain change	100	100	100	73.8	28.9	27.1	10.3	3.8	4.2	15.7	67.4	68.7
Shandong												
1936	7,104	1,359	9,650	3,449	1,030	3,551	465	1,029	478	3,190	1,762	5,621
1937	6,467	1,217	7,872	2,893	998	2,887	403	1,007	406	3,171	1,444	4,579
Percentage changes	-9.0	-10.4	-18.4	-16.1	-2.9	-18.7	-13.3	-1.9	-15.1	-0.6	-18.0	-18.5
A and Y shares in Q change	46.4	53.6	100	84.7	15.3	100	87.5	12.5	100	3.2	97.3	100
Absolute change	-637	-142	-1,778	-556	-32	-664	-62	-22	-72	-19	-318	-1,042
Crop shares in all-grain change	100	100	100	87.3	40.3	37.3	9.7	3.3	4.1	3.0	56.5	58.6

TABLE 4.2. *cont.*

| | All grain | | | Winter grain crops | | | | | | Summer grain crops | | |
| | Area (A) (1,000 ha.) | Yield (Y) (kg./ha.) | Output (Q) (1,000 tonnes) | Wheat | | | Other crops | | | Area (1,000 ha.) | Yield (kg./ha.) | Output (1,000 tonnes) |
				Area (1,000 ha.)	Yield (kg./ha.)	Output (1,000 tonnes)	Area (1,000 ha.)	Yield (kg./ha.)	Output (1,000 tonnes)			
Henan												
1936	8,324	1,210	10,068	4,095	1,290	5,283	1,180	1,314	1,551	3,048	1,065	3,246
1937	7,537	874	6,584	3,271	570	1,872	961	566	544	3,302	1,262	4,168
Percentage change	-9.5	-27.8	-34.6	-20.1	-55.8	-64.6	-18.6	-56.9	-64.9	+8.3	+18.5	+28.4
A and Y shares in Q change	25.5	74.5	100	26.5	73.5	100	24.6	75.4	100	25.2	65.1	100
Absolute change	-789	-336	-3,484	-824	-720	-3,411	-219	-748	-1,007	+254	+197	+922
Crop share in all-grain change	100	100	100	104.4	76.4	97.9	24.3	25.2	28.9	-32.2	-0.7	-26.5

Notes: The relative contributions of area and yield to output are taken to be the ratio of the respective percentage changes to output loss percentage. The percentage shares of the three different crop categories in the all-grain yield changes are estimated with the formula

$$\delta x = (x_1 \delta a_1 + a_1 \delta x_1) + (x_2 \delta a_2 + a_2 \delta x_2) + (x_3 \delta a_3 + a_3 \delta x_3),$$

where δx denotes all-grain yield (kg./ha.) change from 1936 to 1937; x_1, x_2, and x_3 the yields (kg./ha.) of the three crop categories in 1936; δx_1, δx_2 and δx the differences between the yields (kg./ha.) in 1936 and 1937; a_1, a_2, and a_3 the average shares of sown area in all-grain totals for 1936 and 1937; and δa_1, δa_2 and δa_3 the difference between the sown-area shares in 1936 and 1937. In some of the cases the estimated shares do not add up to 100 per cent. They are then proportionately adjusted to make up for the total.

Sources: As Tables AB.5 to AB.7.

Research Bureau attributed the reductions to catastrophic droughts which had begun in autumn 1936 and rendered wide stretches of the wheat region totally unsowable.[15] Deteriorating market prices following the bumper harvests of 1936 were cited as another factor which reduced peasant incentives to maintain the existing scale of sowing and use of current inputs.

For Hebei and Shandong, the situation in 1937 was similar to that of Henan. Reductions in the 1936/7 winter-crops areas already accounted for 84 and 97 per cent of the respective yearly provincial sown-area declines (see Table 4.2). The only difference is that in Henan there was still some increase in both sown area and yields for the summer crops (harvested in autumn 1937) which helped offset, albeit marginally, the losses of earlier winter crops. By contrast, the summer crops in Shandong, and especially Hebei, suffered substantial losses in terms of yield per sown hectare. The yield losses in these two provinces amounted to 18 and 27 per cent based on the 1936 records. They occurred on top of the losses sustained by the winter crops.

A possible explanation of the summer crop yield losses sustained in Hebei and Shandong during 1937 is that the peasants, in anticipation of the Japanese advances, sought to withhold necessary current inputs after the summer sowing. However, in the second half of 1937 the North China plain was dominated by widespread natural disasters, floods in particular. Whatever the relative importance of the war and the weather factors, the data in Table 4.2 suggest that for Hebei and Shandong, the combined sown-area and yield losses sustained by the summer crops in 1937, as a proportion of the annual all-grain output losses, were still considerably less than the comparable contribution of the winter crops. The latter was not at all affected by the war.

What conclusions may be drawn from the observed interprovincial variations in farm instability within the North? One is clearly the long-run improvement in sown-area stability from the 1930s to the 1980s. This consistently characterizes the three major provinces under study. It underlines once again the comparative edge of collectivized rural control, relative to the marketized rural economy of the 1930s. Such collective coercion was strongly reinforced by improvements in farm technology by the early 1980s. The upshot is that large-scale fluctuations in sown areas which in the past were typical of the North China plain, whether because of changing market conditions or peasant emigrations prompted by famines, have now been eliminated.

Yet it is no less clear that yield per sown hectare remains in the North, perhaps more than anywhere else in China, as vulnerable to weather changes as was the case in the 1930s. Yield fluctuations in Shandong in both 1925–7 and 1979–84 even outweighed the sown-area stabilization effects to result in greater output instability compared with 1931–7 (Fig. 4.5). Both Hebei and Henan saw a similar amplitude of yield and output fluctuations in the post-war years. Had it not been for the dramatic fluctuations of 1936/7, both provinces would in fact have been more stable in 1931–7 taken as a whole, compared with the post-war periods.

Finally, while the North China plain was by no means weather-proof by the early 1980s, the increased per capita grain availability brought about by the persistent grain-self-sufficiency drive had, nevertheless, given the peasants a surplus to hedge against bad years. It also helped to increase state procurement and centralized allocation of relief grain to areas devastated by natural disasters. Such relief supplies are a prerequisite of any attempt to prevent forced rural emigrations. This has probably been the most visible long-run change in the North China plain.

The Centre and East

Most of the provinces in these two regions lack a complete set of sown-area data for 1952–7. We can only consistently compare the provincial differences in grain output instability trends. The output instability indices plotted in Fig. 4.4 will serve as a reference point for our discussion.

During 1931–7 grain output in the Centre and East was extremely unstable. In all the provinces concerned, the degree of output instability in this pre-war period was comparable with that of provinces in the North-west and North. However, the two latter regions face a more unstable agro-climatic environment compared with the Centre and East. It can easily be shown that the high provincial output instability reflects the combined impact of the great 1934 summer droughts and the 1931 Yangzi floods, both of which were especially serious in these two regions.

Most interestingly, amidst the declining output instability trends from 1931–7 to 1952–7, interprovincial disparities widened substantially, disfavouring Hubei in the Centre, and neighbouring Anhui province in the East region (Fig. 4.4). In both these provinces, grain output in 1952–7 (instability index 8.42 and 7.45 respectively) was basically as unstable as in 1931–7 (index 7.86 and 9.49). The reason for this is readily apparent. Both provinces, especially Hubei, were at the centre of the 1954 Yangzi floods whereas other provinces in the Centre and East were less seriously affected. They could still take advantage of the newly emerging collective institutional framework, hence the widening interprovincial discrepancies in output instability.

Similar intraregional discrepancies existed in 1979–84. Grain output instability in Hubei (index 6.95) remained very high relative to that of Hunan (index 3.64) and Jiangxi (index 2.87) in the same Centre region

(Fig. 4.4). This can be explained by the 1980 Yangzi floods which were concentrated heavily in Hubei. Meanwhile, in Zhejiang province a declining output instability trend was abruptly reversed between 1952–7 (index 1.10) and 1979–84 (index 6.83). Grain output in Zhejiang fluctuated most sharply during 1979–84, while sown area remained very stable (index 0.56). For example, it fell by 11 per cent in 1980, amidst 'severe natural disasters rarely encountered in the past several decades'.[16] The impact was apparent through 1981, with a further loss of 1.2 per cent. It then recovered sharply by 21 per cent in 1982, when there occurred 'a full-fledged harvest . . . characterized by increases in output, continuing from one season to another for all the spring, summer, and autumn crops'.[17]

It would be premature to assess the relative contributions of good and bad weather to observed grain output fluctuations in Zhejiang in 1979–84. However, it should be clear that the harvest fluctuations occurred at a time, when peasants in Zhejiang, as elsewhere in China, showed no signs of declining incentives in a buoyant rural environment conditioned by the decollectivization drive.

The South-west and South

For lack of consistent sown-area and yield data we shall again confine our remarks to grain output trends. As shown in Fig. 4.4, grain output in 1952–7 was very stable in both the South-west as a whole (index 0.94), and its three major provinces, Sichuan (index 1.09), Guizhou (index 1.32), and Yunnan (index 1.06). It was very stable, not only in relation to 1931–7, but compared with the Centre and East regions during the same period 1952–7. However, while provincial output instability trends within the Centre and East tended to converge during 1979–84, the three South-west provinces began to diverge (see Fig. 4.4). The South, comprising Fujian, Guangdong, and Guangxi provinces, also exhibited similar divergent trends in provincial output instability from 1952–7 to 1979–84, though not as strongly as in the South-west.

These long-run interregional changes in output instability again reflect the random nature of regional shifts in weather disturbance. As in the Centre and East in 1952–7, intraregional discrepancies in output instability in both the South-west and South in 1979–84 are clearly associated with known weather events. The Sichuan basin floods of 1981, for example, resulted in a yield loss of 3.30 per cent (from the trend) which was the largest during 1979–84. Provincial grain output in that year in turn declined by 1.34 per cent (also from the trend), despite a comparable increase of nearly 2 per cent in grain-sown area, which was the largest recorded in 1979–84.[18]

In contrast to Sichuan, Guizhou province saw its grain output instability increase to a much greater extent between 1952–7 (index 1.32) and 1979–84 (4.66). Unimportant as the province was in terms of grain production, it was disastrously affected by prolonged drought and unusually high temperatures in 1981.[19] Its grain output fell by an alarming 13 per cent.

Similar weather factors could explain the observed reversal in grain output instability in Guangdong in the South from 1952–7 to 1979–84 (see Fig. 4.4). This need not, however, detain us. A more important concluding point is that, despite increased output instability, from 1952–7 to 1979–84, the interprovincial discrepancies for both the South-west and South remained, as in most other regions, limited in 1979–84, compared with 1931–7.

Some General Observations

First, the random incidence of weather disturbance can be clearly seen within the relatively confined provincial context, and to a lesser extent within broader administrative regions. Thus, there were clear cases of chance reversals in provincial instability trends, especially in terms of grain yield and output, within the long-run context of improving agricultural stability from the 1930s to the early 1980s.

Second, the consistently declining scale of interprovincial disparities in yield and output instability for most of the regions underline the basic finding of our national aggregate analysis that the weather as a factor in agricultural fluctuations has, over the long-run, increasingly become less important compared with the institutional and technological hedges available in the post-war periods. The stabilization effect is especially obvious with respect to sown-area fluctuations. There has been a most marked intraregional as well as interregional convergence in sown-area instability towards near-perfect stability in 1979–84.

Third, much as the comparative weather influence has been reduced in the post-war years, our regional or provincial verifications of the national aggregate instabilities, reveal that bad or good weather was still clearly associated with major short-run (i.e. year-to-year) fluctuations in yield and output in 1952–7, as well as 1979–84, given the prevalent institutional and technological milieu. It remains to be determined to what extent the degree of weather influence has changed from one period to another, relative to institutional and technological factors.

Fourth, within each of the three periods, 1931–7, 1952–7, and 1979–84, the degree of grain yield and output instability generally varied in ascending order from the South-west and South to the Centre and East, and further to the North-west and North. This is in line with known regional differences in agro-meteorological parameters, in particular the absolute volume of rainfall and its relative annual variability.

Finally, large-scale disturbances in one or two major regions or provinces could, even if offset by good harvests elsewhere, readily translate into major national fluctuations in grain output. A good case in point is the 1954 Yangzi floods. These affected mainly the Centre and East regions. For the country as a whole, however, grain output declined mainly because of the floods by nearly 3 per cent, in relation to the potential or trend output for that year. This percentage loss was the largest recorded for any single year in 1952–7. In the following chapter we shall examine some notable episodes of major instability from the national and longer-term historical perspectives.

NOTES

1. *JJNJ 1982*, VI–56 and *JJNJ 1983*, V–52.
2. Xue Weimin and Zhang Jijia (eds.), *Dangdai Zhougguo de Qixiang Sheye* (Meteorological Works in Contemporary China) (Beijing: Zhongguo Shehui Kexue Chubanshe, 1984), 203–4. See also Ce Jianjun 'Heilongjiang Sheng Nongye Jishu Gaizao de Mubiao Yi Fangxiang' (Goals and directions of agricultural-technological renovation in Heilongjiang province), in *XXYTS* 6 (Nov. 1984), 85–9; and Du Youlin, 'Sanjian pingyuan di qu nongye kaifa shexiang' (Thoughts on agricultural development in the Sanjiang plains), in *XXYTS* 4 (July 1984), 71–8, for a more extensive discussion of the ecological disturbances in the North-east as a consequence of excessive practices of cultivation. Cf. also Guan Yuzuan 'Guanyi Sanjiang pingyuan diqu kaifa jianshe fangxiang he tujin de tantao' (An exploration of the direction and approach for development and construction of the Sanjian plain area), in *NYJJWT* 12 (Dec. 1980), 8–17; see p. 10 in particular for a specific example of how the oversubstitution of maize crops (which are vulnerable to both deficiency in solar radiation and low temperature) for soya bean led to increased overall yield instability.
3. See *RMRB*, 28 Apr. 1953 for Hebei; and *DZRB*, 19 and 23 Jan. 1957 for Shandong.
4. *DZRB*, 11 Feb. 1958; the 1957 disasters in Shandong were also prominently reported in *RMRB*, 11 Feb. 1958. For Henan, see *HNRB*, 14, 23, and 25 Aug. 1957. These sources are cited in detail in Table AB.16, together with all the other similar reports in 1952–61. Henceforth we shall avoid repeating the same citations in the main text.
5. This is especially the case in 1956; see Kenneth R. Walker, *Foodgrain Procurement and Consumption in China*, 222.
6. *NYNJ 1981*, 22.
7. *NYNJ 1983*, 37 and *NYNJ 1982*, 34.

8. See Table 7.4 for the sources of these and other similar figures cited in the text of grain yield fluctuations.

9. *RMRB*, 17 Mar. 1957 and 23 Oct. 1959; more than 15 million tonnes of relief grain were dispatched by the State to the province as a result of the flood losses.

10. The scale of the 1957 disasters in Shandong can be gauged by press reports at the time, esp. *DZRB*, 6 and 18 Jan. 1958: total grain output only 129.68 million tonnes, fulfilling only 84.1 and 82.8% of the targets set by the central and provincial governments respectively. It represented a reduction of 12% from the 1956 output. A total of 74 *xian* were covered by floods, 3.1 million houses collapsed, and 9 million population were affected. The floods were followed by autumn droughts which affected more than 60 *xian*, with 2,667 thousand hectares of late autumn crops most severely affected. Earlier the provincial authority repeatedly appealed to peasants in the worst-affected localities to resort to all possible means for self-survival including eating wild vegetables, etc.

11. *JJNJ 1983*, V–105.

12. *SBNJ 1934*, 1009–10.

13. *The Chinese Yearbook 1937* (in English) published by the Nanjing-based Council for International Affairs reported that 'The year 1936 was by far the best year for the Chinese farmers as a whole' (p. 778).

14. Available in Su Zhenshen, *Zhongri Guanxi Shi Shijian Biao* (A chronology of Sino-Japanese Relations, AD 57–AD 1970 (Taipei: Huagan Chubanshe, 1977), 387–407. See also Morius B. Jansen, *Japan and China from war to peace, 1894–1972* (Chicago: Rand McNally College Publishing Co. 1975), 407 (map).

15. *NARBCR* 5/8 (1937), 251 and 254, and 11 (1937), 311.

16. *NYNJ 1981*, 22 and *JJNJ 1981*, IV–263.

17. *JJNJ 1983*, V–72.

18. *NYNJ 1981*, 22 and *NYNJ 1982*, 34.

19. *JJNJ 1982*, 153–4; see also Zhang Yuhuan, Qin Shaode, and Zheng Chengjiu, 'An enquiry into the problem of Guizhou's grain production', in *GZSHKX* 6 (Nov. 1983), 31–9, for a specific account of the weather and the extent of sown area affected and output losses.

5

Notable Episodes of Great Instability

In our analysis of national trends of agricultural instability (Chapter 3), and their regional variations (Chapter 4) from the 1930s to post-war periods (1952–8, 1970–7, and 1978–84), we frequently referred to specific instances of large-scale fluctuations in grain production since the early 1930s. They were all closely associated with a known weather disturbance of unusual magnitude. The most familiar cases can be tabulated as follows:

1931   The Yangzi and Huai River basin floods, embracing the Centre, East, and part of the North regions (South Henan).

1934   The Yangzi and Huai River basin droughts, more widely, distributed within the same region as the 1931 floods.

1937   The North China plain drought, affecting Henan most seriously (droughts of similar scale occurred in the Sichuan basin).

1954   The Yangzi River and Huai River basin floods, affecting most parts of the Centre and East regions.

1972   The great North China droughts, covering the North-west, North, and parts of the Centre and East regions.

1980   The Yangzi River basin floods, affecting the Centre and the East (also widespread summer droughts in the North, North-east, and part of the North-west).

1981   The Sichuan basin floods (also widespread spring and autumn drought in the North, and summer and autumn droughts in the Centre).

The above list omits, however, the most notorious episode of 1959–61. This omission is, first of all, due to the fact that unlike other periods under study, we do not have a complete set of provincial sown-area, yield, and output data for these three years on which to base a comprehensive verification of the regional sources of the changing agricultural instability at the national level. More importantly, by virtue of the scale of the disturbance, the 1959–61 episode deserves to be treated separately.

This chapter attempts to place these three tumultuous years, 1959–61, into proper historical context, by making use of available national aggregate data in order to compare the magnitude of fluctuations in grain output during this period with that of other notable episodes of great instability. Before doing so, however, we look more closely at other known examples of major regional output fluctuations, in order to con-

sider how they might have translated into national-scale disturbances.
This facilitates a verification of the possible regional sources of instability
during 1959–61.

Regional Disturbance and National-Scale Instability

Figs. 5.1 to 5.3 show the yearly fluctuation in grain output for each of the
seven administrative regions, demarcated in exactly the same way as in
our instability analysis, for the three periods, 1931–7, 1952–7, and 1979–
84, for which a complete set of provincial data are available for deriving

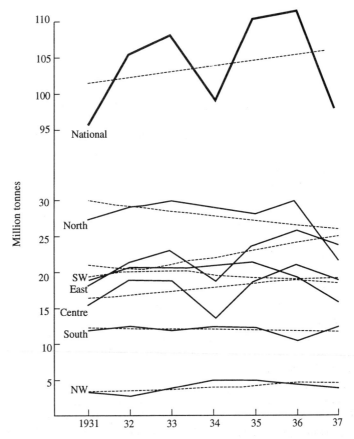

FIG. 5.1. Yearly fluctuations in regional and national grain output in China,
1931–1937

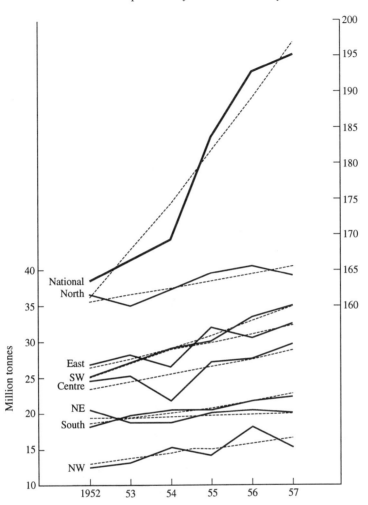

Fɪɢ. 5.2. Yearly fluctuations in regional and national grain output in China, 1952–1957

regional totals. The yearly fluctuations are plotted against the estimated trend values to show the relative magnitude of fluctuations. The national grain output totals, which comprise the regional totals for the three periods, are also shown in order to facilitate corroboration of the regional and national variations.

Before we proceed, a number of methodological points must be made. First, the overall grain output figures include different crops (wheat and rice, for example), and, as annual aggregates, they may conceal or

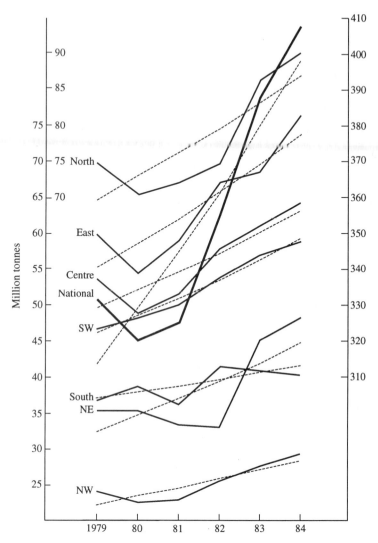

Fig. 5.3.   Yearly fluctuations in regional and national grain output in China, 1979–1984

Sources: Table AB.2 and as Table AB.11.

blur possible seasonal fluctuations. A bad winter wheat harvest resulting from a spring drought, for example, may not be adequately reflected in the diagrams if it was offset by a good autumn harvest of high-yielding coarse grains.

Second, the vertical output scale adopted in the figures is identical throughout (except for the expanded scale for national output in 1979–84). This facilitates a visual comparison of the relative output weights of the different regions and thus the possible quantitative implications of any output losses incurred in one region compared with another. Likewise, for any one region, the level of output and changes in different periods can be compared with one other. The relative size of implied output losses or gains in any one year can be read from the figures by relating the vertical distance between the actual and trend output values to the respective baseline output, say the trend or the actual values. Thus, while the absolute level of losses for the Centre region in 1980 was approximately the same as in 1954, as the diagrams show, the percentage loss for 1980 was smaller, in view of its much higher output base, compared with 1954.

On the basis of this methodology, the regional output variations shown in the three diagrams may be interpreted with relative ease. For most of the years during 1931–7 and in the two post-war subperiods, observed sharp output fluctuations can be readily identified and correlated with the important weather disturbances, negative and positive. More importantly, most of the large-scale regional disturbances are clearly reflected in national grain output fluctuations. Thus, the great Yangzi and Huai River basin droughts of 1934, which embraced both the East and Centre regions resulted in a dramatic downturn in national grain output in the same year, as shown in Fig. 5.1. The same can be said of the great Yangzi River floods of 1931 and the disastrous droughts in both the North China plain in 1936/7 and Sichuan basin in 1937. Likewise, the disastrous Yangzi River floods of 1954 are clearly reflected in the sharp grain output losses in national terms, in relation to the potential output trend of that year.

Even during 1979–84, when after three decades of technological progress one would have expected fluctuations in farm production to have become substantially muted, the yearly variations in grain output were still very much in evidence, as shown in Fig. 5.3. Thus, the remarkable downturn of the national grain output in 1980, as shown in Fig. 5.3, undoubtedly derived from losses in the East and Centre regions, which were both affected by the disastrous Yangzi River floods.

The same conclusion applies to other notable episodes of great instability. But the most important point is that, while good and bad weather conditions in different regions tend in the main to cancel each other out, and so leave national grain output relatively stable, in some years the incidence of bad weather in critical grain-producing regions may lead to a marked downturn in aggregate output. Alternatively, most or all of the important regions may enjoy good weather in the same year and so bring about a bumper harvest, good examples being 1979, 1982, 1983, and 1984.

The other side of the coin is that within the same agricultural policy framework, in any particular year the grain output of different regions may fluctuate greatly in opposite directions. This happened in almost all the periods under study, and it implies that the weather has played and still plays a substantially greater role than either policy or institutional arrangements in influencing agricultural output and yields whether negatively or positively. This is clearly illustrated by the grain output variations for 1980 and 1981. Despite the much improved incentive schemes offered since 1979, it was still impossible for the peasants effectively to stave off the disastrous effects of the 1980–1 Yangzi River floods.

The Provincial Scene

As a component of regional output, fluctuations in output at the provincial level are bound to be more substantial than those of individual regions. Thus, as shown in Table 5.1, Hubei province, which was at the centre of the Yangzi River floods in both 1954 and 1980, suffered a startlingly high grain output loss of 23.18 per cent in 1954 and 12.97 per cent in 1980, measured against the respective trend outputs, whereas for the Centre as a whole the comparable losses amounted to only 13.76 and 6.70 per cent respectively.

The same holds for Anhui province, in relation to the East region (Table 5.1), which was also at the centre of the same Yangzi River floods in the same two years. Note also that in Hubei and Anhui, as well as their associated regions, Centre and East, grain output losses in relation to the much higher output bases of 1980 were consistently lower in percentage terms compared with 1954, although the absolute level of losses in 1980 was comparable to, or even higher (as in the case of the East region) than that sustained in 1954 (Table 5.1).

The year 1980 was also marked by sharp provincial output fluctuations in other regions, notably the North which is the largest grain-producer among the seven regions under study. Again measured against the trend output, grain output in Hebei, Shanxi, and Henan, which together accounted for 62 per cent of the regional output total, was reduced respectively by 9.56, 7.17, and 2.50 per cent in 1980 (Table 5.2).

Taken together, the Centre, East, and North regions, which, with a combined total grain output accounting for around 55 per cent of the national total of 320.56 m. tonnes in 1980, suffered a total loss of 14.4 m. tonnes, measured against the actual grain output of 332.12 m. tonnes obtained in 1979. This was equivalent to a loss of 4.3 per cent, excluding compensatory gains from other regions, which helped only marginally to reduce the national loss rate to 4.1 per cent.

TABLE 5.1. Relative provincial grain loss rates in the Centre and East regions in China, 1954 and 1980

	1954 Actual output (million tonnes)	1954 Losses as measured against			
		trend output		1953 output	
		(million tonnes)	(%)	(million tonnes)	(%)
	(1)	(2)	(3)	(4)	(5)
Centre					
Jiangxi	5.74	−0.308	−5.09	−0.01	−0.14
Hubei	6.85	−2.065	−23.18	−2.22	−24.51
Hunan	9.28	−1.068	−10.32	−1.06	−10.21
Regional	21.87	3.49	−13.76	−3.29	−13.08
East					
Shanghai	0.04	−0.003	−7.32	−0.01	−13.64
Jiangsu	11.50	−0.230	−1.96	−0.25	−2.13
Zhejiang	7.09	−0.215	−2.95	−0.07	−0.98
Anhui	7.77	−1.810	−18.89	−1.32	−14.47
Regional	26.39	−2.310	−8.05	−1.64	−5.85

	1980 Actual output (million tonnes)	1980 Losses as measured against			
		trend output		1979 output	
		(million tonnes)	(%)	(million tonnes)	(%)
	(1)	(2)	(3)	(4)	(5)
Centre					
Jiangxi	12.40	−0.27	−2.13	−0.57	−4.39
Hubei	15.36	−2.29	−12.97	−3.14	−16.92
Hunan	21.24	−0.92	−4.15	−0.94	−4.24
Regional	49.00	−3.52	−6.70	−4.65	−8.67
East					
Shanghai	1.87	−0.34	−15.38	−0.72	−27.80
Jiangsu	23.57	−1.53	−6.09	−1.57	−6.20
Zhejiang	14.35	−0.75	−4.96	−1.77	−10.92
Anhui	14.54	−1.91	−11.61	−1.55	−9.69
Regional	54.33	−4.47	−7.58	−5.41	−9.03

Source: As Table AB.11.

TABLE 5.2.   Provincial grain loss rates in the North and North-west regions in China, 1980 and 1981

	1980 Actual output (million tonnes)	1980 Losses as measured against			
		trend output		1979 output	
		(million tonnes)	(%)	(million tonnes)	(%)
	(1)	(2)	(3)	(4)	(5)
North-west					
Neimenggu	3.97	−0.94	−19.14	−1.13	−22.16
Shaanxi	7.57	−0.78	−9.34	−1.53	−16.81
Gansu	4.93	0.20	4.23	0.31	6.71
Qinghai	0.96	0.09	10.34	0.14	17.07
Ningxia	1.20	0.01	0.84	0.14	13.21
Xinjiang	3.89	−0.05	−1.27	−0.05	−1.27
Regional Total	22.52	−1.39	−5.81	−1.52	−6.31
North					
Beijing	1.86	0.02	1.09	0.14	7.51
Tianjin	1.38	0.12	9.52	0.01	−0.72
Hebei	15.22	−1.61	−9.56	−2.58	−14.44
Shanxi	6.86	−0.53	−7.17	−1.15	−14.36
Shandong	23.84	−0.10	−0.42	−0.88	−3.56
Henan	21.48	−0.55	−2.50	0.14	0.66
Regional Total	70.63	−2.75	−3.74	−4.34	−5.79

	1981 Actual output (million tonnes)	1981 Losses as measured against			
		trend output		1980 output	
		(million tonnes)	(%)	(million tonnes)	(%)
	(1)	(2)	(3)	(4)	(5)
North-west					
Neimenggu	5.10	−0.05	−0.99	1.13	28.46
Shaanxi	7.50	−1.09	−12.69	−0.07	−0.92
Gansu	4.35	−0.46	−9.56	−9.58	−11.76
Qinghai	0.80	−0.08	−9.09	−0.16	−16.67
Ningxia	1.27	0.03	2.42	0.07	5.83
Xinjiang	3.90	−0.17	−4.20	0.01	0.26
Regional Total	22.92	−1.68	−6.83	0.40	1.76

TABLE 5.2. *cont.*

	1981 Actual output (million tonnes)	1981 Losses as measured against			
		trend output		1980 output	
		(million tonnes)	(%)	(million tonnes)	(%)
	(1)	(2)	(3)	(4)	(5)
North					
Beijing	1.81	−0.06	−3.21	−0.05	−2.69
Tianjin	1.07	−0.14	−11.57	−0.31	−22.46
Hebei	15.75	−1.37	−8.00	0.52	3.42
Shanxi	7.25	−0.37	−4.86	0.39	5.69
Shandong	23.13	−1.29	−5.28	−0.71	−2.98
Henan	23.15	−0.22	−0.94	1.66	7.72
Regional Total	72.15	−3.54	−4.68	1.51	2.14

Source: As Table AB.11.

Turning to 1981, another recent year of major instability, when the Sichuan basin was extensively flooded, the losses in that province may not have been adequately reflected in our diagram (Fig. 5.3). Grain output fell by only 1.3 per cent from the trend, compared with 6.7 and 7.58 per cent losses sustained in the Centre and East regions in the wake of the equally disastrous Yangzi River basin floods which had occurred a year earlier.

There are, however, two points which deserve to be made. The first is that there was a significant 2 per cent increase in grain-sown area in Sichuan in 1981 to compensate for losses in grain yield per sown hectare, which amounted to 3.30 per cent (from the trend). The second is that the previous year, 1980, was itself marked by a mediocre harvest in Sichuan. More importantly, unlike the North, East, and Centre regions which all recorded bumper harvests in 1979, Sichuan had a relatively bad year, as is revealed in Fig. 5.3.[1] This helped to depress the trend output for the province in subsequent years and to explain its comparatively low rate of grain output losses in 1981, measured against the trend value.

At any rate, 'depressed' grain output for Sichuan in 1981, coupled with slow recovery in the Centre and East regions from the 1980 floods, together with a drastic downturn in output in the South (due to a disastrous flood in Guangdong province), and continuous drought in the

major northern provinces (Table 5.2), all helped to make the year 1981 a
poor harvest year.

It is also instructive to consider the year 1982, which introduced three
consecutive years of exceptionally good harvests for the country as a
whole (Fig. 5.3). There is little doubt that accelerated growth in grain
output in 1982–4 was due to both improved current inputs (chemical
fertilizers, in particular) and peasants' incentives during the decollectivization process. But the effect of the weather is nevertheless also very
evident.

This will be shown in detail in Part III, especially Chapter 8. Suffice it
to say here that the sharp upswing in national grain output in 1982,
accompanied, as shown in Fig. 5.3, rapid output recovery from the trough
of 1980–1 in such critically important regions as the East, Centre, South-
west, and South, but not the North-east, where extensive drought in both
Heilongjiang and Jilin during 1982 continued to keep grain output below
the 1981 trough (due to disastrous floods).

Perhaps the most important point to be made about the relative
significance of the weather and policy (or institutional) changes as a grain
output determinant is that the accelerated output growth in 1982–4 (given
the buoyant mood of decollectivization) was actually no more striking than the major increases in grain output in 1955–7, following the
disastrous Yangzi River floods of 1954 (cf. Figs. 5.2 and 5.3). In particular, the years 1955 and 1956 were, in contrast to 1982–4, characterized by accelerated collectivization. Equally noteworthy is that 1957,
which saw a marked slow-down in grain output growth, witnessed the first
retrenchment ever made from the 'high tide' of rural socialization in the
1950s.

Before we embark on a rigorous analysis of the possible relationships
between the weather and grain yield changes in Part III, we shall attempt,
although it is difficult, to ascertain exactly the relative significance of the
human and non-human factors underlying the observed episodes of great
instability in Chinese agriculture.

The 1959–1961 Episode

Fig. 5.4 shows the historical significance of events during 1959–61 relative
to other episodes of major instability. The comparative scale of fluctuations is measured in terms of grain output, sown area, and yield per sown
hectare, with breakdowns for rice and wheat. Note also that the post-war
years have been divided into just two periods, i.e. 1951–66 and 1970–84,
in order to estimate the year-to-year magnitude of fluctuations around the
trend values. These two periods are of equal length, and are also longer

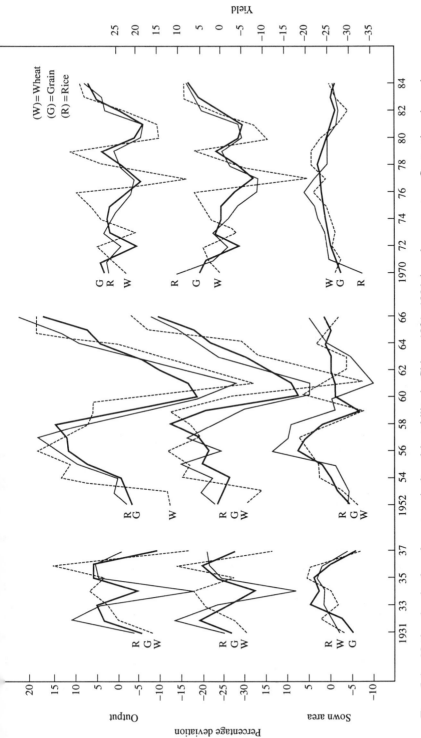

FIG. 5.4. National episodes of great agricultural instability in China, 1931–1984 (yearly percentage fluctuations in grain-sown area, yield, and output from the trend values)

Sources: As Tables AB.3 and AB.4.

than any of the short subperiods hitherto used for comparing intertemporal differences in agricultural instability. This helps to place the
scale of the 1959–61 disasters into proper historical context. Several
points emerge from Fig. 5.4.

First is the enormity of the grain yield and output losses, especially in
1960–1, as well as of the sown-area losses in 1959. Against the trend
values, the grain output losses were 19 and 17 per cent in 1960 and 1961
respectively. These are much greater losses than those of the other
episodes of major instability during 1970–84 or 1931–7. In 1970–84, the
greatest annual loss was 5.9 per cent in 1981. In the pre-war period, the
maximum loss was 7.6 per cent in 1937. The grain-sown area loss in 1959
was 7 per cent, compared with 3.5 and 5.3 per cent in 1931 and 1937, and
a mere 2 per cent in 1982, the year which saw the most dramatic cropping
shifts in favour of non-grain crops under the early post-Mao rural reforms.
More importantly, the spectacular fluctuation in the sown area under
wheat and rice in 1959 abruptly ended the pervasive and forceful sown-
area stabilization trends across the whole of China during 1952–7, discussed in Chapter 3.

Second, in the light of similar experiences in other major grain-producer
countries of the world (cf. Fig. 1.1), the extent of grain output fluctuations in 1960 and 1961 observed in China may not be totally incompre-
hensible. It is, however, interesting that for both rice and wheat crops,
outputs and yields per sown hectare were biased consistently downward
in 1960 and in 1961. The losses in output were at an extraordinarily high
level—29 per cent for rice and 32 per cent for wheat from the respective
trend outputs. This stands in sharp contrast to other years of major
fluctuations, which, as shown in Fig. 5.4, often saw rice and wheat output
deviating from the trend in opposite directions. In other words, in other
episodes of large-scale disturbances good harvests in other regions or
other crop seasons often served partially to offset the losses incurred. The
1959–61 episode was, by contrast, characterized by an absence of such
compensatory adjustments.

The third point to emerge from Fig. 5.4 is the slow recovery from the
trough of 1960–1. It was not until 1964–5 that grain yield and total
output had recovered their previous levels and resumed the pattern of
development observed prior to the 1959–61 episode.

The comparative scale of the 1959–61 disturbance can be further
gauged by the summary statistical measure of the instability index. The
grain output instability index for the period 1952–66 taken as a whole
amounted to 8.92.[2] This is extremely high and reflects the dramatic
fluctuations of 1959–61. Note that the comparable index for 1970–84 was
only 3.3. For the pre-war period 1931–7, which should be more comparable with 1952–66 than with 1970–84 (in terms of technological
setting), the index was no more than 5.35. As a matter of fact, by

virtually all the output, yield, and sown-area measures, the instability index for 1952–66, is a multiple of those for the other periods (see Table AB.3). While technological advances since the early 1970s may have made Chinese agriculture more weather-resistant than in the 1950s or 1960s, a comparison of the degree of instability in 1952–66 with that of 1931–7 and 1952–8, suggests that the powerful institutional hedge associated with the collectivization drive of the mid-1950s, was rendered entirely inoperative during the 1959–61 episode.

Clearly the notorious Great Leap Forward strategy and the concomitant institutional upheavals are implicated in the 1959–61 disasters. There is also no doubt that the three years happened to be affected by large-scale weather anomalies. The question is whether, barring any major human errors, the magnitude of grain output and yield fluctuations in 1959–61 were comparable with those of other episodes of great instability in relation to the scale of weather disturbance. We address this complicated issue in some detail in Chapters 8 and 10.

In contrast to 1959–61, all the other episodes of disturbance fit in well with the historical pattern of fluctuations and also with known weather events, save perhaps for one obscure case. That is the year 1977 which is not even listed in the introduction to this Chapter as one of the seven years of mjor weather disturbance, but which nevertheless saw exceptionally large fluctuations in grain output and yields. This special case merits an explanation.

As shown in Fig. 5.4, grain output and yield per sown hectare in 1977 were reduced by 5.7 and 7.2 per cent respectively from the trend value. The comparable losses in wheat output and yield appear to have been even more striking, amounting to 17 and 20 per cent. Such losses are unmatched in any other historical period since 1931, except for the year 1961. Rice output and yield shared similar though less pronounced trends in 1977.

Two factors may resolve this puzzle. The first reflects a statistical bias. It arises from the fact that the entire period 1970–84 served as a basis for estimating the log-linear trend values which in turn can be used to calculate the given loss rates. A more appropriate approach would be to estimate the trend values separately for the two subperiods 1970–7 and 1978–84. This is because these subperiods are distinctive in terms of trends of grain output and yield changes. This can be seen visually by examining the absolute grain output and yield estimates shown in Fig. 2.2 for the years 1970 to 1984.

The earlier period, 1970–7, was characterized by quite steady rates of increase while the latter period, 1978–84, was associated with rather dramatic upsurges in yield and output as a result of increased and widespread applications of chemical fertilizers and new seed hybrids.[3] Note that wheat yield increased at an annual rate of 8.3 per cent in 1978–84, compared with only 3.5 per cent in 1970–7. The comparable rates of increase for rice yield were 5.2 and 0.9 per cent, and for all-grain yield 6.1 and 2.2 per cent respectively.[4]

By combining the two subperiods 1970–7 and 1978–84, however, the estimated trend line inevitably tends to exaggerate the trend yield (in relation to the actual yield) for the years linking the two periods. This is especially true for wheat yields which grew much faster than rice and all-grain yields during 1978–84. Thus, the exceedingly large loss rates for 1977, shown in Fig. 5.4, are, compared with say, 1972 (i.e. the year of the Great North China drought), largely a matter of statistical bias. If the two subperiods, 1970–7 and 1978–84, are viewed separately in order to estimate the trend yield, then the all-grain yield loss rate for 1977 falls dramatically from 7.2 to 3.1 per cent, while for 1972 it rises from 3.9 to 6.0 per cent. Given that grain-sown area remained relatively stable in both 1972 and 1977, the same conclusion can be drawn with respect to total grain output. These comparative loss rates for 1972 and 1977 seem to be more in line with the known weather situation in the two years.

Granted that most of the overestimates in grain losses in 1977 can be plausibly explained by a statistical readjustment, the remaining 3.1 per cent all-grain yield loss still appears to be substantial. It is higher than the grain yield loss rate of 2.7 per cent for 1980, though still far below the 5.1 per cent loss sustained in 1981, measured against the trend values estimated for the subperiod 1978–84. The question now is whether the 3.1 per cent yield loss in 1977 reflects any real weather disturbance, as in, say, 1954, 1972, 1980, and 1981 in the post-war years, or 1931, 1934, and 1937 in the pre-war period?

Perhaps the dramatic downturn in wheat yield and output in 1977, shown in Fig. 5.4, is partly associated with rural unrest in North China associated with the arrest of the 'Gang of Four' in October 1976, i.e. the time of sowing for cross-winter wheat. We really do not know for sure. It is true that the 'weather story' in 1977 seems to have been overshadowed in the press by reports of political events. What we generally know about the weather in that year is that it was by no means a good year.[5]

Quite a number of major grain-producer provinces did indeed suffer substantial grain output losses in 1977 against the record obtained in 1976, notably Anhui and Jiangsu by 16.3 per cent, Hubei 6.1 per cent, Henan 3.3 per cent, Hebei 2.7 per cent, Shandong 2.2 per cent, and Shaanxi an exceptional 25.1 per cent.[6] There were also important provinces which showed grain output gains in 1977 over 1976. These included Hunan by 5.8 per cent, Jiangxi 10.0 per cent, Fujian 8.0 per cent, Guangdong 10.3 per cent, Sichuan 10.7 per cent, and Zhejiang with a most impressive gain of 37.7 per cent.[7] What is important in this respect is that it seems to have been the weather, rather than any political event in

1977, which helped to differentiate the grain output performances of the various provinces.

At any rate, with or without the somewhat dubious case of 1977, the 1959–61 episode clearly displayed a degree of agricultural instability disproportionately large compared with that of any other notable episode of large-scale weather disturbance.

A Long-Term View, 1931–1984

Setting aside the three years of 1959–61, we must ask how other episodes of major instability stand in relation to the changing institutional and technological settings from the 1930s to 1980s. A glance at Fig. 5.4 shows, first, that all these episodes, namely 1931, 1934, 1937, 1954, 1972, 1980, and 1981, or even 1977, are at once recognizable, showing sharp fluctuations from the estimated trend line, especially in terms of grain yield and output, but also during the 1930s in grain-sown area. The disastrous Yangzi River flood of 1954 may not be adequately reflected in Fig. 5.4, because, being part of the fifteen-year period 1952–66 (which was used as a whole in order to estimate the trend line), its relative magnitude of grain losses is exceeded by that of the three extraordinary years of 1959–61. The diagram in Fig. 5.2, which is based on the shorter period 1952–7, shows however that the fluctuation in 1954 in relation to the trend was dramatic.

No matter how dramatic the fluctuations in grain production may appear to have been in the post-war years, there has nevertheless been a long-term trend of declining instability in terms of the relative magnitude of losses. As shown in Table 5.3, grain output fluctuations in the pre-war episodes of instability were caused either by sharp losses in sown area (as in 1931 and 1937), or reductions in yield per sown hectare (in 1934). The background to this pattern is that large-scale natural disasters either rendered wide stretches of farm area unsowable altogether, as during the Yangzi River floods in 1931 (sown area curtailed by 3.5 per cent), and the 1936–7 North China droughts, which deprived vast areas of water necessary for the sowing of the cross-winter wheat in late 1936 (down by 5.3 per cent); or, if they persisted after sowing was completed, they were reflected in serious reductions of yield per sown hectare, as in 1934 during the great Yangzi and Huai River basin droughts in the summer and autumn (average grain yield down by 7.24 per cent).

The comparative sown-area and grain yield loss rates cited in Table 5.3 for the three years, 1931, 1934, and 1937 thus help to underline the earlier contention that sown area which survived the impact of floods (1931) and droughts (1936/7) tended to have a relatively higher yield (or

TABLE 5.3. Relative magnitude of losses in grain-sown area, yield, and output, as measured against the trend values for notable episodes of great instability in China, 1931–1984 (%)

	1931–7			1952–8	1970–7		1978–84	
	1931	1934	1937	1954	1972	1977	1980	1981
Area	−3.53	+2.85	−5.34	+0.49	+0.52	−0.49	−0.11	−0.92
Yield	−2.51	−7.24	−2.40	−3.46	−6.01	−3.10	−2.70	−5.10
Output	−5.99	−4.58	−7.64	−2.91	−5.30	−3.74	−2.93	−5.97

Source: Table AB.2.

lower loss) rate, compared with areas which after being sown were overwhelmed by natural disaster on a similar scale. This is because a given large-scale weather disturbance is unlikely to persist through the entire cropping season.

At any rate, both weather-bound sources of grain output fluctuations, i.e. sown-area and yield losses, observed in the 1930s, seem to have been significantly curtailed during post-war episodes of major instability. Thus, in none of the years, 1954, 1972, 1977, 1980, and 1981, was the scale of sown-area losses (in relation to the trend) as severe as that sustained in 1931 and 1936–7, as shown in Table 5.3. Note that this includes the year 1981, in which the estimated sown-area curtailment of 0.9 per cent reflects deliberate policy-induced reductions in grain-sown area in favour of cash crops, more than weather-forced readjustments, as happened in both 1931 and 1937.

In fact, as discussed in Chapter 3, there was a strong process of 'area-for-yield' substitution at work in the 1950s, including 1954, which helped, in the wake of yield-impairing bad weather, to reduce substantially grain output losses by increasing the sown area or minimizing possible area losses. All these point to the existence of a powerful 'institutional hedge' associated with collective coercion and mobilization, which helped to stabilize grain production. This has of course been reinforced by increasingly effective harnessing of the major river streams and improved irrigation and drainage capacity in the 1970s and beyond, at least until the decollectivization drive.

As a result, with grain-sown area remaining remarkably stable until the recent past, the 'technological hedge' also helped to stabilize grain yield fluctuations. Thus grain yield losses (in relation to the estimated trend values) during the years of major weather disturbance, 1972 (loss of 6.0 per cent), 1977 (3.1 per cent), 1980 (2.7 per cent), and 1981 (5.1 per cent), were consistently lower than in 1934 (loss of 7.2 per cent). Note that as in post-war episodes of instability, 1934 did not experience any sown-area losses (Table 5.3), although the background is dissimilar: in 1934 the natural disasters struck after sowing was completed, rather than as a result of the collective and technological hedge.

Taken together, the combined effect of sown-area stabilization and reduced losses in grain yield per sown hectare in post-war episodes of large-scale weather disturbance helped to lessen the relative scale of overall grain losses, making them considerably smaller, compared with the pre-war years of 1931, 1934, and 1937.

To conclude, it should be noted that despite the particular institutional and technological hedge available in the post-war years, widespread and prolonged drought (e.g. in 1972), or a basin-wide flood (e.g. 1980–1), was still capable of exhausting existing irrigation and drainage capacity and so causing extensive crop losses over a large region. In short, in China, as well as in advanced Western countries of comparable size, large-scale weather anomalies often remain beyond human control and contribute to observed episodes of major instability.

NOTES

1. Grain output in Sichun province increased by only 0.67% in 1979. This is very much lower than the impressive gains in Hebei (5.6%), Henan (6.7%), Shandong (8.1%), Hunan (6.6%), Hubei (7.2%), Anhui (8.5%), Jiangsu (10.6%), and Jiangxi (15.3%). See K. R. Walker, *Foodgrain Procurement and Consumption in China*, 319.
2. See Table AB.3.
3. Bruce Stone, 'Basic agricultural technology under reform', in Y. Y. Kueh and Robert Ash, *Economic Trends in Chinese Agriculture'*; and Y. Y. Kueh, 'Fertilizer supplies and foodgrain production in China', in *Food Policy*, 9/3 (Aug. 1984), 219–31.
4. Table AB.3.
5. Cf. the computed weather index shown in Table AA.8. The index shows a substantial increase in farm areas covered by natural disasters from 1976 to 1977.
6. The percentage figures are derived from the absolute grain output figures given in Walker, *Foodgrain Procurement and Consumption in China*, 319.
7. Ibid.

III
Major Factors in Chinese
Agricultural Instability since 1931

6

Methods and Problems of Measurement

In our analysis of the history of agricultural instability in China, weather and technological changes have frequently emerged as crucial factors shaping both short-run fluctuations and the long-run stability of agricultural production. Rural collectivization in the 1950s, and the subsequent organizational variations in the different periods of socialization, also seem to have played an important role in this respect, especially when compared with the 1930s, which was dominated by market relations and private ownership. However, our foregoing historical survey has been mainly descriptive, being concerned with the relative impact of the three factors, weather, technology, and institutional changes in *specifically qualitative* terms. We now attempt to determine, in more specific terms, to what extent each of these factors may have contributed to the observed degrees of agricultural instability in the different periods, and their changes over time.

To pursue such analysis involves the difficult problems of defining and measuring each of the three factors. They are therefore considered at some length before embarking on rigorous analysis designed to relate them to measures of agricultural instability in China.

The 'Weather Index'

As Professor Anthony Tang observed, the 'weather [as it impinges on the farmer] is a complex multidimensional phenomenon that defies quantification that is meaningful and operational'.[1] For Professor Kang Chao, it is not meaningful to use 'hydro-thermal indexes and cumulative temperature indexes, even if these indexes could be constructed for China, as a measurement of weather impact on a country with such a vast territory and such great climatic variations'.[2] As a result, Western authors, as well as Chinese scholars, normally resort to the 'categorical weather index' in gauging the weather implications of the yearly fluctuations in harvests.[3] For our purposes, however, the different qualitative 'weather categories', viz. good, average, and bad years, are excessively aggregated classifications. It is not appropriate to relate them to the highly diverse yearly variations in such weather-sensitive measures as grain yield per hectare or total grain output.

To break the impasse, I have made use, in my previous studies, of the official Chinese statistics on the extent of arable land affected by different kinds of natural calamity (notably floods and drought) in order to construct a 'weather index' for the years 1952–81.[4] This is now extended to the year 1984. A similar 'weather index' is also constructed for the prewar period 1931–7. The methodology and major problems involved are discussed in detail in the Appendix, and are merely summarized here.

A distinction is made between farm areas 'calamitously affected' (*chengzai*) and those 'covered' (*shouzai*) by natural disasters. The *chengzai* category is therefore included in the *shouzai* statistics, and it refers specifically to farmland on which a crop loss of more than 30 per cent below the 'normal' yield has been monitored on account of natural disasters. The balance (i.e. *shouzai* net of the *chengzai* area) may be referred to as the *non-chengzai* area which sustained a crop loss of less than 30 per cent. These two area categories are weighted by our assumed average of losses of 60 and 15 per cent respectively, in order to obtain a weighted *shouzai* area index. This has proven to be a good proxy for a 'national weather index' as a basis for analysing the weather influence on agricultural instability in China.[5]

As shown in Fig. 6.1, the 'weather index' is presented as yearly per-

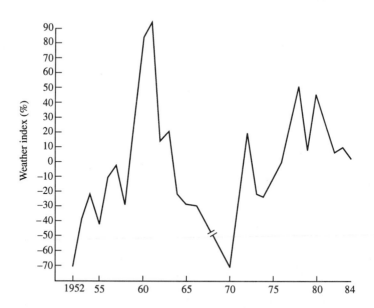

FIG. 6.1. The weather index of China: yearly percentage fluctuations of natural-disasters-covered sown areas from the long-term mean, 1952–1984

Source: Table AA.8.

centage deviations of the weighted *shouzai* area from the 1952–84 mean, although it can also be expressed in terms of absolute hectare size, or as a proportion of total sown area. A positive/negative percentage deviation represents, therefore, weather worse/better than the average weather year. Declining/increasing positive deviations or increasing/declining negative deviations both mean improving/deteriorating weather conditions from year to year.

For the pre-war period 1931–7, *shouzai* area statistics are also available, but no breakdown for the two categories of severity in terms of crop losses is given. Moreover, data for farmland affected by natural calamities other than floods and droughts are incomplete. We can therefore only compile an 'unweighted' *shouzai* area index with respect to these two major weather hazards. As shown in Fig. 6.2, the constructed index for 1931–7 is expressed as the percentage share of *shouzai* area in the total sown area. This is different from the 1952–84 index. The adjustment is made to account for the fact that, unlike the post-war years, the sown-area during 1931–7 itself fluctuated greatly from year to year. Besides, the 1931–7 mean of sown area affected by natural calamities covers too short a period to be representative of the average weather year for judging the relative scale of yearly weather fluctuations.

A number of crucial problems are involved in using the *shouzai* area as

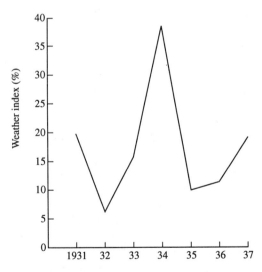

FIG. 6.2. The weather index of China, 1931–1937: sown area covered by floods and drought as a proportion of total sown area

Source: Table AA.8.

a weather proxy. First of all, we do not know how exactly the *shouzai* area, including the *chengzai* statistics, are collected by Chinese officials. For example, the statistics might include losses caused by human errors rather than adverse weather. There seems, however, to have been powerful administrative pressures, in both the post-war and pre-war contexts, against falsifications in this respect.

Secondly, since the *shouzai* area is derived from crop loss statistics, we might have in effect used the 'dependent' variable as proxy for the 'explanatory' variable. This potential 'methodological circularity' (or tautology) can be avoided, however, if we apply the two average crop loss rates (60 and 15 per cent respectively for the *chengzai* and *non-chengzai* areas) indiscriminately to all the relevant years. In this way, the two standard loss rates may be regarded as a set of constant 'weather weights' used for qualifying the changing weather conditions as captured by the non-weighted *shouzai* area series. Note especially that the loss rate of 60 per cent which assumes so much weight in our formula, is derived from the effective losses in the 1950s when progress in agricultural technology had not yet gained momentum. It may therefore be taken to reflect the 'real' weather impact.

Thirdly, some may ask why, after all, we should have preferred the *shouzai* area statistics to the meteorological indices which can be compiled from the monumental contribution of the State Meteorological Bureau (SMB), entitled *Yearly Charts of Dryness and Wetness in China for the Last 500 Year Period 1470–1979*.[6] Briefly, the SMB contribution is basically a summer record of weather events, omitting entirely cross-winter crops. A detailed reconciliation of our *shouzai* area index with the SMB's meteorological records is given in the Appendix to verify the pro-summer bias of the SMB study. The exercise also helps to check the reliability of the *shouzai* area statistics against the SMB records.

The Technological Variables

Agricultural technology embraces a myriad of variables. They include irrigation and drainage capacity, extent of farm mechanization, the multiple-cropping index, advances in scientific seed breeding, and the application of chemical fertilizers per sown hectare.[7] For China, natural fertilizers, which have been collected systematically and with increasing efficiency since agricultural collectivization in the early 1950s, until at least the very recent past, should also be counted.

All these technological elements share, generally speaking, common goals of simultaneously raising and stabilizing farm output per sown hectare. A possible exception is the practice of multiple-cropping. Its excessive expansion in the North without due regard for temperature constraints, for example, may render both early and late crops vulnerable to possible frost damages. As a result, average yield per sown hectare for the year concerned may be reduced, although by virtue of the 'risk-aversion function' of multiple-cropping, there does seem to be a built-in hedge against sharp yield contraction, as was suggested in Chapter 1.

For our analytical purposes, it is important to choose a composite indicator which can embrace the various aspects of farm technological progress. Fertilizers, chemical and natural included, seem to be a good choice in this regard, for their increased application is normally related to, or presupposes greater availability of other technological components, irrigation water in particular.

Another important point concerns the spatial spread of technological progress. In terms of our aggregate analysis, such progress should be fairly widely dispersed to be representative of general trends. In this respect, our previous studies show that there has indeed been marked interprovincial equalizing trends in irrigation technology, and the application of chemical fertilizers since the early 1950s.[8] The supply of natural fertilizers poses fewer problems, for it is basically subject to the natural growth of the generating sources. These are predominantly the human population, pigs, and draught animals, which as part of the agricultural system exist wherever land is under cultivation.

A complete set of official statistics on chemical fertilizers is available for the post-war periods. For natural fertilizers we have established, in a separate study, an independent time series for 1952–81.[9] This is combined with chemical fertilizer data to form an integrated series of fertilizer inputs as a comprehensive measurement of farm technological changes.[10] Fig. 6.3 shows the combined chemical and natural fertilizers input for 1952–81 in relation to grain yields in kilograms per sown hectare. There are two aspects of the fertilizer–yield relationship which are important in any consideration of the relative influence of technology in Chinese agriculture.

The first is the yield-raising effect of advances in agricultural technology. Fig. 6.3 shows that increases in grain yields are closely related to increased application of fertilizer technology. There are upward and downward deviations from the long-run rising trend relationship between the two variables (fertilizer input and grain yield), but the deviations seem to be marginal in relation to the upward trend.[11] The second aspect of the fertilizer–yield relationship concerns the magnitude of observed oscillations around the trend line. As shown in Fig. 6.3, this has tended to diminish over time in relation to the ever-increasing grain yield bases.[12]

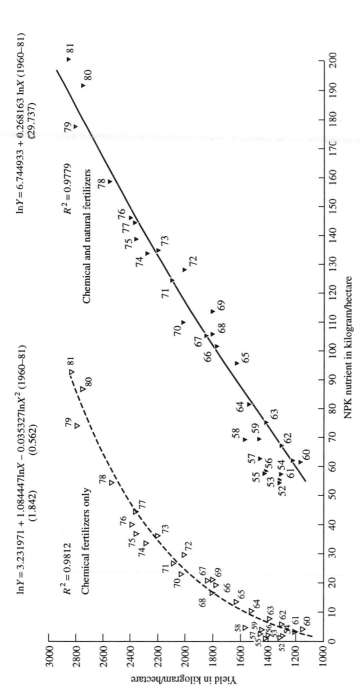

$$\ln Y = 6.744933 + 0.268163 \ln X \ (1960-81)$$
$$(29.737)$$

$$R^2 = 0.9779$$

Chemical and natural fertilizers

$$\ln Y = 3.231971 + 1.084447 \ln X - 0.035327 \ln X^2 \ (1960-81)$$
$$(1.842) \qquad\qquad (0.562)$$

$$R^2 = 0.9812$$

Chemical fertilizers only

NPK nutrient in kilogram/hectare

Yield in kilogram/hectare

FIG. 6.3. Average grain yield in kg./sown ha. in relation to the application of fertilizers in nutrient (N, P_2O_5, K_2O) weight kg./sown ha. in China, 1952–1981

Notes: The per hectare application of fertilizers is based on total sown area, including non-grain crops. It is impossible to make a separate estimate for grain crops alone, especially as it pertains to the supply of natural fertilizers. But this should not pose too serious a problem, given the predominant weight of the grain crops. Moreover, as we are primarily concerned with changes in agricultural technology in general, the per hectare figures as used here should be adequate to serve as an indicator. No effort was made to extend the estimate of natural fertilizer supply to 1984 (the bench-mark year for this study). Time and resource constraints aside, the disintegration of the commune system since 1982 renders it increasingly difficult to measure the exact size of the relevant rural population which is the single most important source of organic fertilizer supplies. At any rate, the period covered here, 1952–81, is long enough to illustrate the post-war agricultural–technological trend in China.

This reflects the yield-stabilizing effect of improved farm technology, including improved flood prevention and modernized irrigation facilities.

Thus, over the long-run, we may treat the rising trend relationship between fertilizer applications and grain yields as a 'secular trend line' for isolating the technological influence. Any short-run or yearly deviations from the trend may therefore be attributed to the weather or policy factors; while the changing relative scale of such fluctuations over time may reflect the weather or policy influence against the comparative technological or institutional settings.[13] The sharp downturn in grain yield observed in both 1980 (the Yangzi River flood) and 1981 (Sichuan basin flood), as well as in 1972 (the great North China droughts) are already clearly visible in Fig. 6.3. Note that the losses (measured against the trend or the record of the previous year) occurred, despite continuous increases in fertilizer input (chemical and natural combined) in these three years.

A word or two should be said about the pre-war years 1931–7. Irrigation practice was widespread in the 1930s, as had been the case in China's previous history. However, in the words of Buck, 'The water supply is anything but constant'.[14] Drainage capacity was also limited, with most major rivers and large streams remaining unharnessed. Again according to Buck, 'until the main drainage arteries are made effective, local drainage systems will be of limited value'.[15] Hardly any farm machines, mechanical or electrical, were in use in the 1930s, and there were virtually no chemical fertilizers.[16] As a result, we can treat the pre-war period 1931–7 as characteristic of 'traditional agriculture' and limit our analysis of technological influence to post-war Chinese agriculture.

Policy and Institutional Factors

The contribution of policy and institutional changes as a factor in agricultural fluctuations cannot easily be quantified. Their role has, however, been subject to considerable speculation in the post-war context, in particular its impact on periodic or annual variations in Chinese agricultural performance.

Our approach to this issue is twofold. First, it is necessary to define the weather and technological variables as precisely as possible in order to isolate their possible influence from that of policy and institutional variations. Second, a clear distinction must be made between the different policy periods after agricultural socialization in order that a comparative evaluation of the various periods, in relation to given weather and technological variables, can help to identify possible policy influences.

Specifically, our methodology is to relate, by way of standard regression analysis, the yearly percentage deviations of grain yield per sown hectare from the technological trend line (Y), to the computed 'weather index', i.e. the yearly percentage deviations of the weighted *shouzai* area from the long-term mean (W). We estimate, therefore, the regression equation,

$$Y = \alpha + \beta W + \mu$$

where Y and W are as defined, and μ is a disturbance term, α a constant, and β is the parameter to be estimated which promises to show the relative scale of weather influence. Any fluctuations in grain yields which cannot be accounted for by the weather index are thus captured by the 'disturbance term', to be explained by possible policy or institutional factors.

The regression estimates are made separately for different periods which have already been defined as having distinct institutional (1931–7 versus the different post-war periods) policy (between the different post-war periods) and technological (1952–66 versus 1970–84) characteristics. A comparison of the differences between the estimated disturbance terms through the various periods may thus help to reveal the possible long-term implications of human factors relative to weather influence. Likewise, a juxtaposition of yearly fluctuations in grain yields and the 'weather index' (*shouzai* area) may show the possible impact of any short-run policy readjustments *vis-à-vis* that of the weather.

Since the 'secular technological trend' as shown in Fig. 6.3 is rather similar to the log-linear trend line used in Part II for estimating the yearly grain yield deviations, we can conveniently apply the estimated deviations for the different periods to the computed 'weather index' in order to examine the weather and yield relationships. In other words, we shall specify to what extent the observed trends and patterns of agricultural instability discussed in Part II are determined by the weather relative to the non-weather factors.

The post-war rural context of China seems to be especially amenable to the proposed regression analysis. This is because agricultural collectivization, control over migration and rural occupation, compulsory physical output procurement, and, until recently, strict limitations on rural non-farm economic activities, have all contributed to minimize other potential non-weather sources of agricultural instability experienced during the pre-war period. This makes the simple weather–yield regression models very appealing for our analytical purposes.

Four different statistical measures have been defined (in Chapter 2), and used (Chapters 3–5) for our historical analysis of agricultural instability in China from the early 1930s. These are grain output, grain-sown area, and grain yield in kilograms per hectare, as well as the broad measure of GVAO. But in our subsequent analysis of the relative influence of the weather, we focus mainly on the grain yield series, although the GVAO series will also be used to show how the broader rural economy may have been affected. The reasons for this choice are clear. The physical grain yield series eliminates possible price influences, and allows abstraction from policy-determined changes in sown area. Of course, annual and regional variations in grain-sown area through geographical shifts—for example, reductions in the dry and less fertile North-west—may still affect average grain yield and therefore yield instability.

The aggregate *shouzai* area index which is used as a proxy for weather variation is also by no means perfect for it ignores differences in the yields of the different crops; differences between regions in the capacity to withstand drought and flood; and regional differences in the impact of natural disasters. Such considerations suggest that our weather and yield regression analysis may underestimate the possible scale of weather influence. This crucial qualification must be borne in mind in interpreting the relative importance of the weather and human factors in influencing agricultural stability in China.

NOTES

1. Anthony M. Tang, *An Analytical and Empirical Investigation of Agriculture in Mainland China, 1952–1980*, 122.
2. Kang Chao, *Agricultural Production in Communist China* (Madison, Wis.: University of Wisconsin Press, 1970), 240.
3. See Table AA.9 and Tang, *Agriculture in Mainland China*, 123.
4. Y. Y. Kueh, 'A weather index for analysing grain yield instability in China, 1952–1981'.
5. Cf. Kueh, 'Weather cycles and agricultural instability in China', *Journal of Agricultural Economics*, 37/1 (1986), 101–4. A similar approach of using the sown areas affected by natural disasters has also been adopted for some Western countries for identifying the weather sources of fluctuations. See e.g. K. Takahashi and M. Yoshino (eds.), *International Symposium on Recent Climatic Change and Food Production*, chapter by E. Maruyama on Indochina.
6. State Meteorological Bureau (Institute of Meteorological Science), *Zhongguo Jin Wubainian Hanlao Fenbu Tuji* (Beijing: Ditu Chubanshe, 1981). Considerable pre-publication publicity was given in the West (see *New York Times*, 15 Feb. 1979, B-11, and *DGBHK*, 20 Oct. 1978, 9), but this major work seems not to have been made much use of since its publication in 1981.
7. For a good survey of the Chinese achievements in this respect, see Bruce Stone, 'Basic agricultural technology under reform', in Kueh and Ash, *Economic Trends in Chinese Agriculture*.
8. Kueh, 'Technology and agricultural development in China: regional spread and inequality,' 553–60.

9. Kueh, 'Fertilizer supplies and foodgrain production in China, 1952–1982', *Food Policy*, 9/3 (1984), 219–31. Our series is substantially different from similar earlier estimates made by other Western scholars, including O. L. Dawson, 'Fertilizer supply and food requirements', in J. L. Buck, O. L. Dawson, and Y. L. Wu (eds.), *Food and Agriculture in Communist China* (New York: Praeger for the Hoover Institution, 1966), 101–48; Kang Chao, *Agricultural Production in Communist China*, 311. Shigeru Ishikawa, 'Changes in the structure of agricultural production in mainland China', in W. A. Douglas Jackson (ed.), *Agrarian Policies and Problems in Communist and Non-Communist Countries* (Tokyo: University of Tokyo Press, 1971), 368–70; and Tang, 'Food and agriculture in China: trends and projection, 1952–77 and 2000', in A. M. Tang and Bruce Stone (eds.), *Food Production in the People's Republic of China* (Washington, DC: International Food Policy Research Institute, 1980), 60–4.

10. For the combination, the estimated amount of natural fertilizers supplies in gross nutrient weights was converted into chemical fertilizer equivalent based on the comparative plant absorption rates of natural and chemical fertilizer. See Kueh, 'Fertilizer supplies and foodgrain production', for details.

11. As shown in Fig. 6.3, this is especially so with the combined series of natural and chemical fertilizers, as opposed to the series of chemical fertilizers only. The former series displays a consistent trend of constant returns to scale over time, while the latter exhibits diminishing response rates from around the mid-1970s. This is clearly caused by the accelerated application of chemical fertilizers to make up for the relative lack of natural fertilizers, in order to maintain a desired rate of increase in crop yield and output to cope with increased demand from an ever-increasing population. The observed diminishing returns of grain yield to chemical fertilizer input should therefore be interpreted as more a matter of statistical correlation, than a technical necessity. Cf. Kueh, 'Fertilizer supplies and foodgrain production', 219.

12. The absolute magnitude of downward fluctuations of grain yield from the trend line in 1980 and 1981, as shown in Fig. 6.3, may be comparable to that in, say, 1972, but in view of the much higher yield base in 1980–1 the percentage losses are bound to be lower than in the earlier year.

13. This is a common procedure which many Western analysts have adopted in order to determine the weather influence relative to the technological trend; see e.g. Louis M. Thompson, 'Weather and technology in the production of corn in the US corn belt', *Agronomy Journal*, 61 (1969) and 'Weather and technology in the production of soybean in the Central United States', *Agronomy Journal*, 62 (1970). Cf. also Douglas B. Diamond and W. Lee Davis, 'Comparative growth in output and productivity in US and USSR agriculture', in *Soviet Economy in a Time of Change*, ii, JEC, US Congress (Washington, DC: US Government Printing Office, 1979), 19–54.

14. J. L. Buck, *Land Utilization in China*, 189.
15. Ibid. 190.
16. Ibid. 191; and T. H. Shen, *Agricultural Resources of China* (Ithaca, NY: Council University Press, 1951), 38.

7

The Long-Run Interplay of Weather, Technology, and Institutions

The main purpose of this chapter is to compare the impact of weather in different periods, which, each taken as a whole, represent either different institutional characteristics or technological settings. A comparison of two periods, 1931–7 and 1952–9, which are similar technologically, is used in an attempt to confirm the contention made in Part II that rural collectivization significantly helped to stablize Chinese agriculture relative to the weather influence. Likewise, by contrasting the two longer periods, 1952–66 and 1970–84, which display the same basic socialist collective arrangements, we seek to prove that advances in agricultural technology since the early 1970s have lessened the negative weather impact.

In addition to institutional and technological factors, we also examine how policy variations between different subperiods, i.e. from 1952–8 (remunerative approach) to 1970–7 (coercive approach), and 1978–84 (decollectivization drive) may have impinged upon agricultural stability in China within the broad socialist context. This chapter is, therefore, an analysis of the sources of long-term trends of agricultural instability. It is different from the problem of short-term or year-to-year fluctuations within an unchanging technological, institutional, or policy framework. This problem will be analysed in Chapter 8 with respect to the possible implications of weather disturbance as against that of discrete readjustments in agricultural policy.

The present chapter is divided into two parts. The first attempts to separate the relative influence of weather and technology on grain yield changes over the entire post-war period, and determines their combined impact in terms of possible policy implications. The second part analyses the comparative weather and policy influence in relation to the institutional or technological trend. In other words, we abstract from the long-term yield and output-raising effect of technological progress, and look at the changing relationships between weather disturbance and agricultural fluctuations in order to examine the relevance of policy changes from period to period. The analysis is preceded by a discussion of the radical institutional transformation from the pre-war years 1931–7 to 1952–9, and the technological transition from 1952–66 to 1970–84.

Weather versus Technological Progress

We take the application of fertilizers (chemical and natural combined) as a proxy for agricultural technology. The annual level of application in terms of kilograms per sown hectare in nutrient weight which shows a long-term (1952–81) upward trend, as revealed earlier in Fig. 6.3, is taken, together with our weather index, as the explanatory variables for the annual increases and fluctuations in grain yields (also in terms of kilograms per sown hectare). The regression estimates shown in Table 7.1 suggest the following major conclusions.

First, the combined input of chemical and natural fertilizers and the weather are capable together of explaining around 85 per cent of grain yield variations in China during 1952–66 and 1970–81. Both factors—weather and fertilizer technology—are highly significant explanatory variables from a statistical point of view.

Second, fertilizer input is a far more important factor than the weather in explaining year-to-year grain yield variations.[1] This is not surprising, since it is the increased use of fertilizers which has consistently brought about the upward movement of grain yield over time. In other words, the yield-raising effect of farm technology is so marked that it has clearly overriden the importance of weather-caused, yearly yield fluctuations, from a long-term perspective.

Third, the importance of fertilizer technology increased substantially from the period 1952–66 to 1970–81 relative to that of the weather. This implies that with a low technological level prior to the mid-1960s, Chinese agriculture was more vulnerable to weather disturbance than in the later period. However, the increased influence of fertilizers can also be explained by the fact that, at prevailing high levels of application, any variations in their application can easily result in yield and output changes. Moreover, the increase in fertilizer use has taken place on a basis of improved irrigation and drainage technology, which helps to insulate the effects of adverse weather, hence declining weather influence at a time of increased fertilizer use.

Perhaps the most important inference is that frequent changes in agricultural policy in the post-war context seem to have had comparatively little impact on the year-to-year changes in grain yield relative to the dominating roles of the weather and technology. This coincides with the conclusion drawn by James McQuigg, Director of the National Climate Assessment Center, on the basis of the US experience, to the effect that 'if one has a regression model in which the components of variability attributed to technology and the component attributed to climatic variability are expressed quantitatively, one can explain from 85 to 90 per cent of the year-to-year variability of yields'.[2]

TABLE 7.1. Weather (W) and fertilizer technology (F) as determinants of yearly variations in grain yield (Y) in China, 1952–1981 and subperiods (estimated regression equations)

	Intercept	Regression coefficients		r^2	t-significance	
		Fertilizer	Weather		F	W
1952–66	733.05534	9.9137100	−1.477422	0.8362	0.0001	0.0051
		(5.727)	(3.484)			
1970–81	745.306428	10.901191	−2.005726	0.8571	0.0001	0.0189
		(6.911)	(2.933)			
1952–81	637.698656	11.541151	−1.727660	0.9336	0.0001	0.0001
		(17.635)	(5.018)			

Notes: Fertilizer technology (F) is taken to be the combined application of chemical and natural fertilizers in kilograms per sown hectare in nutrient weight. See notes to Fig. 6.3 for details. The weather variables (W) are percentage deviations of the weighted *shouzai* area from the 1952–84 mean, excluding the three years 1967–9 for which data are not available. Variations in grain yield (Y) refer to the actual annual yields in kilograms per sown hectare.

The regression equation used for estimating the relative influence of weather and fertilizer technology is, thus:

$$Y = \alpha + \beta_1 F + \beta_2 W + \mu_1$$

where, in addition to the defined symbols, Y, F, and W, μ denotes the residual term, and β_1 and β_2 are the parameters to be estimated. All the estimated regression equations are corrected for 'first-order autocorrelation' by using the maximum likelihood procedure. That is to say, the estimated fertilizer–yield or weather–yield relationships, shown by the respective regression coefficients, are net of possible cumulative year-to-year impact which is a phenomenon common to most economic time-series. See Y. Y. Kueh, 'Weather cycles and agricultural instability in China', *Journal of Agricultural Economics*, 37/1 (Jan. 1986), 102–3, for a discussion of the factors operating in China to cause 'autocorrelation'. Note that the signs for all the regression coefficients are correct, in the sense that, as expected, fertilizer inputs and the weather variables are positively and negatively correlated with the changes in grain yields. The estimates (with the t-value given in parenthesis), are all statistically highly significant.

Sources: AB.2 for grain yield; AB.13 for the fertilizer data; AA.8 for the weather variables.

If anything, the Chinese model is probably a more fitting one than that of the USA. This is not surprising, for within the fully marketized and monetized US context, farm decisions are influenced by price changes and the government's agricultural policy (e.g. farm-aid programme and stockpile policy). In China, however, at least until the very recent past, rural collectivization and bureaucratic physical output control, have helped to contain potentially destabilizing market behaviour.

Institutional and Technological Transformation

Table 7.2 shows the regression estimates of the relationship between the weather and the yearly fluctuations in grain yields and GVAO (both measures net of the secular trend influence) for various periods from 1931 to 1984. The grain yield series is used as a proxy for changes in the farm sector proper, while GVAO relates to the broader rural economy as a whole.

We first comment on the radical transition from pre-war conditions during 1931–7 to rural collectivization in 1952–9. Note again that both periods were characterized by an absence of advanced agricultural technology. Any reduction in the weather influence in the latter period attests to our hypothesis about the stabilizing effect of collectivization. The two periods are not only of nearly equal length, but also displayed similar weather patterns. Each period experienced both a large-scale flood (in 1931 and 1954) and widespread drought (in 1934 and 1959), affecting almost identical regions.

As shown in Table 7.2, weather conditions can explain virtually the whole of the yearly fluctuations in grain yields during 1931–7. This is not quite the case in 1952–9, although about three-quarters of the yearly variations in grain yields are associated with the weather factor. This picture of reduced weather and yield correlation is not basically altered, even if we omit the Great Leap Forward year of 1959, for which the relative importance of the weather and policy factors needs to be clarified in interpreting the sizeable fluctuation in grain production.

Viewed against the background of sharply reduced grain yield instability between 1931–7 (index 3.57) and 1952–8 (index 2.02),[3] the 'institutional hedge' seems to have worked powerfully in the 1950s to reduce the potential weather influence. In addition to collective coercion as a hedge against the impact of floods and droughts, there was a specific policy factor which also helped to stabilize grain yields during 1952–9. This was the remarkable substitution of 'area-for-yield' in 1956 (discretionary sown-area expansion to compensate for possible yield losses), and 'yield-for-area' in 1958 (sown-area contraction in favour of the fertile land against the marginal area).[4] The large-scale sown-area contraction (by 9 per cent) in 1959 under the misconceived 'three-three system'[5] of arable land utilization must have also contributed to yield stabilization. Not only did such contraction involve only infertile marginal land, but against the background of deteriorating weather conditions peasants' 'subsistence urge' seems to have made itself felt in an attempt to raise average yields,

TABLE 7.2. The relative scale of weather (W) influence on the yearly variations in grain yield (Y) and GVAO (V) in China, 1931–1984 and subperiods (estimated regression equations)

	Intercept	Regression Coefficient	t-value	r^2	t-significance
Institutional transformation without technological progress					
1931–7 (Y)	7.55802	−0.41717W	26.312	0.9943	0.0001
1952–9 (V)	9.27796	−0.58986W	4.081	0.7692	0.0095
Technological changes under socialist setting					
1952–66 (Y)	−0.07232	−0.15666W	3.725	0.5524	0.0029
(V)	0.55270	−0.12655W	2.246	0.3324	0.0320
1970–84 (Y)	0.87110	−0.10182W	3.584	0.5174	0.0038
(V)	3.98668	−0.08140W	2.247	0.2943	0.0442
Between the remunerative and coercive approaches					
1952–8 (Y)	−3.67630	−0.12313W	3.972	0.7972	0.0165
(V)	−0.11621	−0.00346W	0.102	0.0034	0.9237
1970–7 (Y)	−0.9765	−0.06423W	1.897	0.4759	0.1162
(V)	−0.8716	−0.05127W	2.814	0.6597	0.0374
1978–84 (Y)	2.63122	−0.09551W	1.454	0.3900	0.2197
(V)	1.95440	−0.04579W	0.551	0.0833	0.6111
The entire post-war context					
1952–84 (Y)	3.82154	−0.13483W	5.476	0.5273	0.0001
(V)	9.86932	−0.09503W	2.904	0.2380	0.0073

Notes: Except for the periods 1952–9 and 1931–7, the weather variables (W) are percentage deviations of the weighted *shouzai* area from the 1952–84 mean, omitting the three years 1967–9 for which data are not available. For 1952–9, W represents *shouzai* area as a proportion of total sown area. This makes 1952–9 comparable to 1931–7, for which W is similarly defined. Moreover, both series cover flood and drought only, as *shouzai* data for other types of natural disasters are incomplete for 1931–7. Unlike the 1931–7 series, however, the 1952–9 series is a weighted *shouzai* area index derived in the same way as that for the other post-war periods. Thus, the regression coefficients for these two periods are also not quite comparable, and neither is directly comparable with those estimated for the post-war periods.

The variations in grain yield (Y) and GVAO (V) refer to the yearly percentage deviations of the actual values from the log-linear trend values.

All the estimated regression equations shown here are corrected for 'first order autocorrelation' by using the Maximum Likelihood procedure. See notes to Table 7.1 for an explanation. Note also that all the regression coefficients show a negative sign, implying that the weather is correctly correlated with the changes in grain yield and GVAO. The estimates show that the weather is a highly significant explanatory variable for the yearly variations in both grain yield and GVAO. The only notable exceptions are, for good reasons, the two periods 1952–8 and 1978–84, as explained in the text.

Sources: The weather variables: Table AA.8 for 1952–84 and 1931–7, and Table AB.15 and *TJNJ 1983*, 154 for 1952–9 (total sown-area base). Grain yield and GVAO: Tables AB.2 and AB.12.

for fear that the disproportionate area contraction should precipitate an output and food crisis.

The sown-area losses resulting from the 1954 Yangzi River flood had a similar yield-raising effect, in that unaffected farmland which could still be effectively sown later that year tended to have higher yields (it was less likely for natural disasters to recur within the same cropping season). A similar 'stabilization' factor was also present in the 1931 Yangzi River flood. In this context, there is another factor which may account for the relative stability of grain yield in 1952–9. That is, in contrast to 1931–7, total sown area in 1952–9 also included the entire North-east and a greater part of the North-west. The broader geographical base may have rendered overall yields less susceptible to the random effects of natural disasters. It is impossible to verify such effects. But the scale of weather disturbance, captured by the *shouzai* area ratio (i.e. the proportion of sown area covered by floods and droughts) was very similar in 1952–9 and 1931–7, averaging respectively 16.2 and 16.7 per cent.[6]

In short, it is probably more the powerful 'collective hedge', rather than any lack of weather disturbance, that helped to stabilize grain yield between 1931–7 and 1952–9. This is bound up with discretionary policy manipulations (made possible by collectivization) of the sown-area and the cropping patterns, designed to improve stability, especially in 1956 and 1958, when a marked sown-area shift towards high-yielding crops took place.

The institutional hedge has clearly been reinforced by advances in agricultural technology to bring about further reductions in agricultural instability during the post-war context. This can be seen in the transition from 1952–66 to 1970–84, both periods defined in terms of broad differences in their technological setting. As Table 7.2 shows, there was a consistent decline in the relative scale of weather influence from the first to the second period with respect to both the grain yield and GVAO measures. For 1952–66, the estimated regression coefficient shows that an average variation in the weather condition of 1 per cent, whether positive or negative, as defined for our estimates, is correlated with shifts of 0.16 per cent in grain yield and 0.13 per cent in GVAO in exactly opposite directions in terms of percentage deviations from the estimated trend values. For 1970–84, however, the relative scale of weather influence was reduced by well over one-third to only 0.10 per cent and 0.08 per cent for the grain yield and GVAO series. And statistically speaking, the weather was no longer as powerful an explanatory variable in 1970–84 as it had been in 1952–66.

Note also that in both periods 1952–66 and 1970–84, the relative scale of weather influence on GVAO was smaller than that on the grain yield series. This is to be expected, for GVAO is a highly aggregated value measure comprising a number of less weather-sensitive components compared with the physical yield series. Nevertheless, from 1952–66 to 1970–84 the estimated weather influence on GVAO declined in a manner perfectly consistent with that operating on grain yields. This reflects the dominant position of grain output in GVAO, and the fact that inter-industrial linkages within the rural context were heavily weighted by agricultural output for the two broad periods.

The Effect of Periodic Policy Variations

We first look at the transition from the subperiods 1952–8 to 1970–7. Consistent with the changes between the two broad periods, 1952–66 and 1970–84, the regression coefficients for the grain yield series given in Table 7.2 show that the relative scale of weather influence was reduced by nearly half from 1952–8 (−0.123) to 1970–7 (−0.064). In both periods, especially the earlier one, the weather sufficiently explains the observed grain yield variations.[7] Curiously, however, while the GVAO regression for 1970–7 reveals a regression coefficient (−0.051) consistent with that of grain yields (−0.064) during the same period, there is a striking absence of correlation between the weather and GVAO during 1952–8 (Table 7.2).

This discrepancy seems puzzling, given the close economic and statistical relationships between variations in grain yield and GVAO. It must be resolved before we can proceed further.

A careful scrutiny of grain yield, output, and GVAO data, presented in Fig. 7.1, immediately points to the years 1956 and 1958 as the source of the 'inconsistency'. In both of these years GVAO deviates sharply (up in 1956, but down in 1958) from the trend values and in the opposite direction to the grain yield values. The reason is that in 1956 the grain-sown area increased by a record 5 per cent and rice-sown area by 14 per cent, as shown in Table 7.3. These increases overcompensated for the reduction in yield per sown hectare, so that total grain output and hence GVAO continued to grow impressively in 1956, following the bumper harvest of 1955. This 'area-for-yield' substitution produces the positive percentage deviation of 1956 GVAO from the trend value, in contrast to the negative deviation of grain yield in the same year.

In 1958 exactly the opposite occurred. The average grain yield per sown hectare increased by 7.4 per cent, partly reflecting the switch to high-yielding grain crops in 1958, potatoes in particular. However, total grain output and hence GVAO did not rise proportionately. Their growth (by only 2.5 and 2.4 per cent respectively) was in fact considerably slowed down by a massive 4.5 per cent reduction in total grain-sown area, a

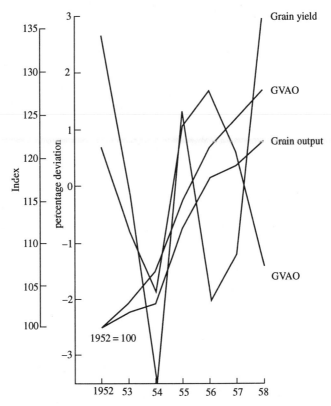

FIG. 7.1.   Yearly fluctuations of the national average of grain yields per sown hectare and gross value of agricultural output (GVAO), in comparison with the index of grain output and GVAO changes in China, 1952–1958

Source: As Tables AB.2 and AB.12.

reduction second only to the dramatic contraction under the 'three-three system' in 1959.

This modest GVAO growth in 1958 (by 2.4 per cent) was preceded by sharp increases in both 1955 (by 7.6 per cent) and 1956 (by 5.1 per cent), and to a lesser extent in 1957 (by 3.6 per cent). The outcome, as can be seen in Fig. 7.1, was a negative deviation of 1958 GVAO from the trend value, contrary to the behaviour of the grain yield series. Had it not been for the yield-raising 'yield-for-area' substitution, the reduction in sown area in 1958 would certainly have caused an even greater negative GVAO deviation, via its effect on the curtailment of grain output in that year.

In short, it is the peculiar policy-induced expansion (in 1956) and contraction (in 1958) in the sown area (and hence in total grain output)

TABLE 7.3. Year-to-year percentage changes in grain- and rice-sown areas in China, 1953–1984

Year	Negative changes		Year	Positive changes	
	Grain	Rice		Grain	Rice
1953		−0.21	1953	2.10	
1957	−1.98	−3.20	1954	1.90	1.42
1958	−4.50	−1.01	1955	0.70	1.57
1959	−9.10	−9.03	1956	5.00	14.19
1961	−0.81	−11.25	1960	5.50	1.98
1963	−0.72		1962	0.15	2.51
1965	−2.03		1963		2.90
1967	−1.45	−0.30	1964	1.10	6.83
1968	−2.24	−1.78	1965		0.74
1973	−0.04	−0.15	1966	1.10	2.36
1974	−0.15		1969	2.09	1.80
1976	−0.26		1970	1.40	
1977	−0.28	−1.91	1971	1.30	7.91
1978		−3.11	1972	0.30	0.64
1979	−1.18	−1.59	1974		1.20
1980	−1.70		1975	0.07	0.61
1981	−1.94	−1.72	1976		1.37
1982	−1.30	−0.67	1978	0.16	
1984	−1.02	−0.13	1980		0.02
			1983	0.52	0.20

Sources: Table AB.3 and *TJNJ 1983*, 154–5.

which made the closely related GVAO measures 'dilute' the weather influence in these years, as well as for the period 1952–8 as a whole. In subsequent periods, however, there was much less room for such policy manipulation of the sown area. Against the background of a basically constant sown-area size, GVAO tended to vary with grain output, which in turn fluctuated closely with yield per sown hectare and the weather as a source of disturbance.

We may now offer some comment on the reduced weather influence from 1952–8 to 1970–7, and during the subsequent period 1978–84. It is not surprising that, in addition to the improved technological hedge, greater collective control and massive concentration of rural resources in favour of grain production during the Cultural Revolution of 1970–7 helped to further stabilize grain yields per sown hectare against the weather impact. In fact, the difference in the relative scale of weather

influence between 1952–8 and 1970–7 would certainly have been even
larger had it not been for the 'yield-for-area' substitution process of
1952–8, which helped to 'dilute' the weather impact in those years
by stabilizing national average grain yield per sown hectare. With the
pro-grain strategy and associated Maoist ideology reducing less weather-
sensitive non-farm economic activities to a minimum, GVAO in 1970–7
also fluctuated closely with grain yield, as shown in Table 7.2. In these
circumstances, any residual variations in grain yields were bound mainly
to reflect weather disturbance.

After 1978, however, the weather seems increasingly to have gained
momentum again as a grain yield determinant. This is shown, in Table
7.2, by the reversal in the estimated relative scale of weather influence for
1978–84 (regression coefficient −0.09551) from the minimum reached in
1970–7 (−0.06423). Nevertheless, the observed weather–yield correlation during 1978–84 is, statistically speaking, no longer as coherent as
in the earlier periods. As for GVAO, not only was the scale of weather
influence reduced to an all-time low (regression coefficient −0.04579),
but the weather itself was of little relevance in explaining GVAO fluctuations in 1978–84. A number of reasons can be given for these estimated
changes.

First, Chinese peasants responded actively to the sharp improvements
in farm prices and accelerated decollectivization measures, by taking
advantage of the good weather, especially in 1979 (output prices increased
by an average of 25 per cent) and 1984 (effective reparcellization of
collective farmland) to expand output and so reap personal income
benefits.[8] This naturally tended to exaggerate the positive impact of the
weather in relation to the previous Cultural Revolution period, when
equality in income distribution no doubt dampened the enthusiasm of
peasants in the face of potential output gains.

Second, at times of bad weather, say in 1980 (Yangzi River flood) and
1981 (Sichuan basin flood), peasants in pursuit of household income
maximization might not have found it worth while to reallocate resources
in order to lessen the weather impact. The opportunity costs incurred
were likely to be too high in the light of alternative earning potentials
offered by post-1978 accelerated rural diversification and marketization
campaigns. Such behaviour, coupled with increased rural decontrol,
certainly made the weather seem worse than in the previous period, when
strict limitations on locational and occupational mobility helped propel
subsistence-orientated Chinese peasants towards self-survival measures in
the face of threatening flood and drought.[9]

Third, the process of economic diversification greatly enlarged the rural
non-farm sector, thus rendering the GVAO measure increasingly less
sensitive to weather disturbance.

It is probably against this background of increased rural diversification and marketization that both grain yields and GVAO became less stable in 1978–84 compared with 1970–7, and 1952–8 (disregarding the yield and output-stabilizing effect of the 'area-for-yield' substitution (and vice versa), as illustrated in Chapter 3). Yet the weather became increasingly irrelevant in explaining increased agricultural instability in 1978–84. To the extent that the weather was responsible, its influence on Chinese agriculture (as measured by the grain yield series) in the post-Mao era remained substantially less than in 1952–8, despite the observed reversal from the minimum reached during the Cultural Revolution period.

This finding underlines, once again, the contribution of yield-stabilizing agricultural technology, which was available in 1978–84, but not in the 1950s, when reduced yield fluctuations (compared with the 1930s), mainly reflected the collective institutional hedge. In this respect, the 1970–7 period enjoyed a double benefit, in that improved irrigation and drainage technology was reinforced by extreme coercive collective control in minimizing the relative scale of weather influence.

The conclusions drawn here about the changing weather influence on the national aggregates of grain yield and GVAO are generally borne out by cross-regional analysis. Table 7.4 presents regression estimates for the three different periods 1931–7, 1952–7, and 1978–84 (represented by 1983 when for the first time a complete set of official provincial *shouzai* area data were made available). Note that our standard period 1970–7 is omitted from the estimate for lack of comparable data. Two points emerge from Table 7.4.

The first concerns the transition from the pre-war years 1931–7 to 1952–7. The relative weather influence (as reflected in both the regression coefficient and r^2) tended consistently to decline, when each of the two periods is taken as a whole. The conclusion should not be pressed too hard, in view of the nature of the data used (see the notes to Table 7.4)[10]. However, the more interesting point is that in both periods, the years of major weather disturbance (notably 1931, 1934, and 1937 on the one hand, and 1954, 1956, and 1957 on the other) all demonstrate a close association between grain yield deviations and weather conditions. Even more remarkable, in exceptionally good years, viz. 1932 and 1952 as well as 1936 and 1955, the same close relationship is virtually non-existent. This suggests that in time of good weather (or lack of weather interference), Chinese peasants were much more flexible in their response to changing market conditions or policy factors.

The year 1983 was not a bad year but it was less favourable than 1952 and 1955. Yet the weather impact (in terms of grain yield deviations as measured in the regression estimate) was consistently lower than in the

TABLE 7.4. Cross-provincial analysis of the weather (W) influence on grain yield (Y) and GVAO (V) instability in China, 1931–1937, 1952–1957, and 1983 (estimated regression equations)

		Intercept	Regression Coefficient	Obs	t-value	r^2	t-significance
1931–7	(Y)	3.20332	−0.27211W	109	4.9907	0.19	<0.000
1931	(Y)	6.09714	−0.22854W	8	2.7586	0.56	0.033
1932	(Y)	3.97070	−0.19678W	11	0.3846	0.02	0.709
1933	(Y)	0.40493	−0.04448W	19	0.5556	0.02	0.586
1934	(Y)	2.64965	−0.27859W	18	1.6868	0.15	0.111
1935	(Y)	2.77585	−0.30645W	19	2.2651	0.23	0.037
1936	(Y)	4.98928	−0.21617W	19	0.6092	0.02	0.550
1937	(Y)	6.68980	−0.79735W	18	9.1952	0.84	<0.000
1952–7	(Y)	1.28565	−0.09799W	118	3.2530	0.08	0.002
1952	(Y)	−0.18897	0.21786W	20	0.6975	0.03	0.494
1953	(Y)	1.34275	−0.24202W	22	3.0819	0.32	0.006
1954	(Y)	3.25905	−0.33251W	21	3.3666	0.37	0.003
1955	(Y)	2.03779	−0.03935W	15	1.0268	0.08	0.323
1956	(Y)	5.09839	−0.44106W	21	3.4056	0.38	0.003
1957	(Y)	3.66377	−0.15714W	19	2.3882	0.25	0.029
1983	(Y)	12.49716	−0.22236W	29	2.6286	0.20	0.014
	(V_a)	11.31921	−0.17950W	29	2.0789	0.14	0.047
	(V_b)	11.99724	−0.09230W	29	1.4921	0.08	0.147

Notes: For 1931–7 the instability of grain yield is expressed as a percentage deviation from the 1931–7 provincial mean. In both 1952–7 and 1983, the three-year mean yield covering the current, preceding, and following years, is used as the base for calculating the percentage deviations to take into account the steadily rising yield levels. The distinction between (V_a) and (V_b) for 1983 is that the former covers the value of cropping output only, while the latter also includes animal husbandry, forestry, fisheries, and rural industries. The weather variables (W) cover *shouzai* area for floods and droughts only, and are defined as the proportion (percentage) of total sown area affected. For both 1931–7 and 1952–7, W refers to the non-weighted *shouzai* area. Apart from the fact that for 1931–7 no distinction is made between the *chengzai* and *non-chengzai* area in the NARB statistics, the 1952–7 series, collected from scattered sources, is not strictly defined and comprises both *chengzai* and *non-chengzai* areas. As for 1983, W represents the weighted *shouzai* area, derived in the same way as elsewhere in this study. Thus, the estimated regression coefficients for the three different periods, 1931–7, 1952–7, and 1983 (especially between 1983 and the two earlier periods) are not quite comparable.

Sources: Grain yields: for 1931–7, Table AB.7; 1952–7 K. R. Walker, *Foodgrain Procurement and Consumption in China*, 239, supplemented by Chao Kang, *Agricultural Production in Communist China* (Madison, Wis.: University of Wisconsin Press, 1965), 300–1; 1983 and 1982–4 average yields, *TJNJ 1983*, 173, *NYNJ 1984*, 85, and *TJNJ 1985*, 265. GVAO: *TJNJ 1983*, 152, *TJNJ 1984*, 132, and *TJNJ 1985*, 242. The weather variable (W): for 1931–7, Tables AB.14 and AB.5 (for the sown-area base); 1952–7, Tables AB.16, and Walker, ibid. 306 (for the sown-area base); 1983; *TJNJ 1984*, 191 and 84 (for the sown-area base).

1950s. More importantly perhaps, for the three different measurements of the weather impact for 1983 (i.e. grain yield, GVAO narrowly defined to include cropping output only, and broadly defined to cover non-cropping agricultural branches as well), the weather influence differs in a manner that is perfectly consistent with the perceived differences in their degree of weather sensitivity, as shown in the national aggregate time-series analysis. The estimated degree of correlation between the weather and the two measures (grain yield and GVAO), may seem less definite (in terms of r^2) than in the national time-series analysis. But it should be noted that the cross-sectional analysis wholly ignores the wide-ranging regional differences in cropping patterns, agricultural technology, and productivity.

Taking the entire post-war period 1952–84 as a whole, however, the estimated weather influence, shown in the more coherent time-series analysis (Table 7.2), is a good representative average of the estimates for the various subperiods. It embodies stabilizing institutional and technological factors, and compares very favourably with the 1930s, when Chinese agriculture was at the mercy of the weather.

There is yet one more point to make in this context concerning the remarkable discrepancy in the estimated weather influence for the two overlapping periods 1952–66 and 1952–8. As shown in Table 7.2, the subperiod which excludes the Greap Leap Forward years of 1959–61, shows a smaller relative scale of weather influence on grain yield variations (regression coefficient -0.12313), compared with the longer period 1952–66 (-0.15666). Nevertheless, the estimated statistical correlation between weather and yield variables is much more coherent for 1952–8 (r^2 0.7972) than 1952–66 (0.5524). This seems to imply that the disastrous 1959–61 episode was more than a matter of weather disturbance. Perhaps the situation in 1959–61 was somewhat similar to that of 1978–84, when in a changing institutional and policy context, non-weather factors, viz. peasant behaviour, helped to reveal the weather influence. We shall, however, postpone our analysis of the 1959–61 period until subsequent chapters.

A Note on Food Security and Peasant Behaviour

We have so far compared the weather influence in different periods, but without considering how improvements in grain supply, and reductions in its instability over time, could have impinged upon the weather–yield relationship. Our approach was first to isolate, in quantitative terms, the relative weather influence, and then interpret the possible long-run technological and institutional implications with respect to the residual variations in grain yields. The results of this analysis appear quite plausible, although no attempt has been made explicitly to account for the role of changes in non-weather factors. The two issues are interrelated, for peasants may respond differently to changes in government policy, depending on their relative security and affluence in terms of per capita grain supply.

The first question is raised because the estimates of the weather and yield relationship, presented in Table 7.2, do not account for the potential cumulative year-to-year impact of grain yield variations. Specifically, the estimates show only the weather impact during the current year. They were already statistically adjusted to take account of the fact that an initial shortfall in grain output in any year might affect grain yield in the following year, to the extent that peasants under subsistence conditions, for example, were forced to dip into seed reserves, resulting in inadequate sowing, and hence lower yields and output. This is a phenomenon common to most economic time-series which show a cyclical pattern of changes, similar to our grain yield series. Econometricians use the term 'positive autocorrelation' to designate such intertemporary relationships.

Our original regression estimates which are not adjusted for such 'auto-correlation' (and not shown in Table 7.2) consistently reveal its existence, especially in 1952–66, if not as strongly in 1970–84. In addition to the problem of inadequate seed supplies,[11] there were other factors operative in the earlier periods which caused such automatic 'spillover'. One is the possibility of a deterioration in seed quality under the impact of pests which have often followed widespread droughts and floods.[12] Another is the threat of prolonged inundation caused by inadequate drainage facilities and leading to inadequate soil preparation for the next crop year.[13] Thirdly, the absence of summer rains under a prolonged drought prevented, especially in the North China plain, the regular leaching of salts which normally accumulate during the dry months of the preceeding winter and spring. This exposed the new crops to possible salt damage.[14]

Such spillover effects may be prevented or mitigated by increased grain availability for consumption and improved irrigation and drainage capacity, as well as through higher applications of chemical pesticides. This explains the reduced degree of 'autocorrelation' in our estimates for 1970–84 and its subperiods.

Turning to the second question, it is difficult to quantify the possible institutional and policy influence, except by assigning 'dummy' variables.

In a previous study I added to a weather–yield regression model for 1952–82, a 'Maoist' dummy for 1958–60 and 1973–8, based on the 'agricultural policy cycle' discovered by Professors Anthony M. Tang and Cliff J. Huang.[15] The results suggest that the Maoist influence on grain yield was insignificant. Estimates for the period 1952–66 yield a similar result. Only for 1970–82 is the 'Maoist' dummy negatively correlated with grain yield variations, but it is still statistically less significant than the weather as an explanatory variable.

A possible explanation is that in 1952–66, the sheer 'subsistence urge' of Chinese peasants induced a kind of lethargy towards policies of egalitarianism and collective coercion, and drove them to adopt whatever measures might be available to lessen the weather impact. Any residual yield variations may therefore largely be attributed to weather changes, especially prior to the mid-1960s. With more secure grain supplies after the 1970s, peasants were perhaps in a better position to resist Maoist extortions, or may have become more sensitive to policy changes, positive or negative, in terms of incentives. This seems to be especially true of the post-Mao era, for our earlier estimates show a decline in the significance of the weather variable, in relation to institutional factors, and their implications for peasant behaviour.

This strongly suggests that future agricultural instability in China will be increasingly determined by discretionary policy measures (e.g. State farm procurement prices), and institutional adjustments (such as changes in the land-tenancy system). The weather will remain an important factor, but its influence on Chinese agriculture will continue to diminish.

Some General Observations

Several major findings emerge from our aggregate analysis of the interplay of weather, technology, and institutional factors in seeking an explanation of the changing degrees of agricultural instability in China since 1931.

First, rural collectivization in the 1950s greatly helped to stabilize agricultural fluctuations in the face of adverse weather. By contrast, the 1930s enjoyed no such institutional hedge. The collective framework served to minimize sown-area abandonment (by prohibiting rural–urban migration), and yield losses (by appealing to the 'subsistence urge' of collectivized peasants).

Secondly, throughout the post-war period, the impact of weather disturbances on Chinese agriculture has gradually diminished, as is shown by the reduced correlation between weather conditions and instability of grain yield or GVAO. This underlines the long-run stabilizing effect of technological progress in Chinese agriculture.

Thirdly, until the implementation of post-Mao reforms, the weather impact loomed larger than any unpredictable policy and institutional changes, no matter whether stabilizing or destabilizing.

Fourth, by the early 1980s, Chinese agriculture had become largely insulated from weather disturbance on account of improved irrigation and drainage technology. Observed increases in agricultural instability increasingly reflected government policy, as it affected peasant income incentives and economic behaviour in the decollectivized rural framework.

Fifth, in spite of its diminishing impact, weather conditions have continued to contribute significantly to annual fluctuations in Chinese agricultural production. This will become especially clear when we analyse the relationship between the weather and short-run agricultural instability in the next chapter.

Finally, the weather itself has basically remained as changeable as ever.

NOTES

1. This can be more clearly shown by the estimated *beta coefficients*, as follows:

	Beta coefficients		r^2	t-significance	
	Fertilizer	Weather		F	W
1952–66	0.9904845 (22.175)	−0.1107690 (−3.072)	0.9706	0.0001	0.0074
1970–81	1.0178453 (23.126)	−0.0977009 (−1.667)	0.9814	0.0001	0.1264
1952–81	1.0080481 (34.874)	−0.1046961 (−3.622)	0.9786	0.0001	0.0013

Note: For the estimates, the units of measurement of the variables are converted into standard deviation units. This makes the relative contribution of the two explanatory variables more comparable.

2. James D. McQuigg, 'Climatic constraints on food grain production', 388.
3. See the discussion in Ch. 3 in connection with Fig. 3.1.
4. See Ch. 3, 53–4.
5. The system was formulated in late 1958 and advocated for allocating 'a third of the arable land to crops, a third to horticulture, and the rest to lie fallow';

see Christophe Howe and Kenneth Walker, 'The economist', in Dick Wilson (ed.), *Mao Tse-tung in the Scales of History* (Cambridge: Cambridge University Press, 1977), 201. As Walker put it in his *Foodgrain Procurement and Consumption in China*, 'The context of this revolutionary proposal of course, was the claim that the grain problem had been solved by the doubling of output in 1958. Although this revolutionary change in land use was advocated as a long-term measure, the political atmosphere of the time led to an immediate decline in the grain-sown area for 1959' (p. 146). However, most of the farmland withdrawn from cultivation involved cross-winter (1958–9) and early spring-sown crops in 1959. See Y. Y. Kueh, 'A weather index for analysing grain yield instability in China', 80. There is clear evidence that the practice was halted by Mar. or Apr. 1959, at the latest; see Ch. 11 of this study for a detailed discussion.

6. If the non-weighted *shouzai* area is used for 1952–9 to make it comparable to the 1931–7 series, the average ratio is reduced by only 2.6 percentage points to 14.1%. The difference is thus still rather marginal.

7. Note the remarkably good *t*-value and r^2 for the 1952–8 repression.

8. Cf. Kueh, 'China's new agricultural-policy program: major economic consequences, 1979–1983', 353–61.

9. Kueh, 'Weather, technology, and peasants organization as factors in China's food grain production, 1952–1981', 22–3.

10. Apart from the lack of comparability between *shouzai* area data in 1931–7 and in 1952–9 as noted in Table 7.4, figures are not available or are incomplete for a number of provinces in both 1931–7 and 1952–7 and the gap varies from year to year. (see Tables AB.14 and AB.16). Note also that no dummy variables are assigned in the regression to account for the regional differences in geographical and agro-climatic conditions.

11. The 1954 Yangzi floods e.g. resulted in serious seedgrain supply shortages in Hubei province; see *JFRB*, 31 July 1955, and similar report in *DGBTJ*, 18 Apr. 1955.

12. *XHRBNJ*, 8 Aug. 1958: the harvested wheat crops in summer 1958 in Xuzhou and Huaiyin prefectures could not be used as seedgrain, because they were seriously plagued by plant diseases and insects. The provincial Party authority appealed to the peasants to find their own solution to the problems as planned transfers from other provinces would not be sufficient to meet the demand for seedgrain.

13. This has always been the case in many localities in both pre-war and post-war China, up to the present period; see e.g. Yang Xiandong, 'A few questions concerning the orientation of agriculture—observations in the study tour of Hubei, Henan, Jiangsu and Auhui provinces', in *NYJJWT* 6 (June 1984), 3–6.

14. For details see the disussion in Ch. 11 about the problems of salinization in connection with the Great Leap Forward débâcle. To the several factors cited here in explaining the cumulative weather impact, Bruce Stone and Scott Rozelle ('The composition of changes in foodcrop production variability in China, 1931–1985: A Discussion of weather, policy, technology, and market'), add, 'rural dislocation, reduced labour productivity due to mal-

nutrition, and the reflexive economic impact on non-agricultural sectors rebounding in generalized depression upon agriculture' (p. 11) in his review of Kueh's hypothesis.

15. Cf. Kueh 'Weather cycles and agricultural instability in China', and A. M. Tang and Cliff J. Huang 'Changes in input–output relations in the agriculture of the Chinese Mainland, 1952–79' for the policy cycle.

8

Weather, Policy, and Short-Run Instability

In the previous chapter we compared the weather influence on Chinese agriculture between different periods, with each one taken as a whole. The present chapter looks, by contrast, at year-to-year fluctuations in agricultural production in relation to the impact of the weather in each of the periods. Since the periods under study are defined in terms of their basic technological and institutional parameters, our analysis promises to reveal the relative importance of weather and short-run policy factors in influencing agricultural stability in given technological and organizational settings. We now ignore the broad measure of gross value of agricultural output (GVAO) and concentrate exclusively on grain yield per hectare. Note again that being a physical measure, the latter standard eliminates possible price influences. In contrast to the grain output measure, it also allows abstraction from policy-determined changes in the sown area. In short, the grain yield measure seems to be the most appropriate key to gauging the relative impact of weather and policy factors, as they affect peasant efficiency and input incentives.

The first section of the chapter compares the relationships between the year-to-year fluctuations in grain yield per sown hectare and changes in weather conditions across two periods, 1931–7 and 1952–9, which are largely similar in technological terms, but radically different in terms of institutional setting. Section two examines how the annual weather and yield relationships differed in the two periods 1970–84 and 1952–66, which share a socialist setting, but are considerably different in terms of agricultural technology. Section three attempts to isolate the relative weather impact on grain yield from policy factors by reconstructing the magnitude of weather-caused losses for selected years of major weather disturbance.

Collectivization (1952–1959) versus Market Economy (1931–1937)

Fig. 8.1 shows the annual fluctuations of the national average grain yield during 1931–7 and 1952–9 in relation to the changing proportions of

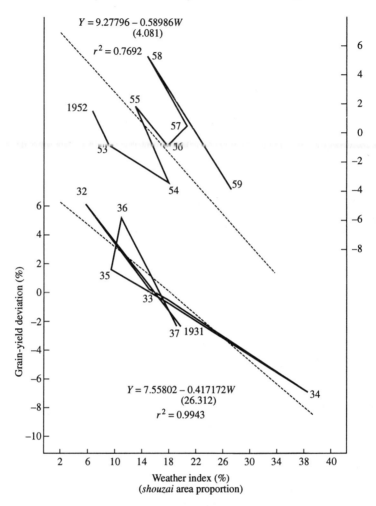

Fig. 8.1. The instability of annual average grain yield per sown hectare (expressed as percentage deviations from the log-linear trend value) in relation to the size of *shouzai* area (expressed as a proportion of total sown area) for China, 1931–1937 and 1952–1959

Source: Table 7.2.

total sown area covered by floods and drought. Virtually all the major negative deviations in grain yields are clearly associated with known large-scale weather disturbances in 1931–7 and 1952–9. Although, as discussed in Chapter 3, overall grain yield instability fell substantially from the earlier to the latter periods (on account of the 'institutional hedge' available in the 1950s against natural disasters), the pattern of weather and grain yield relationship is very similar in the two periods.

Thus, in 1931–7, the disastrous 1931 Yangzi River floods resulted in a negative 2.5 per cent grain yield deviation from the trend value. Likewise, the even more catastrophic 1934 droughts throughout the Yangzi and Huai River basins reduced the yield by 7.2 per cent from the expected harvests. By contrast, the year 1932, which enjoyed very favourable weather in most parts of China saw a gain in average grain yield of 6 per cent above the expected level. Similarly, the sharp deterioration in weather conditions from 1936 to 1937, for which the regional sources were identified in Chapter 4, brought a marked, albeit less dramatic negative shift in grain yields, as shown in Fig. 8.1.

The year-to-year changes in weather and grain yield correlation during 1952–9 seem more complex and merit more lengthy discussion. We note, first, the striking similarity in Fig. 8.1 between the Yangzi floods in 1931 and 1954 on the one hand, and the disastrous droughts in the Yangzi and Huai River basins in 1934 and 1959 on the other hand, in terms of their respective impacts on grain yield.

The relationship between weather disturbances and national average grain yields was not the same, nor is there *a priori* reason why it should have been the same during the 1930s and 1950s. Measured roughly as the ratio of the negative percentage yield deviations and the *shouzai* proportion of sown area, the impact of the 1954 Yangzi floods (ratio −0.19) was greater than that of the 1931 Yangzi floods (ratio −0.13). By contrast, however, the 1959 drought appears to have been less catastrophic (ratio −0.15) than that of 1934 (ratio −0.19).

The observed discrepancies in the relative impact of weather conditions on grain yield seem to suggest nevertheless that, in the absence of any newly-created major capacity in flood control during the 1950s, labour mobilization was not on its own capable of containing major natural disasters (although admittedly by 1954 Chinese peasants were far from being properly organized into collectives). By contrast, the 'institutional hedge' did seem to fare better against drought, than against floods, as the experience of 1959 *vis-à-vis* that of 1934 suggests. The milder impact on grain yield of the 1959 drought could, of course, also reflect the peasants' 'subsistence urge', which emerged strongly, as noted in Chapter 7, in the wake of the unwarranted sharp contraction of sown area in that year, in an effort to protect grain yield and output and avoid a potential food crisis.

In this respect, we should look more closely at other years in the 1950s in an attempt to explain why the yearly weather and yield relationships in 1952–9 were less consistent than those in 1931–7. As shown in Fig. 8.1, annual grain yield fluctuations were frequently quite large, notably from 1953 to 1958. Yet all these years display a relatively small range of *shouzai* area variations.

Consider, first, the fluctuation from 1954 to 1955. The percentage deviation of grain yield from the trend moved rapidly from a negative value of 3.46 per cent in 1954 (Yangzi River flood) to a positive one of 1.94 per cent in 1955. This change was associated with exceptionally favourable weather conditions in 1955. However, although conditions in 1956 appear to have been similar to those of 1954, the impact on grain yield was considerably smaller. The year 1956 displayed only a minor negative deviation of grain yield, by 0.91 per cent, from the trend. This compares favourably with the negative deviation of 3.46 per cent in 1954, but not with the positive gain of nearly 2 per cent in 1955. While the yield change from 1955 to 1956 can be attributed to the deterioration in weather conditions, one wonders why, between 1954 and 1956, this was not the case. Three explanations can be offered.

The first has to do with the possible impact of agricultural collectiviz-ation, which was in full swing during 1955–6. Perhaps the 'institutional hedge' overcompensated for possible peasant disincentives to help stabilize yield fluctuations. A second possibility is that the yield and output stabilizing 'area-for-yield' substitution in 1956 was coupled with a sown-area shift to high-yielding rice crops in the northern provinces. A third explanation is that the main focus of the floods shifted from the Yangzi River in 1954 to farther North in 1956.[1]

A similar explanation can be given for the year 1957, which experienced a further, albeit marginal increase in the *shouzai* area (worsening weather conditions) compared with 1956, but displayed a greater positive deviation of grain yield from the trend, as shown in Fig. 8.1. In contrast to 1956 which was dominated by floods, 1957 was overwhelmingly a year of drought, affecting mainly Shanxi, Hebei, Shandong, and Henan, but only for a short duration.[2] Thus its impact on grain yield was not as catastro-phic as the 1956 flood in these northern provinces.

The even more contradictory weather–yield relationship between the years 1953 and 1958, as shown in Fig. 8.1, can similarly be traced to the shifting regional distribution of floods and droughts, as well as to other known human factors. Unlike 1958, the year 1953 was dominated by floods rather than by drought. The generally more harmful effect of floods apart, the 1953 floods mainly affected Shandong, Hebei, Hubei, Zhejiang, and Jiangxi provinces. These areas are much more fertile than the semi-arid North-west, which was the main locus of the 1958 flood. In contrast, the 1958 droughts mainly affected the wheat crops in the North-east and, perhaps, the North. Judging by the extremely small *chengzai* proportion (7.4 per cent), their impact on the wheat yield per sown hectare seems to have been quite mild in 1958. Besides, the autumn crop, which enjoyed a higher average yield than wheat, was hardly affected. But, the most important factor contributing to the exceedingly high positive yield deviation in 1958 (in relation to the *shouzai* area), was probably the dramatic cropping shifts towards such high-yielding crops as potato and maize (see Chapter 3). This tended to reinforce the yield-raising effects of the 'yield-for-area' substitution process which took place so impressively in 1958. Such policy factors all helped to make observed weather and yield relationships during 1952–8 less consistent than in 1931–7.

Nevertheless, the weather influence in the 1950s was too strong to be wholly diluted by policy factors. This can be shown by a more systematic comparison of the weather–yield co-ordinates displayed in Fig. 8.1. The relative strength of the weather influence in 1931–7 and 1952–9 can also be gauged in another way.

Note first that the average *shouzai* ratio for each period was virtually the same, amounting to 16.7 per cent of the total sown area in 1931–7, and 16.2 in 1952–9. During both periods in almost every year a positive/negative percentage deviation of the grain yield from the trend value was consistently associated with a below-/above-average *shouzai* area ratio.

This point can be demonstrated by moving (in Fig. 8.1) the abscissa rightwards and the ordinate upwards to cross at the point where the average *shouzai* area proportion and the zero-percentage yield deviation meet. The two diagrams are then each divided into four quadrants. The upper-left quadrant includes all the co-ordinates relating the positive yield deviations to the below-average *shouzai* area ratios. This is exactly where all the known good-weather years are located. The lower-right quadrant includes analogously all the co-ordinates relating negagive yield deviations to above-average *shouzai* area ratios. Here we can find all the known bad-weather years. The only exceptions to this classification are 1953 and 1957, which fall, respectively, in the lower-left and upper-right quadrants; their co-ordinates, however, lie not far from the quadrants—respectively the upper-left and the lower-right ones, in which they should have been located.

The two weather–yield co-ordinates for 1934 and 1959, i.e. the two most severe drought years, assume a strikingly similar position in the lower-right quadrant. The same is true for the two notable flood years of 1931 and 1954. This implies that the extent to which grain yield was curtailed by adverse weather was very similar in 1931–7 and 1952–9. It can be more precisely measured and compared by drawing straight lines linking the co-ordinates for the years concerned (1931, 1934, 1954, and 1959 in this case) to the origin (i.e. the crossing-point of the redrawn abscissa and ordinate) of the respective diagrams. The slope of the two lines, so drawn, for the two drought years 1934 and 1959 is almost identical, as is that drawn for the two flood years 1931 and 1954. The only notable difference is that the slope for the two flood-year lines is consistently greater than that for the two drought-year lines, implying the more severe impact of floods than of drought in both 1931–7 and 1952–9 in relation to the average weather impact and grain yield losses during these periods.

In the context of severe natural disasters, the 'institutional hedge' provided by the collective framework ceased to be a decisive mitigating factor. This was especially so in the face of serious floods. In contrast to droughts, serious floods are often beyond human control. In this respect, the situation in the 1950s was not very different from that of the 1930s, because the new regime had only just begun to harness the major river systems. The 1954 flood actually had an even greater impact than its 1931 counterpart when viewed from the standpoint of the average weather–yield relationship. This underlines the observation made in Chapter 3 that increased overall grain output stability in the 1950s was more a result of sown-area stabilization (1959 aside) than of improved yield stability.

All the known good years are characterized in Fig. 8.1 by comparatively large positive yield deviations coupled with relatively small *shouzai* proportions (apart from the special case of 1958 already discussed). Both 1952 and 1955, for example, are generally thought of as very good weather years, as was the bumper harvest year of 1932.[3] However, it is more difficult to compare the influence of good rather than that of bad weather. When weather conditions are favourable, other factors may affect farm input decisions and hence output and yield levels. In the 1930s, for example, it was not unusual to find that, following a good summer harvest, market prices for grain declined to such an extent that peasants in richer areas considered it uneconomic to work harder or increase the application of fertilizer for the second crop, especially if the weather remained as good as it had been during the period of the first crop.[4]

Within the framework of collectivization, once obligatory grain deliveries for the summer crops had been fulfilled, the peasants might, similarly, pay less attention to the autumn crops. Alternatively collective controls might offer a hedge against such a reaction. A third possibility is that confiscatory State procurements, coupled with unfavourable terms of trade, would induce peasants to make a greater work effort in order to maintain living standards. This is the familiar phenomenon of a backward-bending supply curve. Such reactions, however, cannot be accurately measured. Moreover, the random spatial incidence of good weather causes the positive yield deviations to vary from year to year, much in the same way as the random distribution of floods and droughts.

Socialist Setting with (1970–1984) and without (1952–1966) Technological Change

We now turn to the different post-war periods to examine the year-to-year fluctuations in weather and grain yield relationships. For simplicity, a distinction is made between the two broad, technologically defined periods of 1952–66 and 1970–84. The latter was characterized by increased irrigation and flood-control facilities of a modern type, while the former was largely based on traditional farming techniques. Both periods were, however, basically under socialist collectivist control, with the remunerative and coercive approaches varying within each period. The chosen periodization may thus help to reveal the impact of policy changes relative to weather fluctuations from year to year.

In Fig. 8.2 the yearly percentage deviations of grain yield from the trend for the periods 1952–66 and 1970–84 are graphically related to the weather index (i.e. the percentage deviations of the weighted *shouzai* area from the long-term 1952–84 mean). The scale of the weather index, shown along the horizontal axis, as well as the vertical yield instability scale are identical in both diagrams. This facilitates a direct visual comparison of the actual amplitude of weather and grain yield fluctuations between the two periods.

Note, first, that the range for the weather index itself is considerably more limited for 1970–84 than for 1952–66. The index for 1970–84 ranges from the maximum negative deviation (by 73.60 per cent) of the *shouzai* area for 1970 (indicating the most favourable weather within this period) to the maximum positive deviation (by 54.39 per cent) for 1978 (the most unfavourable weather). By comparison, while the largest negative deviation (by 71.46 per cent for 1952) for 1952–66 is nearly equal to that for 1970–84, the maximum positive deviation (by 96.77 per cent for 1961) was much larger—almost double that of its counterpart in the later period.

This difference in the extent of *shouzai* area fluctuations between the two periods 1952–66 and 1970–84 does not necessarily imply reduced weather vicissitudes in 1970–84. Rather, it underlines the fact that improvements in irrigation and drainage capacity helped to reduce the average extent of the *chengzai* area (i.e. the area sustaining a crop loss of more than 30 per cent as a result of natural disaster). It will be recalled that in the statistical formula for converting the *chengzai* area and the non-*chengzai* area (crop loss below 30 per cent) into equivalent areas, a much heavier weight (0.60) was assigned to the former category than (0.15) in the case of the latter. Since, as a result of technological progress, the *chengzai* proportion was smaller in 1970–84 than in 1952–66, the com-

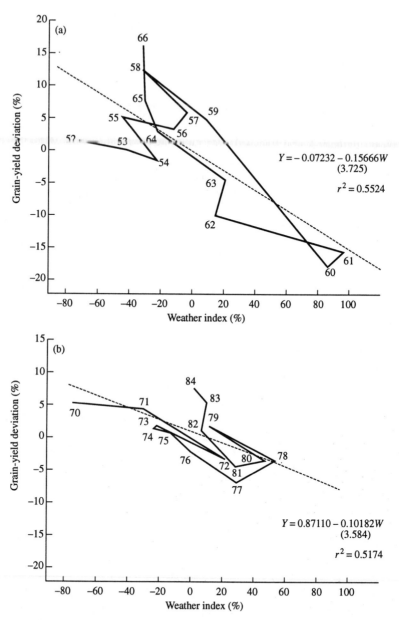

FIG. 8.2. The instability of annual average grain yield per sown hectare (expressed as percentage deviations from the long-linear trend value) in relation to the size of *shouzai* area (expressed as a proportion of total sown area) for China, 1952–1966 and 1970–1984

Source: Table 7.2.

bined weighted *shouzai* area and the extent of its yearly fluctuations are not quite comparable between the two periods.[5] This should be borne in mind when interpreting the two diagrams in Fig. 8.2 as they relate to weather variations.

Both diagrams in Fig. 8.2 point to a cyclical pattern of movements of weather and grain yield. This is especially pronounced for the period 1952–66. Thus, in line with the worsening weather conditions from 1952 to 1960–1, grain yield per hectare, after exhibiting positive percentage deviations from the trend within a comparatively narrow range, moved rapidly into the zone of negative deviations. The recovery from the trough of 1960–1 in the early 1960s and the upward movement into the positive zone until the peak of 1966 are also no less noteworthy than the original downturn. The cyclical weather pattern during 1970–84 may not appear to be as pronounced as during 1952–66, but this is also due to the unavoidable methodological bias which, as just mentioned, tends to underestimate the size of the weighted *shouzai* area in the latter years.

The weather cycles in 1970–84 can be more clearly visualized by expanding the weather index for 1977 and 1978, for example, rightwards parallel to the horizontal axis. Given the amplitude of annual grain yield fluctuations in 1970–84, such a graphic expansion generates a new regression line with a slope less steep than the original one (dotted line in Fig. 8.2) which is already substantially flatter than that drawn for 1952–66.

Such a graphic comparison between 1952–66 and 1970–84 also suggests that both the relative scale of weather influence and the amplitude of annual fluctuations in grain yields were reduced from the earlier to the later periods. However, despite reduced weather impact as a result of advances in agricultural technology, the yearly fluctuations in grain yield in 1970–84 are still rather consistently correlated with the weather changes, much in the same fashion as in 1952–66.

We first look at the period 1952–66. In contrast to the subperiod 1952–9 (already discussed in connection with Fig. 8.1), every year from 1952 to 1959, except 1954, in the context of this longer period 1952–66 assumes *positive* yield deviation values as shown in Fig. 8.2. Likewise, except for 1959, all the other years in the 1950s fall within the range of negative *shouzai* area deviations (good or relatively favourable weather), although in this respect a direct comparison with the earlier 1952–9 diagram (in Fig. 8.1) is not possible, because the weather index for the shorter 1952–9 period is defined as proportions of *shouzai* area in total sown area.

This discrepancy between the 1952–9 and 1952–66 diagrams with respect to the years 1952–9 is readily understandable. It is the extremely large magnitude of both grain yield losses and *shouzai* area in the early 1960s which greatly overwhelmed those of the earlier years in the 'detrending' procedure adopted for the period 1952–66. Nevertheless, the weather and grain yield co-ordinates for the various years 1952–9, as shown in the diagram for this longer period, stand to each other in a relation strikingly similar to those plotted in the diagram for the sub-period 1952–9.

We have already discussed how the interplay of weather and policy factors determined the location of the weather–yield co-ordinates during the years 1952–9. A few words should now be said about the extra-ordinary years of the early 1960s before we embark on a full exploration of the weather and policy background to the collapse of the Great Leap Forward strategy in the next section (and especially in Chapters 10 and 11). From Fig. 8.2, there is no doubt that weather conditions did play a crucial role in bringing about the great débâcle. Grain yield was reduced by an astonishing 18 per cent in 1960 and by a further 16 per cent in 1961 from the respective trend values. Meanwhile, the *shouzai* area reached an all-time record.

In 1962 and 1963 the weather was no worse than in 1959, judging by our weather index in Fig. 8.2. Grain yield was however still exceedingly low. This reflects the lack of momentum of recovery from the 1960–1 crisis, conditioned by inadequate seed and feed supplies, for example. Thereafter, the recovery, encouraged by improved weather conditions, proceeded quite rapidly until it culminated in the bumper harvests of 1966. Interestingly, this occurred amidst intensified rural 'class struggle', advocated by Mao at the Tenth Plenum in September 1962.

During 1970–84, the most notable episode was, as shown in Fig. 8.2, the abrupt downturn in weather conditions in 1972 and in grain yield, following the favourable weather conditions of 1970 and 1971. The background to this was the great North China drought, classified by a senior SMB analyst as one of the six worst droughts since 1470, occurring at a frequency of only one per century.[6] Measured against the trend yields estimated for 1970–84 taken as a whole, the average national grain yield per sown hectare in 1972 was reduced by 3.9 per cent, compared with 5.4 and 4.2 per cent gains in 1970 and 1971 respectively. Based on the trend yields estimated *separately* for the subperiod 1970–7 alone, however, the loss in 1972 amounted to a startling 5.6 per cent, as against gains of around one per cent in both 1970 and 1971. This is likely to be a more realistic assessment of the relative magnitude of yield losses in 1972, because the period 1970–7 is quite different from 1978–84 in terms of both agricultural policy and growth trend. By combining them into a single period, estimated trend yields can easily distort the comparative scale of yield losses for the different years concerned.[7]

In fact, the 5.6 per cent average grain yield loss in 1972, based on the 1970–7 trend, was the largest loss experienced since 1931, except during 1934 (the Yangzi and Huai River basins droughts) and the catastrophic years 1960 and 1961, and 1962, the year of slow recovery. The fact that the 1972 drought mainly affected the North and the relatively infertile North-west, which was not an important grain-producer, points to the severity of the weather impact, especially at the local level. This can be shown by the experience of the celebrated Dazhai brigade.

Dazhai and its parent province, Shanxi, was at the heart of the drought belt which stretched from Shaanxi in the interior to the coastal province of Hebei. Dazhai itself suffered a 14 per cent decline in grain yield and a 13 per cent decrease in output in 1972.[8] The drought in the brigade persisted through 1973 and 1974. Average wheat yield per sown hectare, in 1972–3, fell by a further 55 per cent, from 3,358 kilograms to 1,500 kilograms. From 1972 to 1974 the brigade made a significant shift in cropping pattern from fine grain (wheat) toward such coarse grains as maize and gaoliang, whose average yields per sown hectare were, respectively, 7,127 kilograms (1972) and 9,097 kilograms (1974) during this period.[9] Undoubtedly, such cropping shifts helped partially to stabilize the average yield of grain.

The relative severity of the 1972 drought can also be gauged against the year 1977. No provincial *shouzai* area statistics are available for 1977, but the national aggregates, as well as data indicating the spatial distribution of floods and droughts, show that it was a year dominated by floods, which spanned the latitudes from 20 to 25 °N.[10] This embraced the most productive zone in South-east China, and in terms of the weighted *shouzai* area the situation in 1977 was even more serious than in 1972, as is shown in Fig. 8.2. Nevertheless, the yield loss in 1977, measured against the 1970–7 yield trends, was substantially lower than that of 1972.

The same 'discrepancy' in the weather and yield relationship holds true for 1978, in relation to 1977. In contrast to 1977, 1978 was dominated by droughts, almost the entire Huai River and Yangzi basins being affected. By contrast, flood inundation, which normally has a severe impact on farm production was minimized in 1978, its impact being similar to that of 1952 and 1966.[11] This seems to explain the relatively mild impact on grain yield in 1978, compared with 1977 or 1972.[12] Such arguments about the comparative weather and yield relationships in different years should of course not be pressed too hard, although in general terms the yearly yield fluctuations are visibly correlated with the changing weather. This also holds for the short subperiod 1978–84, despite three or four decades of marked technological progress in farming having occurred in the meantime.

Thus, boosted by a remarkable improvement in weather conditions in 1979, grain yield bounced back strongly to a positive deviation of 4.7 per cent based on the 1978–84 trend value from the low of 0.7 per cent for 1978. The upturn was however followed by an equally powerful downturn to minus 2.7 in 1980 (the Yangzi basin floods) and minus 5.1 per cent in 1981 (the Sichuan basin floods). Note that it was in 1980, in the wake of sharp increases in State farm procurement prices implemented in 1979, that liberal policies designed to increase peasant income and farm output were accelerated.

Likewise, the quick recovery in 1982 and 1983 from the 1980–1 trough was, as shown in Fig. 8.2, due more to weather improvement than due to major policy changes. The rise in yield during 1983, compared with that in 1982, looks large, in relation to weather conditions in these years. Here again, it is tempting to attribute this to the agricultural reforms which, by 1983, tended increasingly to approach quasi-privatization. There is no doubt that there was a strong positive response by peasants to decollectivization. Yet, the weather pattern seems to have been more favourable in 1983 than in 1982, even allowing for the fact that 1982 still awaited recovery from the disastrous losses of 1981.

Provincial statistics for 1983 show that, out of the total sown area covered by flood and drought, nearly 60 per cent was accounted for by drought, of which 65 per cent took place in the North, North-west and North-east.[13] These were less fertile regions in terms of average grain yield per sown hectare (except for Liaoning and perhaps Jilin)—this quite apart from the fact that the impact of drought tends to be smaller than that of floods. Moreover, more than one-tenth of the flooded area in 1983 was in Heilongjiang, which had one of the lowest average grain yields per sown hectare in China. Compared with 1982, the more important grain-producing areas of the South enjoyed generally favourable weather and seem to have been less affected. Unfortunately, the provincial *shouzai* area data available for 1982 are too scattered and limited to facilitate a firmer comparison with 1983 with respect to the shifting nature and impact of natural disasters during the two years.

Turning finally to 1984, the record harvest of this year, in terms of both total output and grain yield, reflects the extension of peasants' leasehold right to land from 3 to over 15 years. This represented a critical first move towards full privatization and no doubt enhanced peasant incentives. Nevertheless, the perceived decollectivization effect was also accompanied by an improvement in the weather index, making 1984 the best-weather year since 1976 (see Fig. 8.2). The inference is that further consideration is needed of the relative impact of weather and policy factors on grain yield. This we attempt to provide in the next section.

Isolating Yield Losses in Major Weather Disturbances

We now seek to assess the relative magnitude of grain yield losses caused by adverse weather and policy factors. We concentrate on the post-war years, because in contrast to the short pre-war period 1931–7, these witnessed periodic dramatic institutional and policy readjustments. To keep the analysis within manageable proportions, we will deal only with some of the more notable episodes of agricultural instability.

Fig. 8.3 presents the quantitative relationship for three sets of grain yield values for the periods 1952–66 and 1970–84. These are: the actual (realized or observed) grain yields, the trend yields derived from the log-linear method, and the 'predicted' yields. The predicted yield series is calculated from the regression equations for 1952–66 and 1970–84 given in Table 7.2. It shows what the grain yield might have been for each of these years, on the basis of the average scale of weather influence as estimated and revealed in the regression coefficient for the weather variable (i.e. the percentage deviation of the *shouzai* area from the 1952–84 mean).

Measured against the trend values, the discrepancy between the weather-predicted and the actual grain yields helps to determine the relative significance of weather and non-weather factors. If, for example, predicted and actual yields are identical, we may conclude that in the year in question the weather was solely responsible for the realized (or observed) yield loss or gain (i.e. the amount of yield below or above the trend value, or yield one would have expected given the prevalent state of farm technology). If, however, the observed yield loss is larger than the weather-predicted yield loss (measured as the difference between the trend value and the predicted yield below it), then non-weather factors are assumed to have accounted for the proportion of the realized loss which is not taken up by the weather-predicted loss.

Estimates of this kind cannot be perfectly accurate. Any weather phenomenon is unique, in terms of its spatial distribution and overall impact intensity. Moreover, the two periods under study, 1952–66 and 1970–84, are relatively long. The latter period especially witnessed accelerated technological improvements. This tends to render the predicted year-to-year yield losses, based on the estimated weather influence for the period as a whole, less than reliable. But the estimates still promise to shed some light on the extent of weather-caused losses *vis-à-vis* those caused by known human factors.

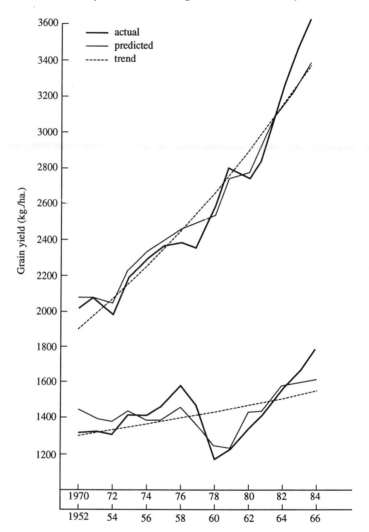

FIG. 8.3.   Actual, predicted, and trend values of the national average of grain yields in China, 1952–1966 and 1970–1984 (kg./ha.)

Source: Table 7.2.

The 1959–1961 episode

We first look at the three most controversial years, 1959–61. Measured against the trend values, weather-induced yield losses, shown in Fig. 8.3, account for 75 and 95 per cent of the observed losses in 1960 and 1961.

Thus, a quarter of realized yield loss in 1960 may be accounted for by policy factors, whereas in 1961 the loss attributable to policy was only 5 per cent. In 1959, weather conditions—again defined in terms of the *shouzai* area—suggest that the average grain yield would have been 2.2 per cent below the trend yield. However, favourable non-weather factors resulted in an actual yield record 3.5 per cent *above* the trend value. These different results for the three years 1959–61 may appear surprising, but they are actually quite in line with contemporary changes in agricultural policy and the scale and pattern of natural disasters. This may be explained as follows.

First, the national *shouzai* area of 1959 was substantially smaller than that of 1960 (Fig. 8.2). The major natural disaster of 1959 was the widespread drought in the Yangzi and Huai River basins, but this did not occur until after mid-July. The extent of floods in 1959 was much more localized than drought, occurring mainly in the Zhujiang River basin in South China, in June 1959, and in Neimenggu and northern Hebei in late July. These summer droughts and floods were part of a weather process which has since become almost a classic meteorological example in many Chinese geographical textbooks. In 1959, the regular South-east Pacific monsoon coincided with an unusually late withdrawal of the Siberian winter cyclone. This resulted in a concentration of rainfall in the Zhujiang area in June. Thereafter, however, 'its frontal rain belt moved abnormally rapidly northwards, leaving a persistent dry spell to prevail for two months in the Yangzi River basin which usually enjoys a comparatively abundant amount of plum rains in the early summer'.[14] Its sudden halt in the far North finally brought about a short but rare series of spectacular rainstorms in the region.[15]

However, 1959 saw other pressures at work which led peasants to hedge strongly against natural disasters. They included widespread fears that the unjustifiable reductions in the grain-sown area, made at the end of 1958 and early in 1959 in response to the 'three-three' system, would seriously reduce total grain output and affect their subsistence level. By late April 1959 or possibly earlier, both the Chinese peasants and the central leadership had become fully aware of the imminent crisis. This is reflected in the urgent personal appeal made by Mao himself on 29 April 1959 in an 'Intra-Party Circular' addressed to the commune cadres, in which he appealed for the stabilization of the sown acreage.[16]

Moreover, the arable land which was withdrawn from spring sowing under the 'three-three' system tended to be either less productive or more vulnerable to weather adversities. This helped to reduce the potential reduction of the average grain yield per hectare in 1959. This 'statistical' gain in terms of yield, together with the 'subsistence urge' already mentioned, probably accounts for the 3.5 per cent positive deviation in the grain yield from the trend, in the face of the weather-predicted loss of 2.2 per cent.

The situation in 1960 was very different. In most areas both the summer and autumn crops were seriously affected by prolonged drought. This reduced the 'risk-aversion' function of multiple-cropping. Worse still, the summer drought was widespread throughout the country[17] and this tended also to render what we may call the 'spatial risk-aversion' function ineffective, in the sense of enabling bumper harvests in a few areas to offset losses incurred elsewhere. There were also severe floods, but almost all of these were confined to the narrow coastal strip, being mainly caused by the eleven typhoons which occurred in 1960—a record number. Instead of taking their normal north-west course through the intervening mountain ranges, the typhoons swept northwards along the Pacific coast, and thereby had no diminishing impact on the severity of the interior drought.[18] Hence the prolonged, catastrophic drought of 1960.

Even if these weather events succeed in explaining 75 per cent or so of the realized yield loss for 1960, can the remaining 25 per cent be plausibly accounted for by policy factors? Rural conditions in 1960 were probably affected by the Anti-Rightist campaign launched in autumn 1959. As Kenneth Walker pointed out, the campaign was associated with a remarkable rise in the government's grain procurement for the grain year 1959–60.[19] In per capita terms, peasants' grain availability was sharply reduced from 288 kilograms, gross of seed and feed, in 1958 to a mere 222 kilograms in 1959.[20] Since most of the State procurements derived from the autumn (coinciding with the Anti-Rightist campaign) harvests, peasants' grain consumption was affected more in 1960 than in 1959. But the real crux of the problem is not so much that of peasants' resistance, but rather that their 'subsistence urge' prompted them to dip deeply into their seedgrain and feedgrain reserves. This was a matter of survival. The consequence can only have been a reduced sowing density, and thus a lower grain yield per sown hectare for 1960. This was then further complicated by the deterioration in the weather conditions. The ensuing crisis obviously could not be averted by concessions in terms of State grain procurements (reduced by 35 per cent compared with a fall of only 15 per cent in grain output), because total output was already at an all-time low of 144 million tonnes in 1960, leaving the peasants with an average of only 209 kilograms per head.[21] In such circumstances, the peasants could only resort to their last lifeline: the seed and feed reserves.

Turning to the year 1961, the relative magnitude of grain yield loss (measured again as the discrepancy between the actual yield and the trend value), which cannot be accounted for by adverse weather, fell from 25 per cent in 1960 to only 5 per cent. Actual grain yield also increased, albeit marginally by 4 per cent. This seems to have been partly attributable to the further reduction in State grain procurements which helped to restore peasants' per capita grain availability (230 kilograms) to a level comparable to that of 1959 (222 kilograms). This certainly helped to replenish the seed and feed stocks; hence the reduced losses incurred on account of the non-weather factor.

Nevertheless, the weather situation in 1961 was still too grave to bring about any quick recovery in grain yield. There was hardly any reduction in the size of sown area affected by natural disasters. As a matter of fact, the weighted *shouzai* area even increased, albeit marginally, in 1961 (Fig. 8.2), for both drought and floods assumed a larger *chengzai* proportion in 1961 than in 1960. This partly reflects the cumulative impact of the prolonged droughts which continued from 1960. Floods were, again, rather localized as in 1960, but areas south of the Yangzi basin were extensively hit by rainstorms in May and June, the most critical farming months.[22] Almost the same frequency of coastal typhoons occurred in the summer as in 1960. The rainstorms and typhoons were then followed by prolonged drizzle which began in autumn in Shaanxi, Sichuan, Guizhou, and Hunan.[23] These helped to lessen the impact of the widespread drought, but the belated turn-around in weather conditions contributed little to raise the average grain yield above the 1960 level.

Several favourable policy factors made themselves felt in 1961. First, the real magnitude of the crisis was recognized by the central leadership by late 1960 and was reflected in the 'Urgent Directive on Rural Work' (the 'Twelve Articles') dispatched in December 1960 in an attempt to ameliorate the situation. Later in March 1961, the 'Sixty Articles' were promulgated, reaffirming the points drafted by Mao two years earlier with a view to redressing the excess of the communes.[24] The commune system was itself radically reorganized, with the production teams, equivalent in size to the co-operatives of 1955, now designated as the basic units for planning, accounting, and distribution. All this certainly helped to improve the rural environment for peasants. However, it seems doubtful that such policy factors contributed more than increased seed and feed stocks, as mentioned before, towards the recovery, albeit a slight one, in grain yield in 1961.

In short, our analysis suggests that weather disturbance may have been the single most important factor in creating grain yield instability during both 1959 and 1961. Only in 1960 can policy failure be held responsible, and even then for only around 25 per cent of the yield loss.

The puzzle of 1958: extraordinary harvest but bad policy

Any discussion of the relative impact of weather conditions and policy factors on grain yield variations in 1959–61 must address the events of 1958, the year which marked the beginning of the Great Leap Forward.

In particular, why, despite the tumultuous socio-economic disruption of that year, did the average grain yield per sown hectare reach a record high in 1958. The communization drive, which began in August 1958 might not have affected the summer crops, but it certainly influenced the autumn crops. The notorious 'backyard furnace' campaign and the mass irrigation movement had, moreover, been in operation for most of 1958.

Fig. 8.3 shows that in 1958 both the actual yield per sown hectare (1,563 kilograms) and the 'weather-predicted' yield (1,460 kilograms) were above the trend value (1,397 kilograms). This implies that favourable weather conditions resulted in an average grain yield higher than might have been expected given the state of farm technology. More important, the actual yield was considerably higher than the weather-predicted yield, implying that non-weather factors even more favourable than weather conditions were at work to bring about the unprecedented yield gain. Could this reflect the buoyant rural mood of the early period of the Leap? It seems unlikely. Moreover, the 'subsistence urge' which had an important influence in 1959, did not operate in 1958, for the sown area did not contract dramatically until spring 1959.

The key to the puzzle is twofold. The first is the particular pattern of regional incidence of drought and floods in 1958 which favoured the important and fertile grain areas. The second is the massive sown-area shifts to high-yielding grain crops in these areas. Above all it should be noted that the year 1958 was not necessarily 'one of unusually favourable weather conditions in most of China'.[25] The SMB Map (8.2) shows excessive concentration of rainfall throughout most of the North-west in the summer. This was preceded by spring drought in Shanxi, and probably Hebei too, although this is not revealed in the SMB map, thanks to its pro-summer measurement bias.[26]

The summer rainstorms in the upper and middle reaches of the Yellow River were so great that the flood flow volume (23,000 m³/second) measured at Huayuankou (near Zhengzhou in Henan) still stands as the highest ever recorded.[27] The flood water was, however, effectively diverted through enhanced dike protection and other rescue operations designed to forestall disaster.[28]

In contrast to the North and the North-west, regions south of the Yellow River generally enjoyed favourable weather in 1958, apart from a localized drought in the lower reaches of the Yangzi. This contrasts sharply with 1959, when in addition to widespread summer drought in the fertile Huai River and Yangzi basins, Sichuan (especially the eastern portion) was also seriously short of rain and began a three-year ordeal of drought.[29]

Grain losses in 1958 in the heavily flooded but less fertile North-west were then probably overcompensated for by the gains obtained else-

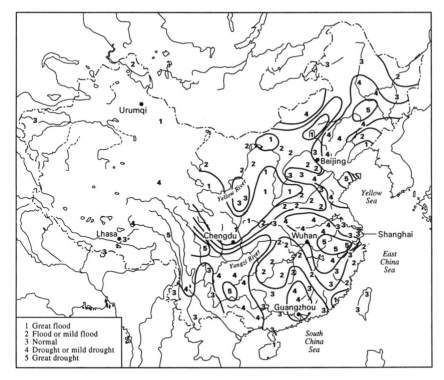

MAP 8.1.   The distribution of floods and drought in China, 1958

Notes: The gradings 1, 2, 3, 4, and 5 refer respectively to 'great flood' (*dalao*), 'flood' or 'mild flood' (*bianlao*), 'normal' (*zhenchang*), 'drought' or 'mild drought' (*bianhan*), and 'great drought' (*dahan*). See Appendix A (pp. 273–4) for details.

Sources: State Meteorological Bureau, *Zhongguo Jin Wubainian Hanlao Fenbu Tuji* (Yearly Charts of Dryness/Wetness in China for the Last 500-year Period) (Beijing: Ditu chubanshe, 1981), 250.

where.[30] More importantly, the North China plain also experienced in the same year a massive sown-area shift from low-yield soya bean to high-yield potato (and probably maize too). This combined with the favourable summer weather to bring about an extraordinary gain in output. The absolute rise in potato output alone was almost double the decline in rice output, rice being the single most important grain crop and accounting for 40 per cent of national grain production in 1958.[31] Thus, it should be clear that the sown-area shift played a major role in generating the impressive yield increases in 1958. None the less this was not achieved without the blessing of the weather.

It is noteworthy, however, that rice yield per sown hectare fell by 5.8

per cent in 1958.[32] Combined with a one per cent decrease in rice-sown area, the yield decline was responsible for the marked reduction in rice output in 1958. This reduction occurred despite comparatively good weather condition in the South, apart from the localized drought in the lower reaches of the Yangzi as mentioned above. I cannot explain this. But any suggestion that the reduction in rice yield was the result of the socio-economic upheaval in 1958 cannot easily be squared with the fact that all the northern grain crops (wheat, potato, and soya bean) simultaneously recorded yield increases. Perhaps the observed yield reduction for rice in 1958 reflected some sort of necessary 'technical adjustment' to its impressive yield increase between 1956 and 1957. This we really do not know.

In this context, reference should perhaps be made to the years 1957, 1956, and especially 1954 which was dominated by the disastrous Yangzi floods. As in 1958, the actual yield per sown hectare in 1957 (1,463 kilograms), and to a lesser extent, 1956 (1,410 kilograms), was higher than the weather-predicted yield (1,385 and 1,387 kilograms respectively). Both actual and predicted values were also above the trend, as shown in Fig. 8.3. Unlike the situation in 1958, however, both these discrepancies (i.e. actual yield higher than the predicted one, and both higher than the trend), can be more easily traced to the geographic incidence of floods and droughts, as has already been shown in the previous section through the relationship between the weather and yield variations.

In 1954, however, the weather-predicted yield (above the trend) was higher than the actual yield (below the trend). The explanation is simple. The 1954 Yangzi flood hit one of the highest-yielding rice belts in China. Floods are, however, more localized than droughts. As a result, the adverse weather may not be 'adequately' reflected in the *shouzai* area compared with the case of droughts, hence the favourable weather-predicted yield for 1954. One might, for example, argue that the lower than predicted yield performance of 1954 is partly accounted for by impaired peasant incentives as a result of the introduction of the compulsory delivery system a year earlier. But it then becomes very difficult to explain the reversal in grain yield in 1956, 1957, and 1958, when, apart from limited institutional retrenchment in the second half of 1957, there was an intensification of collectivization and communization.

In any case, the situation in 1958, or for that matter, 1954, 1956, and 1957 was quite different from that of 1960 and 1961. While the divergent regional incidence of floods and drought can still explain the discrepancy between the actual and the weather-predicted yields (shown in Fig. 8.3) for the earlier years, this is not so during 1960 and 1961. Widespread droughts in both 1960 and 1961, and in particular their prolonged duration, tended to render inoperative the risk-aversion function of both the multiple-cropping practice and the random geographic incidence of natural disasters. The situation in 1960 was made worse by excessive State grain procurements in 1959–60, which drastically reduced per capita grain availability in the countryside and forced peasants to dip deeply into their seed and feed reserves, resulting in a lower sowing density and hence lower grain yield per sown hectare.

The Great North China drought of 1972: bad policy or bad weather?

A methodological remark is in order before we assess the weather impact in 1972 in relation to the policy factor. In Fig. 8.3, the years 1970 to 1984 are taken as a period to make it comparable in length to the earlier period 1952–66. As in 1952–66, the estimated weather-predicted grain yields for the various years in 1970–84 fluctuate, generally speaking, consistently with the realized yields. For all the known bad years, both sets of the values, i.e. realized and weather-predicted yields, are lower than the respective trend yields. This applies to 1972, 1977 (South-east China floods), 1978 (East China drought), 1980 (Yangzi floods), and 1981 (Sichuan basin floods). The reverse holds true for the known good weather years 1970, 1979, 1982, 1983, and especially 1984, in all of which both the weather-predicted and realized yields were higher than (or very near) the trend values.

Nevertheless, the estimated trend-yields line for the whole period 1970–84, shown in Fig. 8.3, tends to distort the comparative scale of yield losses between the different years concerned. The year 1972 is a good case in point. Measured against the estimated trend yield of 2,068 kilograms for that year, the actual yield of 1,988 kilograms represents a yield loss per sown hectare of only 3.9 per cent. This is very low in relation to the comparable loss rate of 7.2 per cent for 1977, given what we know about the comparative scale of weather disturbance in 1972 and 1977. Such 'inconsistency' reflects a statistical bias resulting from the application of the (log-linear) detrending method to the entire 1970–84 period, as discussed at length in Chapter 5 (pp. 103–4).

In fact, if the two periods, 1970–7 and 1978–84, are taken separately, then, as shown in Fig. 8.4, the negative percentage deviation of grain yield from the trend (by 2.8 per cent) for 1977 is only half that of the low rate for 1972 (5.6 per cent). This seems to be a more realistic assessment of the comparative scale of grain yield losses in 1972 and 1977.

Given this more accurate estimate of grain yield loss for 1972, the question is whether it can be entirely explained by the weather factor. Our regression estimates, as given in Fig. 8.4, show that only 2.4 per cent

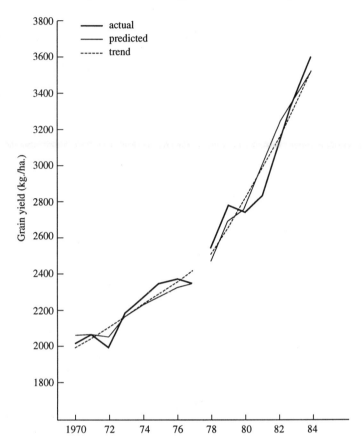

FIG. 8.4.   Actual, predicted, and trend values of the national average of grain yields in China, 1970–1977 and 1978–1984 (kg./ha.)

Source: Table 7.2.

out of the total yield loss of 5.6 per cent (measured against the trend value) can be accounted for by the weather factor. The likely explanation is that the enormous impact of the North China drought of 1972 is not adequately captured by our estimated weather index. This resembles the comparative impact of floods (more disastrous in terms of yield and output losses, yet limited in geographic coverage), as against less visible drought (extensive in coverage, but frequently subject to human control through improved irrigation). In other words, in terms of its meteorological intensity the prolonged 1972 drought more closely resembled a flood.

In the context of the Cultural Revolution, it is also tempting to attribute non-weather-caused yield losses in 1972 to such man-made factors as peasant apathy or resentment in the face of the harsh (or confiscatory) farm siphoning policy and extreme egalitarianism in income distribution. However, this does not easily tally with the fact that in the four subsequent years, 1973 to 1976, realized grain yields were consistently higher than weather predicted yields. These years were characterized by an intensification of the familiar two-line (socialist versus capitalist) class struggle which culminated in the downfall of the 'Gant of Four' in 1976. Perhaps Deng Xiaoping who was then restored as the deputy to Premier Zhou Enlai, was able to help withstand the oppressive policy of the ultra-leftists in creating a more favourable rural environment for the peasants. But we cannot be sure.

1980 and 1981: good policy but bad weather

The years 1980 and 1981 were both marked by disastrous floods. As shown in Fig. 8.4, the magnitude of yield loss in absolute terms and in relation to the trend yield, was substantially larger in 1981 than in 1980. This may be because the main locus of the floods in 1981 was the Sichuan basin, which enjoyed a much higher grain yield per sown hectare (4,013 kg./ha. in 1983) than Hubei province (3,758 kg./ha.) which was the main centre of the 1980 Yangzi floods. Moreover, the contribution of Sichuan to the national grain output was double that of Hubei: 10 compared with 5 per cent.[33]

The more important point in this context is that in 1980 and 1981 both the weather-predicted and the actual yields were below the trend (Fig. 8.3), with the minor exception of 1981 when predicted and trend yields, based on the subperiod 1978–84, were virtually the same. This suggests that the favourable policy and institutional reforms in the early 1980s were unable to avert or mitigate the weather impact. Note too that weather-predicted yields for both years were consistently higher than actual yields, the discrepancy being especially large in 1981. This seems to reflect once again the fact that the scale of floods does not 'adequately' show up in the *shouzai* area, compared with that of droughts.

Another possible explanation is that increased rural decontrol in the early 1980s had deprived the local authority of that collective hedge, which had proved to be so powerful in stabilizing both sown-area and grain-yield fluctuations from the 1930s to the 1950s. This also underlines the point made earlier (in Chapter 7) that in the new rural context of the 1980s increased earnings alternatives made Chinese peasants reluctant to reallocate a substantial amount of resources to cope with the floods. This naturally served to aggravate the flood impact on grain production.

1979 and 1984: good policy or good weather?

As shown in both Figs. 8.3 and 8.4 the years 1979 and 1984, and for that matter 1983 as well, saw considerable gains in yields in relation to the estimated trend values for the period 1978–84, or for 1970–84 taken as a whole. However, the positive yield deviations for these years can also be closely predicted from weather movements. The gains were therefore not obtained independently of weather conditions. By comparison with the years 1980 and 1981 which both saw substantial negative yield deviation (despite increased decollectivization designed to make Chinese peasants more responsive to the purported income incentives provided by the sharp increases in 1979 in State farm procurement prices), the weather seems to have contributed more than any policy factors in bringing about the yield gains in 1979 and 1984.

Note that in contrast to the year 1972, the realized yield for 1979 was higher than the weather-predicted yield. This may have reflected initial enthusiasm by peasants for the offered price benefits after years of imposed austerity. We cannot be sure of this, but what is certain is that the income-distributive structure under the collective framework in 1979 remained highly egalitarian, hence the need to decollectivize in subsequent years.[34]

While in the wake of the disastrous 1980–1 floods increased decollectivization seems to have rendered the collective hedge against yield fluctuations inoperative, consecutive increases in State farm procurement prices in the early 1980s, coupled with the more congenial new institutional rural environment, nevertheless prompted the peasants to take advantage of favourable weather since 1982 and exert more effort to increase yield and output. The situation tended to resemble increasingly that of the 1930s, with, however, one notable difference. That is, at least until 1984, peasants were guaranteed relatively favourable State farm procurement prices, whereas in the market environment of the 1930s increased supply from a good harvest tended to depress farm prices. This, together with continuous improvements in the weather conditions explains the observed upsurges in grain yields since 1982.

Some General Observations

Various observations may be made from the foregoing analysis about the relative importance of weather and policy factors in the observed annual fluctuations in national average grain yield. The most striking finding is that although agricultural technology may have changed, *within* any period of similar technological level, grain yield instability can be more consistently explained by weather fluctuations than by abrupt changes in agricultural policy. Thus, despite the considerable reduction in grain-yield instability from 1952–66 to 1970–84, the observed year-to-year oscillations in yields in the later period can still be clearly traced to the changing weather conditions.

Floods have, in general, a greater impact than drought on the stability of national average grain yield, and for any comparable floods or droughts the impact is greater when they occur in the southern rice provinces than when they do so in the less fertile North-west or North. Similar observations can be made within the same regions, or between regions in the broadly defined North and South which display differentiated yield levels. This is borne out by the contrast between the yield impact of the disastrous floods in Sichuan (1981) and Hubei (1980) for example. Thus, in many cases the known random regional incidence of major floods and droughts can be conveniently used to explain the different degrees of national aggregate yield variations between the years of major weather disturbance.

Nevertheless, at the national level it is difficult, if not impossible, exactly to isolate, in quantitative terms, the weather-caused yield losses from the possible implications of policy influence. This is readily understandable, given the vast size of China and the complexity of its agroclimatic and regional grain-yield variations. As a result, the yearly yield variations cannot always be accurately predicted from our regression estimates about the possible weather impact. In many cases, however, the observed discrepancy between the actual yields and weather-predicted yields in 1952–66 and in 1970–84, can reasonably well be explained by the known institutional or policy parameters and their favourable or unfavourable implications for yield increases.

But in the end, the most important finding is that the actual grain yield for most of the years under study consistently follows the variations in the weather-predicted yields. In short, the weather has been more important than anything else in accounting for observed short-run fluctuations in agricultural yields and output in China from the 1950s until the most recent past.

NOTES

1. Cf. Fig. AA.3 for the latitudinal distribution of droughts and floods and Table AB.16 for provincial statistics of *shouzai* area.
2. With the notable exception of Shandong as noted in detail in Ch. 4.
3. Cf. e.g. Jao Xing, 'Qixiang gongzuo wei nongye jishu gaige fuwu de chubu yijian' (Preliminary views on meteorological work to serve agro-technical work), in *ZGNB* 7 (1963), 12, for 1952 and 1955 in comparison with the

other years. This is also consistent with the *shouzai* area (weighted or non-weighted) index shown in Fig. 6.1 and Table AB.15.

4. Chen Zhenlu and Chen Bangzheng, *Zhongguo Nongcun Jingji Wenti*, ii (Shanghai: Daxue Shudian Chubansbe, 1935), 288–90 in particular.

5. See Appendix A for a detailed discussion of this point, and its possible implications for the computed weather index.

6. Zhang Xiangong, 'Zhongguo dongbanbu jin wubainian ganhan zhishude fengxi' (An analysis of the index of droughts in the eastern half of China in the last 500 years), in State Meteorological Bureau (Research Institute of Climatic Science), *Quanguo Qihou Bianhua Xueshu Tuolunhui Wenji (Pro ceedings of a National Academic Symposium on Climatic Changes)* (Beijing: Kexue Chubanshe, 1981), 47. The other five years involved are 1528, 1640, 1641, 1721, and 1785 (based on a drought index same as that underlying our reconstructed index for 1921–79, as shown in Table AB.19.

7. This point is discussed in greater detail in connection with the great North China droughts of 1972 in the next section.

8. Y. Y. Kueh, 'A weather index for analysing grain yield instability in China', 78.

9. Ibid.

10. See Fig. AA.3.

11. Table AB.15.

12. There is no doubt about the severity of the drought in 1978. In our computed weather index (Fig. 6.1), this is the factor which on its own brings the index to a climax within the 1970–81 period (next to 1960 and 1961 only within the entire 1952–84 period). Nevertheless, the total *Chengzai* area in 1978 (including floods and other natural disasters) was still less than in 1981 (and only marginally higher than in 1981), which were both dominated by floods (in the Yangji and Sichuan basins respectively). As in the catastrophic floods of 1980 and 1981, the drought also received great attention in Chinese press reports and was frequently compared with the equally disastrous droughts of 1934 and 1959 occurring in the same Yangzi and Huai River basins. But with greatly improved irrigation capacity by the late 1970s it was obviously easier to fend off the possible impact of droughts, compared with floods. For major reports on the 1978 droughts see e.g. *RMRB* 4, 13, and 26 Oct.; 1, 4, 11, 12, and 25 Nov. 1978; and 2 and 23 Jan. 1979. There was also spring drought affecting the wheat crops in the North (Shandong, notably); see *RMRB* 27 Apr. and 1 May 1978.

13. *TJNJ 1984*, 191.

14. No author, *Zuguo Jinxiu Heshan* (The Splendid Rivers and Mountains of the Motherland) (Tianjin: Tianjin Renmin Chubanshe, 1975), 131–2. For the disastrous Zhujiang River floods in 1959, see E. Stuart Kirby (ed.), *Contemporary China*, iv (Hong Kong: Hong Kong University Press, 1961), 214: 187 persons killed, 29 missing, 204 injured, 2 million affected, 200,000 houses destroyed, and over 310,000 ha. of land under early rice, and 67,000 ha. under peanuts, sugar cane, jute, and other industrial crops submerged.

15. *Zuguo Jinxiu Heshan*. For more details about the meteorological background

of the floods and droughts occurring in 1959, see the discussion in the Appendix in connection with Map AA.2.

16. *MSSWS* 292–4.
17. This can be easily verified by looking at Map 10.2 (1960); for details see Ch. 10 and Appendix A.
18. *RMRB* 29 Dec. 1960.
19. K. R. Walker, *Foodgrain Procurement and Consumption in China*, 149.
20. For details see Table 11.3 and the discussion in connection therewith in Ch. 11.
21. Ibid.
22. *RMRB* 26 Jan. 1966. This is from a major article 'Woguo qihou yu hanlao zaihai' (Climate and calamities of droughts and floods in our country), by Lu Wu, Deputy Director of the State Meteorological Bureau. Lu gives a lengthy review of the weather in 1959–61.
23. Ibid.
24. Christopher Howe and Kenneth Walker, 'The economist', in Dick Wilson (ed.), *Mao Tse-tung in the Scale of History*, 206 and 209–10.
25. Chao Kang, *Agricultural Production in Communist China*, 133.
26. See Appendix A, p. 274.
27. Ministry of Water Conservancy and Power Generation, *Xiandai Zhongguo Shuili Jianshi* (Water Conservancy Work in Contemporary China) (Beijing: Shuili Dianli Chubanse, 1984), 38–9; and *HNBB* 18 July 1958 and *DZRB* 26 July 1958.
28. Zhou Enlai then rushed to Zhengzhou, capital of Henan, to take charge personally of commanding the rescue operation.
29. Lei Xilu, *Woguode Shuili Jianshe* (Water Conservancy Work in Our Country) (Beijing: Nongye Chubanshe, 1984), 13; cf. also the rainfall charts shown in Fig. 10.1.
30. Cf. Walker, *Foodgrain Procurement and Consumption in China*, 34.
31. Table AB.3.
32. Ibid.
33. *NYNJ 1982*, 34.
34. Kueh, 'China's new agricultural-policy program: Major economic consequences, 1979–1983', 358.

IV
Precipitation and Grain-Yield Variations

9

Precipitation and Yield Relationships in 1931–1935

In our analysis of the relationships between weather and national grain-yield variations in Part III, we relied on the extent of the sown area covered by natural disasters, i.e. *shouzai* area, as a national indicator of weather changes between the different years. This approach is deemed necessary, because the sheer size of China and the complexity of her geoclimatic conditions make it impossible to compile a meteorological weather index for the country as whole based on precipitation or temperature data.

In this chapter, however, an attempt is made to verify grain-yield fluctuations against possible precipitation evidence for selected agricultural areas which are defined as more or less 'homogeneous' in terms of broad agro-climatic and cropping criteria. The precipitation indicator is used, because this is the single most important measure of floods and droughts, which are the most predominant weather hazards for China. We focus on the pre-war years 1931–5, not only because monthly rainfall data are available, but because rural China in those years was, unlike the post-war years, virtually free of direct government policy intervention. This may help to capture more exactly the real magnitude of weather influence on grain yields.

However, an effort is also made to collect monthly precipitation data for the years 1959–61. We make use of the results of our regression estimates of weather and yield relationships of the 1930s to predict the possible magnitude of weather-caused yield losses in 1959–61 based on the precipitation data collected for the related agricultural areas (see Chapter 10).

The present chapter is divided into two sections. The first makes a broad distinction between the wheat and rice regions and uses annual provincial data for 1931–5 in order to make a cross-sectional estimate of the degree of correlation between rainfall and yields for the rice and wheat crops. The second section focuses on smaller agro-geographic entities, namely the Yangzi rice and wheat area and winter wheat and gaoliang area as defined by Buck,[1] and examines the yearly fluctuations in their grain yields from 1931 to 1935 in relation to changes in precipitation. The section concludes with a brief examination of the possible implications of annual yield fluctuations in these two agricultural areas for the annual national average of grain yields in 1931–5.

A Cross-Provincial Perspective

The rice yields

Fig. 9.1 presents the *graphical* correlation between changes in precipitation and rice yields for a pooled (time and cross-sectional) series of data covering the eleven predominantly rice-producing provinces during 1931–5. In both graphs the explanatory rainfall variable is taken to be a percentage of the long-term means. The difference between the two graphs is that, in the first, the 1931–7 mean rice yield is used as a basis for determining the percentage yield deviations of each province in each of the years during the period 1931–5, whereas in the second graph the 'normal' rice yield standard as defined in the 1930s is used. Briefly, the 'normal' yield (as applied by the National Agricultural Research Bureau (NARB) prior to 1934–5 to gauge the yearly agricultural performance) refers to the yield in a 'normal' or 'average' year, as opposed to the 'best' year. The latter, which was also used by NARB, is defined as the maximum yield available or ever obtained.[2] For almost all provinces, the 'normal' yield for both rice and wheat crops is substantially higher than the 1931–7 mean,[3] although strictly speaking it should not be taken as the yield commensurate with the 'potential evapotranspiration rate' for the crops concerned. This background should be borne in mind when we come to interpret the estimated rainfall and yield correlations.

Fig. 9.1 presents the polynomial regression equations for the two alternative measurements of the weather influence. Taking the 1931–7 mean as the standard, over 40 per cent of rice-yield fluctuations may be explained by the fluctuations in rainfall in the main growing season from May to August. By the 'normal yield' standard, the relation between rice yield and rainfall is less close, but the estimated degree of correlation is still remarkable, especially bearing in mind that no differentiation is made among the eleven rice provinces in terms of soil fertility and climatic conditions.[4]

Both graphs show that too little or too much rainfall (drought or flood) is detrimental to rice production: rice yield per sown hectare is depressed in both cases. Most of the provincial samples cluster within the relatively narrow (or normal) range of plus or minus 10 to 15 per cent of rainfall variations. The precipitation-predicted maximum rice yield, which is around 5 per cent above the 1931–7 mean, is associated with a rainfall which is 10 per cent above the long-term mean. This implies that higher rice yields are associated with an amount of rainfall higher than normal, but not so large as to cause floods. This is consistent with the fact that rainwater supplies were normally not sufficient to meet irrigation demands in the southern provinces, since their multi-cropped areas were already quite extensive in the 1930s. This point applies with even greater force to the water-scarce wheat provinces, as is discussed shortly.

Measured against the 'normal yield', the predicted maximum rice yield is, in turn, about 10 per cent *below* the standard. This is in line with the discrepancy noted earlier between the 1931–7 mean yield and the 'normal yield'. In any case, the exact numerical difference is less important than the fact that the two alternative measurements presented in Fig. 9.1 tend to supplement each other. Note especially that, as in the 1931–7 mean-yield series, the precipitation-predicted maximum yield for the 'normal yield' series falls roughly within 5 per cent plus or minus of the long-term mean precipitation.

The wheat yields

Both of the comparable polynomial regressions in Fig. 9.2 for wheat yields provide a less coherent precipitation and yield relationship compared with that for the rice region. This is not surprising, for the ten wheat provinces included in the regressions are far less homogeneous than the rice provinces. If, for example, the ten provinces are grouped into four relatively homogeneous agricultural areas in line with Buck's taxonomy, then through the assignment of 'area dummies' to account for the regional environmental differences, the rainfall and wheat yield correlations turn out to be much closer. This is reflected in the following regression equations,

(1) $Y_1 = -31.827 + 0.42887P - 0.000981P^2 + 2.3188D1 + 15.211D2 + 5.6645D3$
$(-2.7575)\quad(2.0405)\quad(-0.82401)\quad(0.37955)\quad(2.0445)\quad(0.73363)$

$r^2 = 0.3840\quad N = 40,$

(2) $Y_2 = -33.326 + 0.19664P + 0.0003298P^2 + 6.8050D1 - 13.808D2 + 44.033D3$
$(-3.7455)\quad(1.2136)\quad(0.35936)\quad(1.4449)\quad(2.4075)\quad(7.3978)$

$r^2 = 0.8258\quad N = 40$

where Y_1 and Y_2 stand, respectively, for the percentage deviations of wheat yield from the 1931–7 mean and the 'normal' yield of the 1930s, and P the September to May precipitation variations from the long-term mean. The four areas comprise, in addition to the most important winter

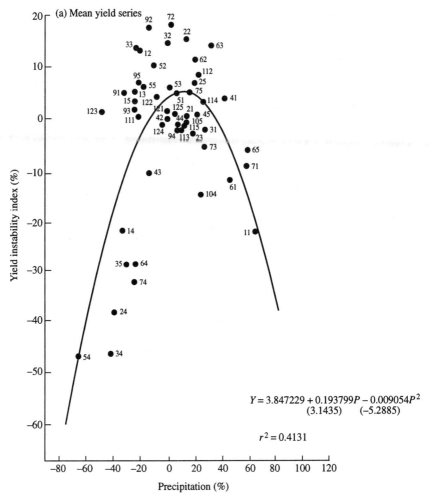

FIG. 9.1. Provincial cross-sectional relationship between precipitation in May to August (expressed as a percentage deviation from the long-term mean) and annual rice yield instability (expressed as a percentage deviation from the 1931–7 mean yield, or alternatively the 'normal' yield of the 1930s) in China, 1931–1935

Provincial and year codes

Provinces		Hubei	6	*Years*
Jiangsu	1	Hunan	7	1931 = 1
Zhejiang	2	Guangdong	8	1932 = 2
Anhui	3	Sichuan	9	1933 = 3
Fujian	4	Guizhou	10	1934 = 4
Jiangxi	5	Yunnan	11	1934 = 5

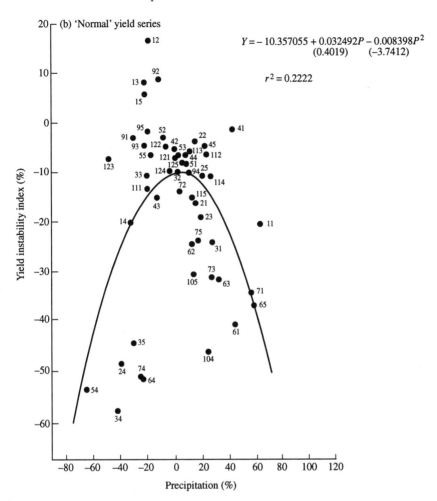

$$Y = -10.357055 + 0.032492P - 0.008398P^2$$
$$(0.4019) \quad (-3.7412)$$

$$r^2 = 0.2222$$

Thus, the code for Jiangsu in 1931 is 11, in 1935 15; Guizhou in 1932 is 102, in 1935 105, etc. The provincial codes used are not the same as those adopted in Fig. 4.4. This helps to avoid using 3-digit codes for the graph.

Notes: For Fujian, Guangdong, and Guangxi provinces, the April to August (instead of May to August) precipitation is applied to take into account the fact that the rice-growing season begins earlier in South China.

Sources: Tables AB.8 and AB.10 for rice yield, and Table AB.17 for precipitation data.

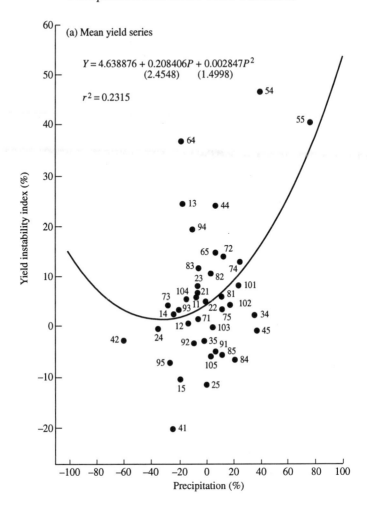

FIG. 9.2. Provincial cross-sectional relationship between precipitation in September to May (expressed as a percentage deviation from the long-term mean) and annual wheat yield instability (expressed as a percentage deviation from the 1931–7 mean yield, or alternatively the 'normal' yield of the 1930s) in China, 1931–1935

Provincial and year codes

Provinces					Years
Hebei	1	Gansu	6		1931 = 1
Shandong	2	Sichuan	7		1932 = 2
Henan	3	Hubei	8		1933 = 3
Shanxi	4	Anhui	9		1934 = 4
Shaanxi	5	Jiangsu	10		1935 = 5

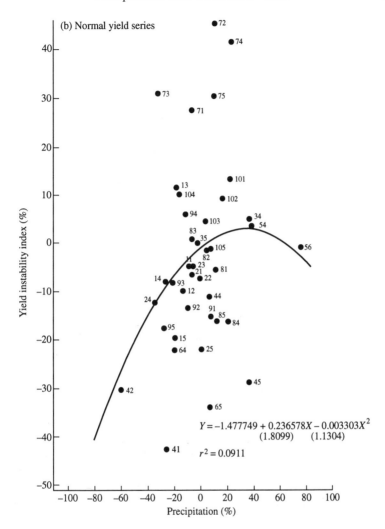

(b) Normal yield series

$$Y = -1.477749 + 0.236578X - 0.003303X^2$$
$$(1.8099) \qquad (1.1304)$$
$$r^2 = 0.0911$$

Thus, the code for Hebei in 1932 is 12, in 1934 14; Jiangsu in 1933 is 103, in 1935 105, etc.

Notes: For Sichuan and Zhejiang provinces the October to April (instead of September to May) precipitation is applied to take into account the fact that the wheat growing season begins later in the South.

Sources: Tables AB.9 and AB.10 for wheat yield, and Table AB.17 for precipitation data.

wheat and gaoliang area (Hebei, Henan, Shandong, Jiangsu, and Anhui) used as the basis for comparison, the winter wheat and millet area D1 (Shanxi, Shaanxi, and Gansu), the Sichuan basin D2 (Sichuan), and the Yangzi rice and wheat area D3.

Analysis of the two sets of regression equations with and without the dummy variables should, again, not focus exclusively on the numerical results, especially those of the equations based on the 'normal yield' series. Taken together, the alternative estimates strongly suggest that the greater the rainfall the higher the wheat yield. This is to be expected because, throughout the wheat region, both winter and spring are the dry seasons and total precipitation normally falls short of the moisture requirements of wheat crops throughout the entire growing season, as shown in Table 9.1 in the case of North Jiangsu.[5]

In terms of the graphical rainfall–yield relationship based on the 'normal yield' series (Fig. 9.2), the implication is that the greater/smaller the positive/negative rainfall deviations, the smaller/larger will be the negative/positive yield deviations from the 'normal' yield standard. Put simply, the greater the amount of rainfall, the higher will be the wheat yield. Excessive rainfall can of course result in reduced wheat yield as is also shown in our estimated regression equations. This relationship is, however, not as pronounced as that in the rice region. Spring flood and waterlogging constitute a rare phenomenon in the North compared with summer floods in the South.[6]

Apart from rainfall, the average yields of wheat (and rice) are also subject to a host of other meteorological anomalies which are not captured in our regression systems. A good example of this is the 'dry-and-hot wind' (*ganre feng*).[7] This phenomenon combines high temperature, low relative and absolute humidity, and high wind speed and occurs over a vast area north of the Huai River in a very short period from late May to early June. Its impact, which is much more severe than that of drought, is virtually unavoidable and its geographical extent is frequently great. It normally lasts for only about three days but, with one broad sweep, the soil moisture can be utterly depleted. Empirical estimates from the North China Plain show that ripening winter wheat crops may be disastrously affected, the harvest falling 30 to 50 per cent below normal.[8]

Another weather disturbance in addition to floods and drought is the unexpected advance of a cold wave (*hanchao*). This may take the form of a 'reversed spring chill' (*daochunhan*), an abrupt temperature decline in the autumn or an exceptionally strong cold spell during the winter months.[9] A 'reversed spring chill' can, for example, result in widespread frost damage to both the rice and wheat crops which have been lulled by the relatively warm early spring into premature tillering or jointing.[10]

TABLE 9.1.   A comparison of moisture requirements (*R*) and moisture supplies (*S*) via precipitation: the case of wheat in North Jiangsu (in mm.)

	Sowing to greening up (early October to mid-Feburary)			Greening up to tillering (late February to mid-April)			Tillering to ripening (late April to late May)			Total growth period (210 days)		
	R	*S*	*S – R*	*R*	*S*	*S – R*	*R*	*S*	*S – R*	*R*	*S*	*S – R*
Xuzhou	150	136	–14	160	67	–93	140	77	–63	450	280	–170
Huaiyin	150	150	0	160	94	–66	140	91	–49	450	335	–115

Source: Writing group, *Nongye Qixiang Zhishe* (Knowledge of Agricultural Meteorology) (Shanghai: Kexue Jishu Chubanshe, 1978), 59.

The Year-to-Year Variations

Comparing the Yangzi rice and wheat area with winter wheat and gaoliang area

The precipitation and yield relationships become all the more pronounced if the regional analysis is conducted at an even more disaggregated level than is implied by the broad distinction between the rice and wheat regions. This is shown in Figs. 9.3 and 9.4 for the two smaller agricultural areas: the winter wheat and gaoliang area for the rainfall–wheat yield relationship; and the Yangzi rice and wheat area for the rainfall–rice yield relationship. The two sets of regressions based on the two different yield bases (i.e. the 1931–7 mean yield and the 'normal' yield) generate nearly identical results.[11] Taken together, the two separate rainfall–yield relationships, for wheat and rice, are consistent with known weather events occurring during 1931–5 in the two agricultural areas under study.

Thus, the disastrous 1934 drought, as shown in Fig. 9.3, was exclusively a summer disaster. It occurred mainly in the Yangzi and Huai River basins, but also extended into areas well north of the Huai River basin, including Henan and Shandong.[12] However, the weather in the spring of 1934 in the winter wheat and gaoliang area (which includes Henan and Shandong) was very favourable, hence the comparatively good wheat harvest shown in Fig. 9.4.

The situation in 1935 was exactly the reverse. The summer and autumn drought of 1934 persisted through the spring of 1935 and spread into the north in a fashion similar to that of 1959–60. As a result, the wheat harvest in 1935 from the winter wheat and gaoliang area was bad, although the yield loss (minus 8 per cent from the 1931–7 mean), was not as disastrous as that of rice (minus 56 per cent) in the Yangzi rice and wheat area in 1934.

The year 1931 is also noteworthy. The average rice yield for the Yangzi rice and wheat area as a whole was reduced by around 15 per cent (from the 1931–7 mean) by the Yangzi flood. But again, in the winter wheat and gaoliang area, the spring weather and hence the wheat harvest were as good as in 1934.

As for 1933, the rainfall–yield relationship also shows a sharp contrast between the two agricultural areas. While for the Yangzi rice and wheat area it was an average year (rice yield 5.6 per cent above the 1931–7 mean), it was a wet spring in the winter wheat and gaoliang area. This greatly helped to balance the potential moisture requirements of the wheat crops, hence the bumper wheat yield (10.5 per cent above the 1931–7 mean) (Fig. 9.4).

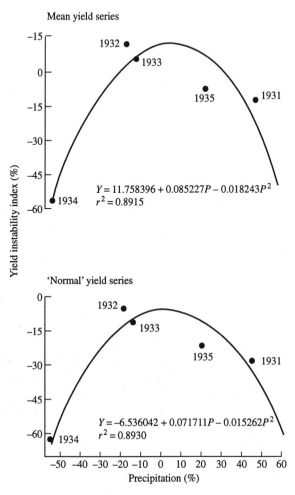

Fig. 9.3. Annual rice yield instability (expressed as a percentage deviation from the 1931–7 mean yield, or alternatively the 'normal' yield of the 1930s), in relation to the fluctuations of precipitation of June to August (expressed as a percentage deviation from the long-term mean) in the Yangzi rice–wheat area in China, 1931–1935

Notes: The area covers the three provinces, Jiangsu, Anhui, and Hubei. It is not identical with, but roughly comparable to the Yangzi rice–wheat area as defined by Buck (see Map 10.1). The precipitation figures are from the meteorological stations in Zhengjiang, Nanjiang, Shanghai, Wuhu, Anqing, Yichang, and Hankou.

Sources: Tables AB.8 and AB.10 for rice yield; and Table AB.17 for the precipitation data.

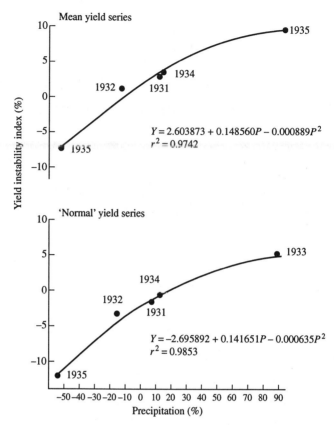

Fig. 9.4. Annual wheat yield instability (expressed as a percentage deviation from the 1931–7 mean yield, or alternatively the 'normal' yield of the 1930s), in relation to the fluctuations of precipitation of March to May (expressed as a percentage deviation from the long-term mean) in the winter wheat–gaoliang area in China, 1931–1935

Notes: The area covers the five provinces Hebei, Shandong, Henan, Jiangsu, and Anhui. It is not identical with, but roughly comparable to the winter wheat–gaoliang area as defined by Buck (see Map 10.1). The precipitation figures are from the meteorological stations in Beijing, Jinan, Kaifeng, and Xuzhou.

Sources: Tables AB.9 and AB.10 for wheat yield; and Table AB.17 for the precipitation data.

Nevertheless, in 1933 the wet spring in the north was apparently reinforced by torrential rainfall in the summer. This culminated in the serious Yellow River flood of 1933 which, together with the 1958 flood, is still used as a yardstick against which to compare similar flood volumes.[13] This flood cannot be captured by either of the rainfall–yield diagrams presented in Figs. 9.3 and 9.4, because it occurred in the north (rather than in the southern rice areas), and came after the wheat harvest (affecting only the autumn crops).

We turn, finally, to the year 1932, which is widely considered to have been one of the most favourable weather years in 1931–7. In our two diagrams (Figs. 9.3 and 9.4), 1932 is not characterized as particularly favourable or unfavourable in terms of rainfall. The reason for this is that unlike in 1933, the weather during the summer of 1932 in the winter wheat and gaoliang area was also favourable, as it was in the spring throughout the Yangzi rice and wheat area[14]. But this is not reflected in the two diagrams which only reflect the spring weather in the wheat area and the summer weather in the rice area.

Nevertheless, during 1931–5, 1932 shows the largest positive yield deviation for rice according to the 1931–7 mean-yield series, or alternatively the smallest negative yield deviation according to the 'normal'— (i.e. maximum) yield series. In short, the year experienced the highest rice yield recorded for 1931–7. The explanation for this is that the 1932 rice yield (Fig. 9.3) refers to the whole year which covers both the summer and autumn rice harvest, but the good 1932 spring weather responsible for the summer rice harvest cannot be reflected in the diagram.

Similarly, both the favourable summer weather in the north and the good autumn harvest to which it gave rise in 1932, are not included in the wheat diagram (Fig. 9.4). This explains why according to this diagram the 1932 yield is not as good as that for 1931, 1934, and especially 1933.

Regional disturbance and national grain yield fluctuations

Both the winter wheat and gaoliang area and the Yangzi rice and wheat area were important in national grain production in the 1930s, and after 1949. One would therefore expect the regional rainfall–yield relationships (Figs. 9.3 and 9.4) between 1931 and 1935 to be reflected nationally (Fig. 8.1) in terms of the relationship of overall grain yield fluctuations to the size of sown area affected by floods and drought. This is the case. The link between the regional and national conclusions is clear enough, not only for the years 1931 (flood) and 1934 (drought), but also for 1932 (the most favourable year in 1931–7). This latter year shows up in the national analysis (Fig. 8.1) as demonstrating the highest positive yield deviation and the smallest *shouzai* area proportion compared with the other years in the 1931–7 period. The background to this is also evident if in addition to the regional rainfall–yield analysis (Figs. 9.3 and 9.4), reference is made to the favourable summer weather in the winter wheat and gaoliang area and good spring weather in the Yangzi rice and wheat area in 1932.

Like 1932, a link between the regional and national aggregate analysis can also be established for 1933 and 1935. In the national 1931–7 aggregate study (Fig. 8.1), 1933 emerges as an average year, judged by both *shouzai* area proportion and the extent of yield deviation (only a very marginal negative percentage deviation from the mean). This is in line with the position 1933 assumes in the regional rainfall–rice yield diagram (Fig. 9.3) for the Yangzi rice and wheat area. Only in the rainfall–wheat yield diagram (Fig. 9.4) for the winter wheat and gaoliang area, does the same year turn out to have been a very good year compared with the other years during 1931–5 included in the diagram.

This may seem to contradict the conclusion drawn from the 1931–7 national aggregate analysis. If so, the contradiction can be easily explained by the fact that the catastrophic Yellow River flood in August 1933 only affected the higher-yielding autumn coarse grains, and not wheat, which had already been harvested. In other words, the bumper wheat yield of 1933 was offset by the reductions in yield of maize, gaoliang, and other autumn crops, resulting in a mediocre average yield for all grain, comparable with that obtained in the Yangzi rice and wheat area. Thus, 1933 turned out to be an average year in terms of national average yield.

As Fig. 8.1 shows, 1935 was also a mediocre year, with the national average grain yield nearly the same as the 1931–7 average. Viewed for the *shouzai* area proportion (10 per cent of the 1931–7 grain-shown area mean), the weather in 1935 was somewhat more favourable than in 1933 (although still not as good as in the bumper harvest year of 1932). This is consistent with the observed grain-yield variations for 1933 and 1935. But how does this national aggregate weather–yield relationship for 1935 stand in relation to the regional rainfall–yield analysis for the two major agricultural areas: the winter wheat and gaoliang area and the Yangzi rice and wheat area?

The rainfall–wheat yield diagram in Fig. 9.4 shows, first, that in 1935 there was serious spring drought in the winter wheat and gaoliang area. However, this mainly affected the winter wheat crop which had a considerably lower yield per sown hectare than rice, gaoliang, and other coarse grains. Another Yellow River flood was reported in summer 1935, but it was not nearly as severe as that of 1933.[15] Thus the more favourable autumn harvests in the winter wheat and gaoliang area and elsewhere seem to have overcompensated for the wheat yield losses in 1935, much in the same way that the bumper wheat harvest in 1933 was neutralized by poor autumn harvests caused by the catastrophic 1933 Yellow River flood.

As for the Yangzi rice and wheat area, our rainfall–rice yield diagram

(Fig. 9.3) reveals that rice yield in 1935 fell by 7 per cent from the 1931–7 mean. The reduction is associated with rainfall which was 20 per cent higher than the long-term mean, but far less than that which caused the 1931 Yangzi flood (by nearly 50 per cent above the mean). The weather factor which generated this 1935 rainfall–rice yield relationship in the Yangzi rice and wheat area can be identified: the July flood of the Han River, the largest tributory of the Yangzi. However, the Han River flood affected mainly Hubei province,[16] in contrast to the 1931 Yangzi flood which affected both the Yangzi and the Huai River basins and areas outside the Yangzi rice and wheat area. The localized nature of the Han River flood explains why the 1935 national *shouzai* area proportion was comparatively limited, as shown in Fig. 8.1. It is likely too that favourable weather in the rice areas outside of the Yangzi rice and wheat area (together with the good autumn harvests achieved in the northern provinces) had overcompensated for the yield losses caused by the Han River flood in 1935, resulting in a national average grain yield which was slightly above the 1931–7 mean.

Taken together, the two sets of rainfall–yield analysis, although confined to two representative areas, can be used to verify consistently the national aggregate yield variations in relation to the changing size of the *shouzai* area in the 1930s. This suggests, once again, that the *shouzai* area index may represent a viable weather proxy, and may even offer the only method of analysing the weather impact on agricultural stability at the national level.

NOTES

1. See Map 10.1 for Buck's demarcation of agricultural areas.
2. See Table AB.10 for details.
3. Compare Table AB.10 with AB.8 and AB.9.
4. A similar analysis is available in Liang Qingchun, 'Zhongguo han yi hanzai zhi fengxi' (Analysis of droughts and drought calamities in China), in *SHKXZZ* 6/1 (Mar. 1935), 1–64. Liang regressed the same 'normal yield' (*jin/mu*) of rice crops on total rainfall (mm.) from Apr. to July (long-term average) for sixteen rice-producing provinces, and the estimated equation is

$$Y = 0.42501 + 0.904787X - 0.0006289X^2, \quad r^2 = 0.684$$

5. The figures given in the table may overstate the extent of moisture deficits since the required rainfall supplies are based on what is necessary to satiate the 'potential evapotranspiration' of the plants (i.e. to sustain the theoretical potential maximum yield). Nevertheless, it should also be remembered that North Jiangsu is a relatively wet coastal area compared with the other wheat areas in the interior.
6. *Nongye Qixiang Zhishi*, 60.

7. Research Institute of Atmospheric Physics (Chinese Academy of Science), *Zaihaixing Tianqide Yice he Yufang* (*Forecast and Prevention of Calamitous Weather*), (Beijing-Kexue Chubanshe, 1981), 156 and 169. See also Chen Kexin, *Zaihaixing Tianqi Jiqi Yufang* (Calamitous Weather and its Prevention) (Shijiazhuang: Hebei Renmin Chubanshe, 1979), 93–105.
8. Ibid. 103 and *Nongye Qixiang Zhishi*, 60. This is also regarded as a major climatic calamity in northern Jiangsu in May and June. See Gao Liangzhi and Zhu Tangsong, 'Xuhuai diqu ganhanfeng de fasheng guilu he fangyu' (The regularity of the happening of dry wind and its prevention in the Xuzhou and Huaiyin areas), in *NYKX* 5 (1961), 31. The Soviet equivalent, *Sukhovey*, of the Chinese *gangrefeng*, is equally, if not even more disastrous; see David W. Carey, Soviet Agriculture: Recent Performance and Future Plans, 581.
9. Research Institute of Atmospheric Physics, *Forecast and Prevention of Calamitous Weather*, 28–49.
10. For a sophisticated analysis of the correlation between wheat yield and solar radiation, precipitation, and temperature in the post-war period, see Wang Shiqi, Cao Yonghua, and Cai Yongci, 'Winter wheat production in Beijing District and the statistical analysis of its meteorological factor', in *NYKX* 1 (1979), 10–18.
11. Note esp. the close comparability in terms of both the regression coefficients and r^2 values which are indeed very high, compared with those for the broader wheat and rice regions.
12. *NARBCR* 2/10 (1934), 640.
13. See Ch. 8 n. 27.
14. For the whole of China the notable exception to the overall good weather were the catastrophic floods in Heilongjiang and Jilin provinces; see *SBNJ 1933*, 974. For the scale of the floods in terms of inundated farm areas, see Table AB.14.
15. Lei Xilu, *Water Conservancy Work in Our Country*, 10; Ministry of Water Conservancy and Power Generation, *Water Conservancy Work in Contemporary China*, 39; and Deng Yunte, *Zhongguo Jiuhuang Shi* (*The Chinese History of Famine Relief*) (Beijing: Sanlian Shudian, 1958), 34–5.
16. Ministry of Water Conservancy and Power Generation, *Water Conservancy in Contemporary China*, 32.

10

Precipitation Evidence of Yield Losses in 1959–1961

The main purpose of this chapter is to compare the relative magnitude of grain-yield losses sustained in 1959–61 in a selected number of provinces (for which both sown area and output data are available) with the comparable loss rates which may be obtained by applying the available precipitation data for 1959–61 to the precipitation–yield regression as estimated in the previous chapter for the related agricultural areas for 1931–5. Such a comparison is based on the fact that the two periods, 1931–5 and 1959–61, are roughly comparable in terms of farm technology (as it bears on farm productivity), but diametrically different in the government's agricultural policy and rural organization. Any observed discrepancy between actual yields and precipitation-predicted yields may thereby help to isolate the possible impact of the policy factors (as they may bear on peasants' incentives or rural resource allocations) in 1959–61.

The accuracy of the predicted yields hinges very much on the precipitation data collected for 1959–61. Thus, in the first section the nature of these data in relation to other comparable sources is discussed. In the following section a few methodological caveats are set out, which need to be borne in mind in interpreting the predicted yield against the realized yield losses. In the third section the two sets of yield values are compared separately for each of the years, 1959 to 1961. The chapter is concluded with a brief evaluation of whether the major findings are in line with the conclusions drawn from the national level analysis, based on the *shouzai* area index, that actual yields were higher in 1959 (due to favourable non-weather factors), but lower than the weather-predicted yield in 1960 (unfavourable policy factor), while by 1961 they had become largely comparable, implying that the losses incurred were mainly a matter of adverse weather.

A Word about the Precipitation Data

Fig. 10.1 presents the seasonal and monthly precipitation data for 1959–61 for the nine agricultural areas of China as classified by Buck (see Map 10.1). The data are from the daily rainfall charts furnished by the Hong

MAP 10.1.   Major agricultural areas of China

Source: Adopted from J. L. Buck, *Land Utilization in China*, 27.

Kong Royal Observatory, which receives daily despatches telegraphed from its Chinese counterparts giving these and other meteorological data. The same data are complete for all the other years but it would be impossible to tabulate them all. Here we concentrate on the three extraordinary years of 1959–61.

Before considering the yield implications of these rainfall data, we relate the five sets of regional rainfall diagrams (Fig. 10.1) to the yearly weather maps furnished by the State Meteorological Bureau (SMB) for 1959–61 (Map 10.2). The SMB's mapping of regional droughts and floods, like the diagram in this study, is based on deviations in rainfall precipitation from the long-term mean. The SMB, however, used only the average of the rainfall from June to September for the North and the North-west and that of May to September for all the other areas in determining the summary seasonal deviations. But the monthly rainfall deviations for all these months are included in our diagrams. This facilitates verification. The monthly figures allow a closer look at the weather situation, compared with the SMB mapping. The latter often tends to conceal the occurrence of drought following a shortfall in rain over a period of a month or so. Similarly, an excessive concentration of rainfall within even a shorter period may cause serious floods, but this cannot be easily shown in the SMB mapping.

It should be noted, however, that the long-term mean precipitation may differ considerably between weather stations in a given agricultural area, which is likely, after all, to embrace a large part of the country. The simple average of rainfall deviations computed for each area as a whole may therefore also conceal possible intra-area differences in the incidence of drought and floods. In this respect, the SMB mapping is a useful check, because it refers to the individual weather stations rather than to broadly defined agricultural or administrative regions.

Another point to bear in mind when comparing the rainfall diagrams for 1959–61 with the SMB study concerns the criteria adopted in our diagrams for the classification of 'drought' (lightly shaded zone) and 'great drought' (darkly shaded zone). These criteria are borrowed from the northern areas, notably Shaanxi.[1] The North and the North-west have much more stringent drought standards than other regions. That is to say, a greater shortfall in rain (measured against the long-term mean precipitation) is needed in the northern regions to constitute a drought than in the southern regions.

The indiscriminate application of the northern drought criteria to the southern agricultural areas, as is done in our rainfall diagrams, implies, therefore, that droughts could have already occurred in many southern areas, although they may not be indicated as such by the shaded drought area in the diagrams. This comparison between the North and South,

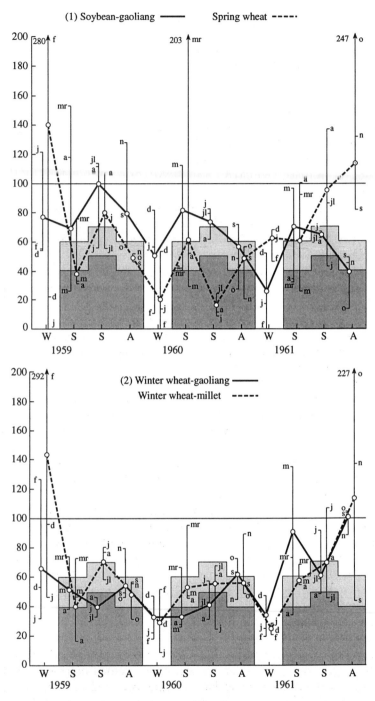

Fig. 10.1. Monthly and seasonal precipitation fluctuations and 'drought' and 'great drought' indicators in the nine agricultural areas in China, 1959–1961 (% of the long-term mean)

Source: Table AB.18.

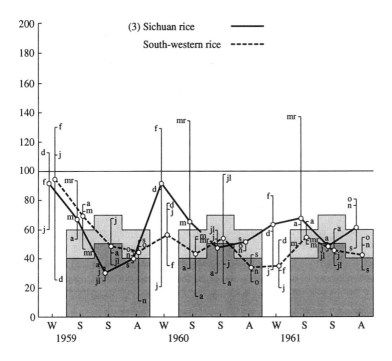

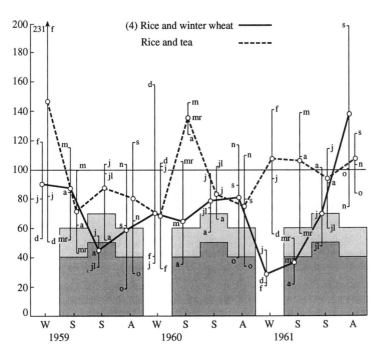

Fig. 10.1. *cont.*

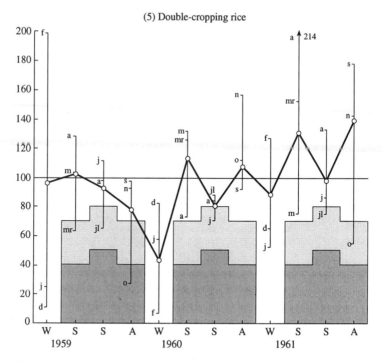

(5) Double-cropping rice

FIG. 10.1. *cont.*

broadly defined, applies to the difference between our drought standards and those used by the SMB. The latter seems to be less stringent in this respect.[2] In other words, if SMB drought criteria were used, then in our 1959–61 rainfall diagrams for all the nine agricultural areas, the shaded drought area (especially that indicating 'great drought'), would become larger. Despite these minor differences, however, the droughts (and floods as well) reflected in our independently compiled rainfall diagrams for all the agricultural areas can be consistently verified by the SMB mappings, except perhaps in a number of minor cases.

Predicting Losses from the 1931–1935 Precipitation–Yield Relationship: A Methodological Note

A number of problems are involved in applying the 1959–61 precipitation data to the estimated 1931–5 precipitation–yield relationship to predict the possible scale of yield losses for purposes of comparison with the actual yield losses in 1959–61. The first is that the daily precipitation data for 1959–61 (as they appear on the daily rainfall charts prepared by the Hong Kong Royal Observatory) were grouped by us according to Buck's nine agricultural areas rather than the provincial demarcation. This helped to keep the tabulations to manageable proportions, although the data could have been classified by province. We have used only the average monthly and seasonal rainfall figures for an 'agricultural area' taken as a whole rather than for the individual provinces, as shown in Fig. 10.1. As a result, the same focus on broader agricultural areas rather than individual provinces has to be used for the pre-war years 1931–5. For simplicity, we focus only on the two most important areas, namely the winter wheat and gaoliang area and Yangzi rice and wheat area, which have assumed special importance in the country's grain output.

However, we lack a complete set of provincial grain-yield data for 1959–61 with which to compile 'agricultural area' averages of rice and wheat yield which might be compared with those predicted by the precipitation–yield relationship of the pre-war years. That is to say, we can only compare the actual overall grain-yield losses in 1959–61 for selected provinces with the predicted rice- and wheat-yield losses for the related agricultural areas. The two sets of yield values are of course not directly comparable, because the different grain crops (rice, wheat, gaoliang, etc.) included in the all-grain yield averages are sown in different crop seasons.

Moreover, estimated precipitation–yield regressions for 1931–5 used for predicting the wheat yields in the winter wheat and gaoliang area and rice yields in the Yangzi rice and wheat area in 1959–61 are based respectively on spring (March to May) and summer (June to August) rainfall. While these are the two principal growing seasons, the two crops are also subject to precipitation in other seasons; for example cross-winter wheat yields are vulnerable to winter moisture. In addition, as a result of increased multiple-cropping practice in the 1950s, the average annual rice yields in the Yangzi rice and wheat area and in other southern provinces also reflect the spring and autumn precipitation. The precipitation-predicted wheat- and rice-yield losses for the two agricultural areas should therefore at best be regarded as only rough guidelines for gauging the realized provincial yield losses in 1959–61.

Where grain-sown area and grain-output data are available, we will also examine the grain-yield losses in 1959–61 for selected provinces outside the Yangzi rice and wheat and winter wheat and gaoliang areas. Their actual losses will be discussed against the background of relevant regional seasonal and monthly precipitation fluctuations, as shown in Fig. 10.1. This may help to verify our comparisons of the actual yield losses for the

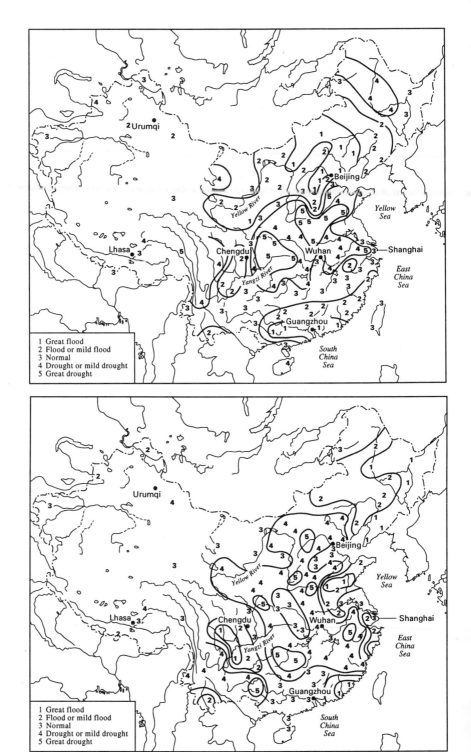

1 Great flood
2 Flood or mild flood
3 Normal
4 Drought or mild drought
5 Great drought

1 Great flood
2 Flood or mild flood
3 Normal
4 Drought or mild drought
5 Great drought

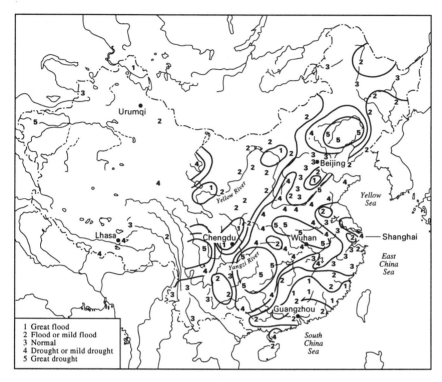

MAP 10.2.   The distribution of floods and drought in China, 1959–1961

Notes: The gradings 1, 2, 3, 4, and 5 refer respectively to 'great flood' (*dalao*), 'flood' or 'mild flood' (*bianlao*), 'normal' (*zhenchang*), 'drought' or 'mild drought' (*bianhan*), and 'great drought' (*dahan*). See Appendix A (pp. 273–4) for details.

Sources: State Meteorological Bureau, *Zhongguo Jin Wubainian Hanlao Fenbu Tuji* (Yearly Charts of Dryness/Wetness in China for the Last 500-year Period) (Beijing: Ditu chubanshe, 1981), 250–1.

other provinces with the precipitation-predicted losses for the respective agricultural areas.

Another problem is that the estimated precipitation–yield regression for 1931–5 used for predicting the rice and wheat yield losses in 1959–61 is the one based on the 'normal yield' rather than the 'mean yield' series for 1931–7.[3] The former is consistently higher than the latter, but in the absence of a comparable 'normal yield' series for the 1950s, we can only use the maximum yield ever attained for the related provinces in 1952–7 to measure the loss rates in 1959–61.[4] This again means that the realized loss rates are not directly comparable with those predicted from the precipitation–yield regression for 1931–5.

The reason why the estimated precipitation–yield regression for 1931–5 based on the 1931–7 mean yield series (as shown in the previous chapter) is not used as a basis for predicting the loss rates in 1959–61, is that this would have required the realized loss rates for 1959–61 to be measured against the average yields of 1952–7. Unlike the years 1931–7, however, the period 1952–7 saw a consistently upward moving trend in grain yields for the various provinces. This makes any mean yields for 1952–7 unsuitable as a basis for measuring the relative loss rates for 1959–61.

These methodological remarks clearly imply that the results of the simulations should be interpreted with great care. Any findings are bound to be speculative. In anticipation of more detailed crop-specific data being made available from China for the 1959–61 period, it is hoped that this initial effort, together with its accompanying valuable rainfall data, will pave the way for more rigorous analysis in the future.

Comparing Actual and Predicted Losses

The results of our computation of the percentage losses for rice and wheat yields in 1959–61, as predicted from the estimated precipitation–yield regressions for 1931–5, are shown in Table 10.1, together with the actual yield loss rates for a selected number of provinces for which yield data are available. We compare the realized and predicted yield loss rates separately for the three years 1959, 1960, and 1961.

1959: Realized yields higher than predicted yields

First, for 1959, the rainfall-predicted loss rate for wheat yield in the winter wheat and gaoliang area amounts to 11.22 per cent, measured against the 'normal yield' of 1931–7. No wheat yield data are available from Hebei, Henan, and Shandong which comprise the major part of the winter wheat and gaoliang area to facilitate verification of this figure. The actual national wheat yield in 1959 increased, however, by 3.6 per cent, above the highest wheat yield attained, in 1956, during the 1952–7 period. This 3.6 per cent rise cannot of course be directly compared with the predicted loss of 11.22 per cent for the winter wheat and gaoliang area, despite the latter's important weight in the country's overall wheat output.

What is certain, however, is that if instead of the 'normal yield', the 1931–7 mean yield were to be applied to the precipitation–yield regression for 1931–5, the predicted wheat yield loss for 1959 would be substantially lower than the 11.22 per cent shown in Table 10.1, even if it did not approach the 3.6 per cent margin above the 1956 record yield level.

TABLE 10.1.   Actual rates of grain losses per sown hectare for selected areas and provinces in 1959–61 in comparison with the rate of losses as predicted from the rainfall–yield regressions for the 1930s (%)

	1959	1960	1961
Predicted losses (1)			
Winter wheat and gaoliang area: wheat only	−11.22	−15.41	−4.44
Yangzi rice and wheat area: rice only	−62.35	−15.62	−24.63
Actual losses (2)			
Area averages: all grain			
Soya beans and gaoliang area			
Heilongjiang	15.89	−33.99	−40.82
Liaoning			−35.38
average			−39.11
Spring wheat area			
Qinghai	−25.23	−44.76	
Winter wheat and gaoliang area			
Hebei		−23.72	
Shandong		−44.17	
Henan		−29.67	
average		−30.20	
Sichuan rice area			
Sichuan	−27.40	−41.74	−39.28
Yangzi rice and wheat area			
Jiangsu	−8.57	−24.02	−21.52
Rice and tea area			
Hunan			−43.45
Double-cropping rice area			
Guangdong	−12.20	−17.97	−15.59
National averages:			
wheat	3.63	−10.57	−38.66
rice	−11.29	−25.10	−24.21
all grain	0.00	−20.03	−16.95

Sources and notes: (1) The predicted losses are calculated from the two estimated regression equations for 1931–5 shown in Figs. 9.3 and 9.4. The 'normal' (maximum) yield, rather than the 1931–7 mean yield series, is used. See text for explanation. The rainfall data for 1959–61 are from the same sources used in Fig. 10.1.

(2) The national average rice and wheat yields are from *TJNJ 1984*, 138 and 141. The provincial data are from K. R. Walker, *Foodgrain Procurement and Consumption in China*, as follows: 239 for the underlying maximum yields in 1952–7, and 150 for yields in 1959. The provincial grain output figures for 1960 and 1961 were kindly supplied by the late Professor Walker. They also appear in his book (p. 159) in per capita terms, where the original sources are also indicated. The only exception is the figure for Hebei for 1960 which was originally given in per capita terms and is derived therefrom by assuming that the population had grown at the same rate after 1957 as the national rate of increase (*TJNJ 1984*, 81).

This seems to be consistent with the pattern of rainfall distribution during the wheat-growing season 1958–9, shown in Fig. 10.1.

Note that while the rainfall in the spring of 1959, in the winter wheat and gaoliang area (but also the winter wheat and millet area) may have fallen short of what was needed to sustain a favourable wheat yield, the moisture carried over from the preceding winter was plentiful not only in these regions, but in virtually all the other major wheat-growing areas. This stands in sharp contrast to winter 1959–60 when, except perhaps for the Yangzi rice and winter wheat area, rainfall was well below the long term average. Moreover, there were also wheat-growing areas in which spring rainfall in 1959 was favourable or at least adequate. These include, for example, the soya bean and gaoliang area which covers Heilongjiang. This province, which produces mainly spring wheat and soya bean, enjoyed exceptionally favourable amounts of rain in March and April 1959, and indeed recorded an increase of 15.89 per cent in its average grain yield in 1959 (Table 10.1).

Sichuan is another example of a province which had adequate rainfall in spring 1959. Its large loss rate of 27.40 per cent for the annual all-grain measure for 1959 was probably mainly due to the fall in rice yield, rather than wheat yield, judging by the severe shortfall in rainfall in the main rice-growing seasons of summer and autumn 1959 (Fig. 10.1). The same holds for the Yangzi rice and wheat area where rainfall in the winter (1958–9) and spring 1959 necessary for wheat growth was even more favourable than in the Sichuan basin (before the area was affected by the disastrous Yangzi and Huai River basin drought in the summer) (Fig. 10.1). It seems likely therefore that the gain in wheat yields helped partly to offset the rice-yield losses in the Yangzi rice and wheat area, resulting in a relatively mild loss rate of 8.57 per cent in terms of all-grain-yield average, for Jiangsu province in 1959 (Table 10.1).

The only province, among those shown in Table 10.1, which suffered from a major wheat-yield loss in 1959 was Qinghai in the spring wheat area. Its substantial loss rate of 25.23 per cent in all-grain yield clearly reflected a declining wheat yield, consequent upon a serious shortage of spring rain in the spring wheat area. A similar rainfall pattern prevailed over the winter wheat and millet area in 1959, but unfortunately provincial yield data are not available to verify the extent of losses. In any case, these two areas were relatively insignificant in terms of their contribution to national wheat output, compared with the other areas discussed.

Thus, the average wheat yield for China as a whole seems not to have been seriously affected by adverse weather in 1959. Moreover, the wheat fields which survived the sown-area contraction under the 'three-three' system were relatively fertile. The 'subsistence urge' which emerged in late spring 1959 in response to the 'sown-area crisis' was yet another factor which helped provide a hedge against wheat-yield losses in 1959. This, supplemented by adequate rainfall during the wheat-growing season, may explain the record wheat yield in 1959.

Turning to the possible extent of rice-yield losses in 1959, the rainfall-predicted loss rate for the Yangzi rice and wheat area was a startling 62.35 per cent (Table 10.1). No rice yield data are available for Jiangsu, Anhui, and Hubei—the core of the Yangzi rice and wheat area—to facilitate verification of this high loss rate. For the country as a whole, however, the average rice-yield loss in 1959 was only 11.29 per cent, measured against the highest yield attained in 1952–7. If the 1931–7 mean yield is substituted for the 'normal yield' in estimating the precipitation–yield regression for 1931–5 in order to predict the 1959 loss rate for the Yangzi rice and wheat area, the relative discrepancy between the two loss rates for rice yields is reduced considerably, but the gap remains large.

It is understandable that the national loss rate should be lower than that incurred at the regional level. Nevertheless, the predicted loss rate in rice yield for the Yangzi rice and wheat area (which experienced the disastrous Yangzi and Huai River basins drought in summer 1959), still looks disproportionately large in relation to the all-grain-yield loss of 8.57 per cent sustained in Jiangsu province, which is mainly located in the same area (Table 10.1). This provincial figure is admittedly measured against the highest annual yield obtained during 1952–7. But even if it is set against some 'normal yield', as defined for the 1930s, it is in no way comparable with the predicted rice yield loss of 62.35 per cent for the Yangzi rice and wheat area as a whole.

Because of lack of provincial yield data, it is not possible to determine whether yield gains from other grain crops in Jiangsu helped to offset the disastrous rice-yield loss in 1959 to result in the relatively mild decline of 8.57 per cent for the province in terms of overall average grain yield. What we do know is that during the wheat-growing seasons, rainfall was quite favourable from February through June, except perhaps for March 1959 (Fig. 10.1). It seems unlikely, however, that the yield gains from wheat or from other crops, could have so substantially compensated for the tremendous loss in rice yield as to reduce average all-grain-yield by a mere 8.57 per cent in 1959.

This leaves us with two possible explanations. The first is that the predicted rice loss rate of 62.35 per cent contains a strong upward statistical bias, deriving from the fact that the prediction is based on an extreme shortfall of rain in the summer (June, July, and August) alone, without taking into account conditions in the spring (quite favourable) and autumn (comparatively adequate). The second explanation is to be found in the 'subsistence urge' of the peasants in hedging against the potential 'food crisis' in the wake of the unwarranted sown-area contraction early in 1959. Note that wide stretches of land for early rice crops were also taken out of cultivation in 1959; the remaining, more fertile sown areas displayed a higher yield per sown hectare.

Such speculative arguments cannot of course easily be corroborated. A firmer verification of the precipitation implications of the yield losses may emerge from a comparison of the all-grain-yield loss rates in the handful of provinces shown in Table 10.1 against the background of their positions *vis-à-vis* spatial and/or seasonal patterns of rainfall distribution in 1959. We note, first of all, that there seems to be no gross inconsistency between the national loss rate of 11.29 per cent for rice on the one hand, and, on the other, provincial all-grain loss rates of 27.40 per cent for Sichuan, 8.57 per cent for Jiangsu, and 12.20 per cent for Guangdong. Note that these were all important rice-producing areas.

In individual provinces, the large discrepancy between the all-grain-yield loss rates of Sichuan (27.4 per cent) and Jiangsu (8.57 per cent) also seems to be consistent with the provincial precipitation patterns in 1959. Virtually throughout all the growing seasons, from winter 1958–9 through spring, summer, and autumn 1959, average seasonal rainfall in Sichuan was consistently lower than in Jiangsu in relation to the respective long-term mean precipitation (Fig. 10.1). Indeed, summer rainfall, which is most critical for the rice crops, was considerably lower in Sichuan, although strangely enough, only the shortfall in the Yangzi and Huai River basins received much publicity outside China at that time.[5]

In the predominantly rice-producing province of Guangdong, the average grain-yield loss rate in 1959 was 12.20 per cent (Table 10.1). But in contrast to the other provinces discussed, average seasonal rainfall in spring and summer in the double-cropping rice area of which Guangdong is a part was close to the long-term mean. It was the same in the autumn, except for an extremely dry October which, however, affected only the late rice crops. The explanation for Guangdong's substantial loss is in fact to be found in the disastrous Zhujiang flood of June 1959, which was described as a 'once-in-a-century' probability.[6] The rainfall which caused the flood is clearly reflected in our monthly rainfall diagram shown in Fig. 10.1, although it is shown to have affected the whole of the double-cropping rice area including faraway Fujian. Note that in June 1959, the excessive rainfall amounted to 20 per cent above the long-term mean.

Two major points emerge from the foregoing discussions about precipitation and grain-yield relationships in 1959. The first is that while it may not be possible precisely to quantify the impact of rainfall fluctuations, observed provincial yield losses in 1959 tended none the less to vary rather consistently with changes in precipitation. The second point is that while it is also impossible to verify the contribution of the peasants' 'subsistance wage' in stabilizing grain-yield fluctuations in 1959, the fact that the realized grain-yield losses were consistently lower than the precipitation-predicted yield losses, seems, nevertheless, to suggest that relatively favourable human factors were at work to help check any aggravation of losses. In both respects, the situation in 1960 was very different from that of 1959.

1960: Predicted losses lower than actual losses

We first comment on the grain-yield situation in the winter wheat and gaoliang area, for which we have a fairly complete set of provincial grain-yield data. The predicted wheat-yield loss rate of 15.41 per cent for the winter wheat and gaoliang area in 1960 is consistently lower than the high actual average grain-yield loss rates either for the area as a whole (30.20 per cent) or for individual provinces (Table 10.1). Although the actual loss rates embrace grains other than wheat, it is doubtful whether the inclusion of gaoliang and other autumn crops would help significantly to reduce the overall grain loss rates, even disregarding the fact that rainfall conditions in the winter wheat and gaoliang area were also very unfavourable for the autumn crops. The likely explanation for the discrepancy is the lack of seed and feed grains for 1960, following excessive State grain procurements of 1959–60 which resulted in a lower sowing density.

A similar discrepancy also exists for other agricultural areas. In Jiangsu, for example, which was a major rice-producer in the Yangzi rice and wheat area, the provincial average grain-yield loss rate of 24.02 per cent for 1960 was much higher than the predicted rice-yield loss of 15.62 per cent for the entire region (Table 10.1). The two figures are of course not directly comparable. Nevertheless, in relation to rainfall conditions, there is no reason why the provincial-yield loss rates should have increased sharply from 8.57 per cent in 1959 to 24.02 per cent in 1960, even allowing for some understatement of the impact of bad weather on the 1959 loss rate.

Both seasonal and monthly summer rainfall in the Yangzi rice and wheat area improved substantially in 1960, compared with the summer of 1959 (Fig. 10.1). September rainfall was indeed not far below the long-term mean, compared with a shortfall of nearly 40 per cent for the same month in 1959. Admittedly, spring rainfall in the area in 1960 was less favourable than in 1959. But this affected wheat more than rice and the former contributed less to total grain output in Jiangsu province.

A similar comparison between 1960 and 1959 can be made for the Sichuan rice basin, suggesting that a man-made shortage of seed and feedgrain was the origin of the accelerated reduction in grain yield in

1960. Unlike Jiangsu, the geographical extent of Sichuan coincides closely with the agricultural area delineated by Buck. This eliminates the uncertainty surrounding the comparison between the provincial and regional yield changes. Sichuan suffered an average grain yield loss of 41.74 per cent in 1960, compared with 27.40 per cent in 1959. Yet our rainfall diagrams reveal that summer precipitation in the Sichuan rice basin as a whole was considerably higher in 1960 than in 1959, although the shortfall was still very substantial, amounting to around 50 per cent of the long-term average rainfall. Autumn rainfall for the area as a whole was also somewhat higher in 1960 than in 1959, whilst the spring rainfall in both years was nearly the same. Thus, as in the Yangzi rice and wheat area and the winter wheat and gaoliang area, the yield loss rates for 1960 and 1959 in the Sichuan rice basin are inconsistent with the comparative rainfall patterns for the two years. This cannot be explained except in terms of a serious shortage of seed and feed in 1960.

The same explanation applies to Guangdong province which suffered an average grain-yield loss rate of 17.97 per cent in 1960, compared with 12.20 per cent in 1959 (Table 10.1) for which the great Zhujiang flood was mainly responsible. Again there was neither serious flood nor drought in 1960 to explain the increased yield loss between the two years. The SMB map also indicates that the summer weather in the province was normal, except for some localized floods in the narrow eastern tip near the border of Fujian province.

As for Heilongjiang, the last remaining province cited in Table 10.1, the realized yield loss in 1960 was no less than 33.99 per cent. This was probably the result of extensive summer floods, which are clearly shown on the SMB map. It is not possible to determine whether these floods could have generated an all-grain-yield loss rate of such a magnitude. The more likely explanation for this is, again, depleted seedgrain stocks reflecting declining grain availability in peasants' hands. It is likely that amidst widespread famine in 1960–1, a greater effort was made by the State to squeeze more grain from such grain-rich provinces as Heilongjiang to improve the lot of the starving millions elsewhere in the country.[7] To the extent that per capita grain availability in Heilongjiang was higher than in other provinces, such forced transfers tended to squeeze its seedgrain supplies to a greater extent, in relation to available sown area, compared with provinces which had a more unfavourable man–land ratio.

Thus, in contrast to the situation in 1959, actual losses in grain yields in 1960 turned out to be consistently higher than the precipitation-predicted losses for the provincial samples shown in Table 10.1. More importantly, for all the provinces, the grain-yield losses in 1960 were consistently higher than those recorded in 1959. The same holds for average wheat and rice yields for the country as a whole. Not only are the observed discrepancies exceedingly large, but they all appear to be totally inconsistent with the rainfall patterns for 1959 and 1960. This strongly points to the negative implications of policy factors which were reflected in serious shortfalls in feed and seedgrain supplies due to excessive State grain-procurements in 1959–60. We will discuss in more specific terms, in the next chapter, how this miscalculation eventually led to the Great Débâcle of the early 1960s. But first we should look at the year 1961 in order to examine the precipitation implications of the observed grain-yield losses.

1961: Predicted yields nearing realized yields

In our analysis of the weather–yield relationship based on the national weather (*shouzai* area) index, we noted that in contrast to the year 1960, the influence of the negative policy factor was reduced rather substantially in 1961 (see Chapter 8). Seedgrain stocks began to be replenished in 1961 as a result of reduced State farm-procurement. Realized yields therefore became increasingly similar to weather-predicted yields in 1961, although the weather situation remained grave, and only a marginal improvement over 1960 was recorded for the national average grain yield. We now examine whether this trend can be corroborated by a precipitation-based regional analysis of weather–yield relationships.

We first look at the situation in the winter wheat and gaoliang area. As shown in Table 10.1, the precipitation-predicted loss rate for wheat yield in this area was only 4.4 per cent in 1961, compared with 15.41 and 11.22 per cent in 1960 and 1959 respectively. If the lower 1931–7 mean yield were substituted for the higher 'normal yield', which is used for estimating the predicted wheat-yield loss in 1959–61, the cited loss rate of 4.4 per cent would certainly become even smaller, if not positive.

Provincial yield statistics for Hebei, Shandong, and Henan (which make up the winter wheat and gaoliang area) are not available for 1961 as a means of verifying actual losses against the predicted loss rate. But for the country as a whole, the average wheat-yield loss was a startling 38.66 per cent in 1961, measured against the highest yield obtained in 1956 during 1952–7 (Table 10.1). It is difficult completely to reconcile the large discrepancy between this figure and the predicted wheat-yield loss of 4.4 per cent for the winter wheat and gaoliang area.

Setting aside the problem of incomparability in coverage between the two figures, a plausible explanation is that the predicted wheat-yield loss was based exclusively on the spring rainfall in 1961, which in contrast to the all-time low of the preceding winter, was quite favourable (Fig. 10.1). In other words, wheat yields in the winter wheat and gaoliang area might have suffered substantially from the extreme shortfall of winter rain. The consequent losses must have contributed significantly to the national loss rate of 38.66 per cent, given the important weight this area assumed in the national wheat output.

In fact, other important wheat-producing provinces also recorded very large loss rates in 1961, although only statistics for loss in terms of all-grain yield are available. These include Heilongjiang (minus 40.82 per cent), Liaoning (minus 35.88 per cent), Sichuan (minus 39.28 per cent), and Jiangsu (minus 21.52 per cent) as shown in Table 10.1. The comparative loss rates for wheat in these four provinces, as distinct from those of other grain crops (gaoliang and soya bean in the two North-east provinces, and rice in Sichuan and Jiangsu), must have been even higher than the declines in all-grain yield bearing in mind that precipitation conditions (Fig. 10.1) in terms of seasonal averages prevailing in the wheat-growing seasons in these four provinces during 1961 were more unfavourable than in the other cropping seasons. It also seems likely that reduced feedgrain and seedgrain stocks had not yet been adequately replenished. This was especially the case in the provinces of the winter wheat and gaoliang area, as well as in the North-west where peasants had a lower per capita grain availability.

The situation in the Yangzi rice and wheat area seems to have been quite different with respect to rice-yield losses. Note, first, that the predicted rice-yield loss for this area in 1961 was 24.63 per cent, again measured against the 'normal yield' as defined for the 1930s (Table 10.1). Yet, the realized yield loss, in terms of all-grain yield, was no more than 21.52 per cent for Jiangsu province. Given the serious shortfall in precipitation for the Yangzi rice and wheat area during the wheat-growing seasons (as compared to more or less adequate summer rainfall and very favourable conditions in the autumn) (Fig. 10.1), there is little doubt that the cited all-grain-yield loss of 21.52 per cent for Jiangsu in 1961 mainly, if not exclusively, reflected losses in wheat output.

The implication of this is not only that the wheat output losses in Jiangsu (or for that matter, the Yangzi rice and wheat area as a whole) substantially contributed to the very high 35.66 per cent decline in average national wheat yield in 1961, as discussed earlier. In addition, effective rice yields must have been much higher than the small discrepancy between the higher precipitation-predicted rice-yield loss rate of 24.63 per cent and the lower realized all-grain-yield loss rate of 21.52 per cent for Jiangsu (Table 10.1) would suggest. This is made clear if we consider that the predicted rice-yield loss of 24.63 per cent for the Yangzi rice and wheat area is even smaller if not lower than the realized all-grain-yield loss of 21.52 per cent for Jiangsu province if the 1931–7 mean yield is substituted for the 'normal yield' as a basis for the 1931–5 precipitation–

yield regression which was used for making the predicted rice–yield loss in 1961.

Thus, the comparatively favourable rice yields in 1961, which benefited from the turn-around in precipitation conditions in the second half of the year, seem to have more or less compensated for the serious yield losses in the summer (wheat) harvests, which resulted, bad weather aside, from the continuous shortage in seedgrain supplies. In contrast to the northern wheat provinces, the South, including the provinces in the Yangzi rice and wheat area, was probably better off in this respect. With a higher level of per capita grain availability, and hence adequate seedgrain supplies, the southern provinces were in a better position to take advantage of the improving weather conditions. If correct, then the better yield performance in the rice areas might partially have offset the losses in the wheat areas to result in an overall national grain-yield average not much below the weather-predicted level in 1961 as shown in our analysis based on the national weather (*shouzai* area) index (Chapter 8).

Looking at the years 1960 and 1961, it is interesting that in those provinces for which data are available, notably Sichuan, Jiangsu, and Guangdong, the all-grain-yield loss rates in 1961 tend to be consistently lower than in 1960. The only exception is Heilongjiang province which suffered a higher all-grain-yield loss in 1961 compared with 1960 (Table 10.1). The sample is of course too small a base from which to draw a definite conclusion. Nevertheless, it is still noteworthy that at least in Sichuan and Jiangsu our rainfall diagrams show hardly any improvement in spring and summer weather conditions (critical for rice growth) between 1960 and 1961 (Fig. 10.1). And if the losses in these two provinces in terms of all-grain average yields in 1961 were in fact heavily weighted by the losses in wheat yields, then the implied (relatively favourable) rice yields in 1961, as compared with 1960, are somewhat inconsistent with the comparative rainfall situation during these two years.

The same inconsistency between 1960 and 1961 may hold for Guangdong province as well. In terms of average seasonal rainfall, weather conditions for the province in 1961 appear to have been much better than those in 1960 (Fig. 10.1). But it seems doubtful whether this really explains the slight improvement in all-grain-yield performance in 1961. On the contrary, the slight gains in grain yield in Guangdong in 1961 might even have been achieved despite tumultuous rainfall (with or without typhoons), as reflected in the extreme monthly fluctuations, especially in the spring and autumn of that year (Fig. 10.1).

Turning to Heilongjiang province, the higher all-grain-yield loss rate in 1961 compared with 1960 appears to be consistent with the variations in rainfall patterns, in that average seasonal rainfall in virtually all the growing seasons in 1961 was lower than in 1960. However, the higher loss rates in 1961 may have also reflected the depletion of seedgrain reserves due, as noted earlier, to forced transfers of grain out of the province to balance desperate food demands elsewhere in the country. If so, the result may have been a greater degree of seedgrain losses in relation to the comparatively more favourable man–land ratio for the province.

Thus, in the light of the precipitation evidence, it seems desirable to distinguish between the wheat-producing North and the rice-producing South, broadly defined, in assessing the contributory factors to grain-yield losses in 1961. Generally speaking, the losses in the North were attributable to a combination of continuous bad weather and slow replenishment of seedgrain stocks due to low per capita grain availability. On the other hand, the South enjoyed not only better weather conditions in 1961, but with a comparatively higher per capita grain availability, peasants there could also take advantage of reduced State farm procurements to restore sowing to a more or less normal level. If correct, then the relatively good (or mediocre) harvests in the South must have partially offset the losses in the North to bring the national average of yield losses in line with the overall (or generalized) weather conditions as measured by our national weather (*shouzai* area) index.

Some General Observations

It should by now be clear that many, if not most, of the observations made in the foregoing discussions are highly speculative. Until a complete set of provincial grain-yield data (with breakdown for rice and wheat yields) is available as a basis on which to conduct a correlation analysis with our rainfall statistics, it is difficult to draw definite conclusions about the impact of weather on grain yields in relation to man-made factors.

What is clear, however, is that our assessment of the weather–yield relationship for the three years 1959–61 shows that 1960 is conspicuously characterized by a set of provincial grain yields (incomplete as it may be) somewhat inconsistent with known precipitation patterns. The inconsistencies unequivocally point to the depletion of essential seedgrain stocks as a consequence of the government's excessive farm procurement quotas imposed during 1959–60 as the prime cause.

In contrast to 1960, the situation in both 1959 and 1961 is somewhat ambiguous, because while the weather was the main cause of the observed fluctuations in grain yield, it is not possible to isolate completely possible policy factors, which were less clear-cut than during 1960. In 1959 the peasants' urge to hedge against a possible food crisis in the wake of unwarranted sown-area contraction helped to arrest weather-caused grain-yield reductions in *both* North and South China. In 1961, however, grain yields seem mainly to have reflected prevalent weather conditions, although in contrast to the South, the North apparently continued to suffer from the continuing shortage of seedgrain supplies.

NOTES

1. *NYKXJS* 25; Original note: 'the drought indicators were determined as the ratio of the rainfall of each season of a year to the respective seasonal average of the past years, whereby the characteristics of rainfall and the water requirement of crops of each season were also taken into consideration.' The drought indicators seem to be primarily for Shaanxi province, but they should be basically applicable to the winter wheat and millet area as a whole, if not to the spring wheat area and the winter wheat and gaoliang area as well. Indeed, scattered evidence also suggests their general applicability to other areas. See e.g. Zhu Yungzhuo and Xie Mingen, 'Jiangsu sheng Ningzhen qiuling diqude ganhan leixing jiqi fangyu' (Types of drought and their prevention in the Ningzhen Highlands in Jiangsu Province), *ZGNP* 5 (1962), 16–19; Zhu Guangzuo, Lu Bing, and Zhao Xiuzhen, 'Mianyang zhuanqu chunhan xiahan quilu zhi fenxi' (An Analysis of the regularities of spring and summer droughts in Mianyang Special District), *ZGNB* 8 (1962), 15–17; and *RMRB*, 21 Dec. 1959 and 10 Feb. 1960. For a more recent classification for the Yangzi River basin, see Sha Wanying and Zhou Yufu, 'Changjiang Liuyu xiaji hanlao huanliu tezheng ji qi changqi yubao' (The characteristics and long-term forecast of the air flows during droughts and floods occurring in the Changjiang basin in the summer), in *DLJK* 11 (Sept. 1979), 66.
2. See the SMB's formula for the classification as discussed in Appendix A. pp. 273–5.
3. Tables AB.9 and AB.10.
4. See notes to Table 7.4 for sources.
5. There is little doubt that drought was widespread and most severe in Sichuan in summer 1959; see *SCRB*, 2, 4, 7, 9, 14, 15, and 17 July; and 2, 4, and 6 Aug. 1959. The drought appeared to continue into 1960; see *SCRB*, 15, 22, and 28 Jan. 1960; 2 Feb. 1960 (32% of sown area in more than 80 *xian* affected); 27 May 1960; and 10 June 1960 (provincial authority appealing for a quick decision to be made to substitute maize and potatoes for rice crops for summer sowing due to the serious shortfall of rainfall supply).
6. Lu Wu, 'Our country's climate and calamities of droughts and floods,' in *RMRB*, 26 Jan. 1966.
7. Heilongjiang ranked with an alarmingly high average per capita grain output of 593 kg. for 1952–7 as the richest province in the 1950s. It is by far the largest grain exporter, second only to Sichuan, responsible for balancing the grain deficits incurred elsewhere in the country. See Kenneth Walker, *Foodgrain Procurement and Consumption in China*, 26, 82, and 85. Cf. also Thomas L. Lyons, *Economic Integration and Planning in Maoist China* (New York:

Columbia University Press, 1987), 49–50; and esp. Zhao Fasheng, *Dangdai Zhongguo de Liangshi Gongzuo* (Grain Work in Contemporary China) (Beijing: Shehui Kexue Chubanshe, 1988), 22, in relation to the situation in 1960–1.

11

An Excursion: Identifying the Non-Weather Factors in the Great Débâcle of the Early 1960s

In our previous analysis of the weather and grain-yield relationship (Chapter 8), we concluded that the weather was the main cause of the enormous grain-yield losses in 1960 and 1961. Measured against the possible trend values, grain yields were reduced by 18 and 16 per cent respectively for the two years. However, while in 1961 the losses seem to have been largely a matter of adverse weather, coupled with the sluggishness in recovery from the disastrous trough of the previous year, in 1960 only about 75 per cent of the total yield losses were attributable to the weather factor. We argued that the remaining 25 per cent loss reflected reduced sowing density as a consequence of excessive forced procurement by the State, which left Chinese peasants with no alternative but to dip into their seed and feedgrain reserves in other to survive.

We have yet to specify how the man-made factors helped to precipitate total chaos in rural China, with millions starving to death, through an examination of the year-to-year changes in the level of peasants' per capita grain consumption from the late 1950s to the early 1960s. We will argue that another equally and potentially more important factor contributing to the Great Débâcle was the large-scale contraction of sown area starting in 1958, and particularly 1959 affecting under the misconceived 'three-three system' of cultivation. The reduced sown-area base, combined with reduced yields per sown hectare because of lowered sowing density and bad weather, helped drastically to bring down overall grain output and per capita grain consumption, leading to widespread famine.

The most important point to emerge from our analysis is that even without bad weather, the peasants could not possibly have survived, purely on account of the unwarranted sown-area contraction and excessive State farm procurement. For students familiar with the Great Leap Forward policy upheavals, it may be difficult to reduce the entire Great Débâcle exclusively to factors influencing the two simple variables of sown-area and grain yield per sown hectare. It seems, therefore, desirable to begin our discussion with a review of the features of the Great Leap

Forward strategy in order to see how they stand in relation to the major factors which we perceive to be most critical in bringing about the disasters of the early 1960s.

The Great Leap Forward Strategy Revisited

The familiar features of the Great Leap Forward strategy include (1) the 'backyard furnace' campaign launched throughout the Chinese country-side, which diverted large amounts of agricultural labour; (2) the creation of the people's communes, with their extremely egalitarian income distribution, impairing peasant incentives; (3) the high compulsory grain delivery quotas which caused widespread peasant resentment and attempts to conceal grain output; (4) mass labour mobilization for large-scale rural infrastructural construction, especially irrigation projects, which were in many cases so badly designed that they resulted in an enormous waste of labour and farmland, and allegedly in the salinization of much arable land; (5) the adoption of deep ploughing and close planting without due consideration to local farming conditions; and (6) the implementation of the 'three-three' system of planting, which was responsible for the withdrawal of farmland from cultivation in late 1958 and early 1959.

A careful chronological scrutiny reveals that these institutional and policy upheavals were almost all phenomena of 1958 rather than 1959 or subsequent years. Take the backyard steel campaign for example. It was launched in the spring of 1958, originally as part of a programme designed to initiate technological revolution in Chinese agriculture.[1] It was accelerated in the summer to assist the fulfilment of the unrealistic steel output target of 10.70 million tonnes set for that year.[2] The mass mobilization involved left wide stretches of farmland either unsown or unattended.[3] Together with the arable area taken up by the nation-wide irrigation campaign (which started in late 1957), this resulted in a marked decrease in the grain-sown area in 1958.[4] By the winter of 1958–9, however, there were extensive calls from the central authorities for the 'technical consolidation' of the backyard furnaces, and early in the spring of 1959 the campaign was virtually halted.[5]

Mobilization for deep ploughing and close planting was also mainly a 1958 phenomenon, although it is not clear to what extent these practices were implemented. Close planting threatened to be technically harmful in many circumstances, while deep ploughing, which was labour-intensive, involved the peasants in very hard work and might therefore have constituted a disincentive. There were, however, few press reports about these practices in 1959.[6]

It is clear that rural conditions in 1959 were comparatively congenial.

In February and March 1959, the widespread resentment and resistance of the peasants in the autumn of 1958 and the winter of 1958–9 to excessive grain deliveries had already filtered up to the central authorities.[7] At the same time, the more extreme egalitarian features of the commune system were recognized and policy correctives were adopted. For example, the production brigade (equivalent to the former agricultural collective) was restored as the basic unit of planning and accounting.[8] The reopening of rural trade fairs and restoration of private plots were also advocated, although it is not known exactly how far these measures had been implemented by the time the anti-Rightist campaign was launched in autumn 1959.[9]

In fact, the Great Leap euphoria had already greatly subsided by early 1959, following, at the latest, the Second Zhengzhou Conference (February through March 1959), which Mao himself convened and at which he called for a series of rectification measures to be adopted against 'communist and commandist styles' and exaggerated output claims.[10] The changes are more or less reflected in the figures given in Table 11.1 which compare the output claims for 1958 and 1959 with the grain and steel output targets for those years.

Thus, both the exorbitant claim made in December 1958 that grain output for that year reached a total of 375 million tonnes and the unten-

TABLE 11.1. A chronology of the changing grain and steel output targets and claims made during the Great Leap Forward years in China, 1958–1960

	Grain output targets (claims) (million tonnes)		Steel output targets (claims) (10,000 tonnes)	
	1958	1959	1958	1959
1958				
January	196			
February	212			
March	212			
April	212			
May				1,100[1]
June	250	487	1,070[2]	
July				
August			1,070[3]	3,000[4]
September	300–350		1,200[5]	
October				
November				1,800–2,000[4]
December	297–349[6]	525[7]	(600–900)[8]	
	(375)[7]		(1,100)[9]	

TABLE 11.1 *cont.*

	Grain output targets (claims) (million tonnes)		Steel output targets (claims) (10,000 tonnes)	
	1958	1959	1958	1959
1959				
January		525[10]		
February				1,800[11]
March				1,650[12]
April	(375)	525		
May				
June				1,300[12]
July				
August	(250)	275		
September				
October				
November				
December				
1960				
January		(270)	(800)[13]	(1,335)[14]

Notes and sources: Unless otherwise noted, the data are from Kung-chia Yeh, 'Agricultural Policies and Performance', in Yuan-Li Wu (ed.), *China: A Handbook* (New York: Praeger Publishers, 1973), 513–19.

[1] *MSSWS* 225: Mao's speech at the Second Conference of the chief delegates of the 5th Party Congress.

[2] *MSSWS* 303.

[3] *RMRB* editorial, 1 Sept. 1958 citing the resolution of the familiar Beidaihe Politburo meeting.

[4] *MSSWS* 252.

[5] *MSSWS* 228: Mao's speech at the supreme state affairs meeting, 5 Sept. 1958.

[6] *MSSWS* 268: Mao gave a target of 730,000 to 860,000 million *jin*, but stated that a quarter would be made up of potatoes. The figures are derived by using the conversion ratios of 1 kg. to 2 *jin* and 1 *jin* of grain to 4 *jin* of potatoes.

[7] E. S. Kirby (ed.), *Contemporary China*, iii. (Hong Kong: Hong Kong University Press), 316, citing the resolution of the 6th Plenum of the 8th Party Congress convened in Wuchang, Nov./Dec. 1958.

[8] *MSSWS* 227–8.

[9] Kirby (ed.), *Contemporary China*, iii. 316.

[10] Kirby (ed.), *Contemporary China*, iv. 193, citing the resolution of the conference of 6,000 delegates of 'outstanding units building socialism in agriculture'.

[11] Kirby (ed.), *Contemporary China*, iv. 197, citing a directive issued jointly by the First Machine Building Industry and the Ministry of Metallurgical Industry on 4 Feb. 1959.

[12] *MSSWS* 396.

[13] Xue Muqiao, Su Xing, and Lin Zili, *The Socialist Transformation of the National Economy in China* (Beijing: Foreign Languages Press, 1960), 265–6. The figure excludes steel produced by indigenous techniques.

[14] Xue Muqiao *et al.*, *The Socialist Transformation*, 265–6.

able target of 525 million tonnes set at the same time for 1959 seem to have been quietly dropped after April 1959 at the latest.[11] Both figures were in fact drastically scaled down to 250 and 275 million tonnes respectively at the Lushan Plenum convened in September 1959 to initiate the anti-Rightist campaign. Note that the adjusted output (for 1958) and target (1959) figures are still considerably in excess of the effective figures of 200 and 170 million tonnes officially published since 1978.

If anything, the anti-Rightist campaign of autumn 1959 may have caused renewed rural uproar in China, although it is difficult to ascertain how the campaign affected the morale and incentives of the peasants. Not only did the Lushan Plenum dramatically revise the grain output target for 1959 from 525 million tonnes down to 275 million tonnes, but it 'also formally declared the brigade to be the basic unit of ownership within the "three-level ownship" structure'[12] of the commune which was subsequently to remain the institutional arrangement until December 1982.

But most important was the excessive State farm procurement which followed, but which was not necessarily a consequence of, the anti-Rightist campaign. Note that the Lushan Plenum was held shortly before the autumn harvests. With total grain output targeted at a high of 275 million tonnes as against an effective level of only 170 million tonnes for 1959, it was likely that State farm procurements would be disastrously excessive in relation to what the peasants could be expected to bear.

Before we turn to the exact statistics of what this implied in terms of the trade-off which faced the peasants between immediate survival and the necessity to preserve seed and foodgrain for future survival, it is necessary to clarify two important points frequently addressed by analysts of the Great Leap Forward period. The first is how the Great Leap policy and institutional upheavals might have impaired peasant incentives, and as a result grain yield per sown hectare; the second is how the large-scale irrigation campaigns conducted in late 1957 and through 1958 may have contributed, as many Western analysts maintain, to the problem of land salinization and alkalization in the early 1960s.

State extortion, over-exertion, and peasant disincentives

It has almost become a consensus among Western scholars that it was the Great Leap Forward strategy, with its radical policy measures, in terms of excessive farm procurement, forced mobilization for the backyard furnace campaign, deep ploughing and close planting, and extreme egalitarianism in income distribution (this last representing another extreme measure of State extortion), which severely depressed peasants' incentives and led to disastrous grain output losses, and subsequently a major food crisis and widespread starvation in the early 1960s.[13] Bad weather is also seen by many Western observers as being implicated, but very often it is regarded as a side issue.

Such arguments obviously imply that Chinese peasants were willing to trade their very survival against State oppression or 'slavery'. This runs contrary to our argument that it was precisely the 'subsistence urge' of Chinese peasants which unfolded strongly to help avert an early food crisis in 1959 in the wake of the unwarranted sown-area contraction of winter 1958–9 and which even more forcefully prompted them to dip into their feed and seedgrain reserves in 1960, resulting in inadequate sowing density and hence grain-yield losses.

But whatever the reality, it is difficult to explain why at the height of the Great Leap Forward in 1958 both total grain output and yield per sown hectare achieved record highs of respectively 200 million tonnes and 1,563 kilograms (an increase of 7.2 per cent over 1957). Had it not been for the reduction in grain-sown area by 4.5 per cent in 1958, total grain output would have been even higher than in 1957.[14]

It is of course true that part of the increase in the national average of grain yield can be accounted for by the massive sown-area shift from such low-yielding crops as soya bean and wheat to high-yielding potato and perhaps maize (but not rice which had a much higher yield than potato). But the most crucial point in this context is that, as revealed in Table 11.2, the per hectare yield of wheat, soya bean, and potato all recorded increases from 1957 to 1958. This points to the fact that the increase in the national grain yield in 1958 was partly brought about by an improvement in yields of individual crops as well as by changes in the cropping pattern.

It is possible that the yield improvement for wheat, soya bean, and potato resulted from the sown-area shift rather than, say, through an improvement in peasant' incentives, or for that matter in weather. It is also possible that the level of peasant incentives was reduced and regular farm-work disturbed by mass labour mobilization in 1958, but that the negative impact on yield was greatly outweighed by the gains from both sown-area shift and the resultant yield improvement. Whatever the truth, there is no evidence of a consistent decline in grain yield across the various crops to suggest that peasants' incentives were drastically impaired by the rural upheaval of 1958.

More interestingly, the national grain yield declined by 7 per cent in 1959 from the record high of 1,568 kilograms per sown hectare in 1958 to resume its 1957 level of 1,463 kilograms per sown hectare. This occurred in spite of the fact that the excessive sown-area shift of 1958 was hardly corrected in 1959 in terms of area distribution among the different

TABLE 11.2. Changes in sown area and yield for the major grain crops in China, 1957–1959

Grain-sown area

	Million ha.			Percentage distribution		
	1957	1958	1959	1957	1958	1959
Rice	32,241	31,920	29,034	24.1	25.0	25.0
Wheat	27,542	25,775	23,575	20.6	20.2	20.3
Maize	14,943	—	—	11.2	—	—
Soya bean	12,748	9,551	9,863	9.5	7.5	8.5
Potato	10,495	15,382	12,289	7.9	12.1	11.0
Others	35,664	—	—	26.7	—	—
TOTAL	133,633	127,613	116,023	100.0	100.0	100.0

Grain yield

	(kg. per sown ha.)			Rate of change (%)		
	1957	1958	1959	1956–7	1957–8	1958–9
Rice	2,693	2,535	2,393	8.79	−5.87	−5.60
Wheat	855	878	938	−5.79	2.69	6.83
Maize	1,433	—	—	9.81	—	—
Soya bean	788	907	888	−7.29	15.10	−2.09
Potato	2,093	2,127	1,938	−4.21	1.87	−8.89
Others	875	—	—	1.16	—	—
TOTAL	1,463	1,568	1,463	3.72	7.18	−6.70

Sources: Tables AB.3 and AB.4.

crops (Table 11.2). More remarkably, the average yields of potato and soya beans were reduced, along with the rice yield. The wheat yield however recorded an impressive increase of 7.4 per cent in 1959, on top of the 2.1 per cent increase in 1958.[15] It is difficult to attribute all these changes in yield between 1958 and 1959 to the much reduced scale of labour mobilization in 1959 compared with 1958.

Irrigation campaign and land salinization

As a number of Western scholars suggest, the rapidly increased salinization of farmland in the early 1960s was a direct consequence of the large-scale irrigation campaign which started in winter 1957–8 and was greatly accelerated in 1958.[16] Chao Kang, for example, has pointed out that numerous irrigation systems built in China in 1958 made no provision for draining irrigated fields. As a result, in many localities 'the soil gradually became saturated with salts and less productive agriculturally'.[17]

Contrary to Chao Kang's claim, the period 1959–61 was, however, not one dominated by 'floods over the whole country for three consecutive years beginning in 1959',[18] so that the 1958 irrigation campaign 'became major flood causes in subsequent years in many localities, especially in the northern plains'.[19] Rather, the three years were overwhelmingly drought years, particularly in the North China plain. Thanks to the monumental work of the SMB, it is known almost exactly where the exceptional (and often localized) floods occurred during these years (Map 10.2). The data obtained from provincial newspapers for the *shouzai* area corroborate the yearly mappings of the SMB of the distribution of both the floods and drought.[20]

Had floods been the major problem, they might have helped to leach away the salts which, in the northern plains, normally accumulate during the dry winter and spring months.[21] Admittedly, in areas lacking adequate drainage for their irrigation facilities, salt might have sedimented when water in the inundated fields eventually evaporated.[22] Unfortunately, however, it was the prolonged absence of the summer rains which deprived peasants in most of North China in 1959–61 of the opportunity for regular natural leaching of their fields to eliminate or lower the level of salt content.[23] Worse still, the persistent dry spell tended to soak away, through capillary action, the underground water, and thus added to increased salt sediments.[24]

There is yet another dimension to the problem of salinization. In many localities of severe drought, peasants were often forced to tap water from inappropriate sources to enable sowing to be carried out or to rescue wilting plants. They did this in the knowledge of the possible threat from salts.[25] It is true, as Chao Kang noted, that in many newly irrigated areas, the malpractice or inexperience in managing irrigation also contributed to salinization, when 'the farmers simply irrigated by a general flooding of the whole plot and held water in the field for a prolonged time, or irrigated with unnecessary frequency'.[26] But this refers to localities or periods in which water was either plentiful or anyway still available.[27]

On balance, the extraordinary drought of 1959–61 can probably be regarded as the major cause of the increased soil salinization in the northern plains. Its prolonged duration explains the slow rehabilitation from the accumulation of salt sediments and the slow growth in grain yield after 1961 in spite of improved weather conditions. Other factors in the slow recovery include insufficient seed supplies and deteriorating seed quality. Salinization is a long-term process and therefore cannot entirely explain the sharp, short-run fluctuations of 1959–61.

The Combined Impact of Confiscatory Farm Procurement and Sown-Area Contraction

It should be clear that it was not a matter of reduced peasant incentives or for that matter large-scale land salinization that brought about the Great Débâcle, and the disastrous food crisis in the early 1960s. The main cause must have been the depletion of the seed and foodgrain reserves (which resulted from excessive State grain procurement) and the unwarranted sown-area contraction, which directly reduced grain output. It is difficult to ascertain how this policy miscalculation occurred. The most probable explanation seems to lie in the disastrous mistakes made in assessing or forecasting the possible grain output situation in late 1958 and 1959.

The grain output for 1959, for example, was forecast to be 275 million tonnes as late as September 1959, i.e. the time of the Lushan Plenum, compared with the actual 170 million tonnes, as mentioned earlier. The effective State grain procurement of 45 million tonnes, net of resale to the peasants engaged in cropping and subsidiary activities other than grain farming, represented a procurement ratio of only 17 per cent, which is comparable with the average ratio of 18 per cent for 1952–7 (Table 11.3). In relation to the lower effective base of 170 million tonnes, the forced procurement implied, however, an alarming rate of 28 per cent.

The same miscalculation seems to have led to the proposal for the gradual adoption of the 'three-three system' of cultivation—an idea which was first articulated by Liu Shaoqi with the personal blessing of Mao at the first Zhengzhou conference held in December 1958.[28] Note that in June 1958 the national total of grain output was targeted to be 250 million tonnes and was upgraded to 300–50 million tonnes in September 1958, and claimed to be realized in December 1958 (Table 11.1). There was then a widespread view that China's grain problem had been solved once and forever and that insurmountable problems were being encountered in grain storage.

It has remained an intellectual myth how, in the first place, such a profound miscalculation about the grain situation came about. Even more puzzling are two facts: first, that there was an enormous lag between the time when the first 'high-yield satellite field' was publicized from Macheng County in April 1958 (early rice yield claimed to be at the astronomical high of 279,969 kilograms per hectare),[29] to the time when the first signs

TABLE 11.3. Actual and hypothetical grain production, State procurements, and rural grain availability from domestic sources in China, 1952–1966

	Mid-year rural population (m.)		Total grain-sown area (m. ha.)		Weather-predicted yield (kg./ha.)
	Actual (1)	Hypothetical (2)	Actual (3)	Hypothetical (4)	(5)
1952	49.99		123.98		1,449
1953	50.64		126.64		1,402
1954	51.49		128.99		1,379
1955	52.60		129.84		1,438
1956	53.41		136.34		1,386
1957	54.17		133.63		1,384
1958	54.99	55.05	127.61	134.99	1,457
1959	55.05	55.94	116.02	134.99	1,382
1960	53.99	56.85	122.43	134.99	1,236
1961	53.14	57.77	121.44	134.99	1,228
1962	54.39	58.70	121.61	134.99	1,426
1963	56.58	59.65	120.74	134.99	1,428
1964	57.54	60.62	122.10	134.99	1,544
1965	58.52	61.60	119.63	134.99	1,579
1966	60.36	62.60	120.99	134.99	1,599

	Total grain production (m. tonnes)		Net State grain procurement (m. tonnes)		Per capita grain availability for rural population		
	Actual (6)	Hypothetical (7)	Actual (8)	Hypothetical (9)	Gross (kg.) (10)	Net (kg.)	Calories (11)
1952	164		28.19		272		
1953	167		35.89		259	(226)	1,842
1954	170		31.59		268	(228)	1,877
1955	184		36.18		281	(245)	1,998
1956	193		28.70		307	(273)	2,248
1957	195		33.87		298	(251)	2,031
1958	200	197	42.73	31.59	288	(244)	1,991
1959	170	187	47.57	29.92	222	(178)	1,452
1960	144	167	30.90	26.72	209	(165)	1,346
1961	148	166	25.81	26.56	229	(185)	1,509
1962	160	192	25.72	30.72	247	(203)	1,656
1963	170	193	28.92	30.88	249	(205)	1,673
1964	188	208	31.85	33.28	271	(227)	1,852

TABLE 11.3. *cont.*

	Total grain production (m. tonnes)		Net State grain procurement (m. tonnes)		Per capita grain availability for rural population		
	Actual	Hypothetical	Actual	Hypothetical	Gross (kg.)	Net (kg.)	Calories
	(6)	(7)	(8)	(9)	(10)		(11)
1965	195	213	33.60	34.08	275	(231)	1,885
1966	214	216	38.24	34.56	291	(247)	2,015
1959–66	1,389	1,542	262.61	246.72	1,993	(1,641)	

	Per capita grain availability for rural population (standard of adequacy)	Potential per capita grain availability for rural population			
		Gross (kg.)	Net (kg.)	Calories	Std. or adequacy
	(12)	(13)		(14)	(15)
1953	adequate/marginal				
1954	adequate/marginal				
1955	high/adequate				
1956	high				
1957	high				
1958	high/adequate	301	(257)	2,097	high
1959	poor	281	(237)	1,934	high/adequate
1960	poor	247	(203)	1,656	marginal/poor
1961	poor	241	(197)	1,607	marginal/poor
1962	marginal/poor	275	(231)	1,885	adequate/marginal
1963	marginal/poor	272	(228)	1,860	adequate/marginal
1964	adequate/marginal	304	(260)	2,121	high
1965	adequate/marginal	306	(262)	2,138	high
1966	high	301	(257)	2,097	high
1959–66		2,227	(1,875)		

Note: All grain figures refer to 'unhusked' grain. The procurement figures refer to the production year (April to March of the following year), rather than calendar year. The population size is taken to be the average of the year-end figures for the preceding and the current years. It may thus be regarded as the mid-year population size. The derived grain availability figures should be interpreted with care. In a way, they may be considered as availability for the following year, rather than the current year, as indicated. This is because procurements are not closed until the following March, and total production net of procurements may thus appear to be the effective balance available. Nevertheless, State procurements do take place as soon as the harvests are completed, starting with the early crops in late spring or early summer. And since grain reserves are minimal, especially in the 1950s, the peasants also rely on current output for consumption.

TABLE 11.3.   *cont.*

Sources: (1) *TJNJ 1984*, 81.
(2) Assumed to grow at annual rate of 1.62%, i.e. the average for the period 1952–7.
(3) Table AB.2.
(4) Assumed to be stabilized at the average level of 1956–7.
(5) From the estimated regression equation shown in Table 7.2.
(6) Table AB.2.
(7) Columns (5) × (4).
(8) *TJNJ 1984*, 370.
(9) Assumed to be 16.1% of total potential output, i.e. the average procurement ratio for 1956–7.
(10) Figures gross of seed and feed are from [columns (6)–(8)]/(1). The net figures were kindly supplied by the late Professor Kenneth Walker for 1953–7 as derived from the provincial figures in his *Foodgrain Procurement and Consumption in China*. The five-year (1953–7) average of 245 kg./head is by 44 kg. lower than the comparable average (289 kg.) for the gross series. The implied absolute seed and feed amount is uniformly applied to all the following years to give the net figures for 1958–66.
(11) Figures for 1953–7 also from Walker. For 1958–66, they are assumed to be related to the net grain figures at the same proportion as the 1953–7 average of 1,999 to the net grain figure of 245 kg. This of course ignores possible influences from changes in the composition of grains consumed.
(12) Walker, 100–1. Where two standards are given, e.g. adequate/marginal as for 1953 they refer respectively to standards for South and North China.
(13) [(7)–(9)]/(2).
(14) and (15) Same source and same conversion procedure as for columns 11 and 12.

of readjustment appeared in early or mid-1959; second, that the total grain output figures published at the time, namely 250 million tonnes (as claimed in August 1959) for 1958, and 270 million tonnes (claimed in January 1960) for 1959, were still much higher than the actual figures of 200 and 170 million tonnes.

Basically, the root cause of the disasters is then to be found in the statistical fiasco of 1958,[30] and hence the breakdown of the communication link between the central policy-makers and the peasants. This deprived the central planners of solid information for their policy decisions. How the black-out could last so long is a matter of further research by mass communication experts. Our concern here is to reconstruct, in quantitative terms, how confiscatory farm procurements and the unwarranted sown-area contraction led to the great depression of the early 1960s.

Draining the pond to catch the fish

Table 11.3 shows the changing volume of State grain procurements in relation to total grain output, per capita grain availability, and the implied calorie intake from 1958 through 1966. Total State procurements net of resales to rural areas for the grain year April 1959 to March 1960 increased by 11 per cent from the already very high record of 41 million tonnes in 1958–9 to nearly 48 million tonnes. As national grain output declined from the record high of 200 million tonnes in 1958 to 170 million tonnes in 1959, per capita grain availability for rural residents fell dramatically from 288 to 222 kilograms per head. These figures are gross of seed and feedgrain requirements. In net terms, they amount to 244 kilograms per head for 1958 and a mere 178 kilograms per head for 1959. As a result, the implied calorie intake per peasant was reduced from a high adequacy of around 2,000 to only 1,452 (which implies malnutrition) as indicated in Table 11.3.

The burden of the massive State grain procurements for 1959–60 probably fell much more heavily on the autumn than on the summer harvests in 1959. This can be explained by the fact that the shortfall in grain output in 1959 overwhelmingly reflected in the autumn harvests, following the sharp deterioration in weather conditions, i.e. widespread drought in the Yangzi and Huai Rivers basins, as well as in Sichuan province. Despite this, average grain output for that year was still being claimed to have attained the unrealistically high level of 270 million tonnes, as against an actual 170 million tonnes (Table 11.3). This inevitably resulted in a disproportionately large level of State procurement from the autumn harvests.

By contrast, the summer harvests in 1959 were relatively good, and probably also subject to comparatively less harsh state procurement, since following widespread peasant resentment against excessive state extortion after the autumn harvests of 1958, readjustments were already being made in this respect in early 1959. More importantly, no matter how harsh the extortion in 1958 had been, it had still left the peasants with a per capita grain availability of 288 kilograms (Table 11.3). This was more than the average of 281 kilograms per head for 1952–7, and was sufficient to allow stocks to be carried over into 1959 to ensure sufficient sowing.

In short, the situation in 1959 still appeared comparatively 'congenial', at least until the summer or autumn. The average per capita grain availability of 222 kilograms for the peasants suggested the need to live with a considerably lower amount of grain starting late 1959 and continuing into 1960. This tended to force them to dip into their seed and feedgrain reserves in 1960.

The depletion of the seed and feedgrain reserves resulted not only in an inadequate sowing density, and hence lower yield per hectare, but also in unavoidable starvation and slaughter of livestock which reduced supplies of organic fertilizers and draught-animal power which was so important for ploughing and other farmwork. In fact, the number of pigs (each of which

Mao described as a small chemical-fertilizer plant), declined by a massive 32 per cent in 1960, and that of draught animals by 12 per cent.[31]

Viewed against this background, the further deterioration in the weather conditions during 1960 was the final blow to the peasants. Probably as a rescue measure State grain procurements were reduced by 35 per cent for the grain year from April 1960 to March 1961 to 31 million tonnes, compared with a reduction of only 15 per cent in total grain output. Even so, output reached an all-time low of 144 million tonnes in 1960. As a result, the State's concessions could not prevent rural per capita grain availability from declining further to a low point of 209 kilograms. The most telling evidence of dire food shortages is the widespread incidence of dropsy and oedema and of actual starvation. There was a dramatic rise in the rural death rate and a sharp downturn in the birth rate, resulting in a negative growth rate of 3.1 per cent for the rural population in 1960.[32]

The catastrophe of 1960 naturally spilled over into 1961, with the weather showing no sign of real improvement until late in the year. Further rescue measures were adopted, including the dispatch of the 'Urgent directives on rural work' (commonly known as the Twelve Articles) to the peasants in November 1960 and the promulgation of the Sixty Articles in April 1961.[33] The new regulations essentially brought rural organizations back to the pre-1956 situation, with the production team now designated as the basic unit of planning, accounting, and distribution.

It is likely that the most important measure that had an immediate effect towards refilling the 'pond' in order to replenish the 'fish' was a further reduction in State grain-procurement to an all-time low of 26 million tonnes in 1961.[34] At the same time, grain imports rose almost overnight from virtually none in 1959–60 to 5.81 million tonnes in 1961,[35] mainly in order to compensate for the 5.09 million tonnes shortfall in State procurements, and to feed the urban population. In addition, grain exports, predominately rice, were curtailed by 1.37 million tonnes, from 2.72 million tonnes in 1960 to 1.36 million tonnes in 1961.[36] All this led to per capita grain availability for the peasants rising to 230 kilograms in 1961, higher than the average of 222 kilograms of 1959. In this way, seed and feed stocks began to be gradually replenished again in 1961.

Withdrawing the firewood from beneath the cooking-pot

The deliberate contraction in sown areas in 1958–9 was tantamount, in effect, to the Chinese saying of 'withdrawing the firewood from beneath the pot' (*fudi chouxin*), and so halting the cooking (i.e. cropping activities) altogether. It is difficult to quantify the magnitude of grain output losses due to the area contractions. One way of doing so is to apply, say, the average grain yield per sown hectare for 1952–7 to the total sown area withdrawn from cultivation.[37] This will, however, overstate the extent of losses, since the sown area that was withdrawn involved mostly infertile marginal farmland.

What is clear, however, is that the reduction in national grain output by 15 per cent from 200 million tonnes in 1958 to 170 million tonnes in 1959, partly reflected losses due to the sown area curtailment by 9.1 per cent from 127.61 to 116.02 million hectares (Table 11.3). In fact, sown area had already started to decline in 1958 by 4.5 per cent from the 1957 total of 133.63 million hectares. This was a result of the irrigation campaign which took up farmland, as well perhaps as of the backyard furnace movement which diverted labour from agriculture.

Against this background, it therefore seems appropriate that we take, say, the average sown area for 1956 and 1957 as a basis on which to estimate losses due to sown-area contraction, in order to capture the full scale of losses due to the entire Great Leap Forward strategy which had begun in 1958. Our discussion is mainly directed to the question of whether, without the sown-area curtailment, or indeed, excessive State grain procurements, Chinese peasants would have been able to survive the unusually bad weather of 1959–61.

For this purpose, we first assume total sown area to be stabilized at the average level for 1956–7 of 134.99 million hectares through the years 1958 to 1966. We then ask what total grain output would have been, given the prevalent weather conditions for these years. This necessitates the application of our weather-predicted grain yield per sown hectare to the assumed sown-area base in order to obtain the hypothetical series of grain outputs shown in Table 11.3 (column 7). To estimate the potential per capita grain availability for the peasants we have established a hypothetical series of rural population for 1958–66, by assuming that the yearly natural rate of increase for these years was the same as the average of 1.62 per cent during the period 1952–7 (Table 11.3).

The population estimates are consistently and indeed considerably higher than the actual series. The reason for not using the actual series for our simulation is that it in itself reflects the impact of the great agricultural depression and food crisis. Note that after consecutive years of increases from 1952 to 1959, total rural population declined sharply in absolute terms in both 1960 and 1961, and did not recover until 1963.

Moreover, we also assume that net State grain procurements out of hypothetical grain output in 1958–66 remained at a ratio comparable with the realized average, i.e. 16.1 per cent for 1956–7. This gives a hypothetical series of per capita grain availability which, compared with the actual series, may also help to reveal the impact on peasant grain consumption and calory intake. Several points emerge from our estimates which are shown in Table 11.3.

First, if both the 1959 total grain-sown area and the net procurement ratio had been stabilized at the 1956–7 level, then despite the bad weather, peasants would still have been left with 281 kilograms per head in that year. This is comparable with the bumper harvests of 1955 and 1958. Such a level of availability would have enabled them to surmount the crisis of 1960–1 with relative ease.

Second, even without the benefit of grain reserves carried over from 1959, the peasants could still have survived the weather extremes of 1960 and 1961 had the grain-sown area and the level of net State procurements for both years remained at the 1956–7 level. The calculations made in Table 11.3 show a potential per capita grain availability for the peasants of 247 kilograms for 1960–1 and 241 kilograms for 1961–2. Both these hypothetical figures are substantially greater than the per capita amount of only 222 kilograms of grain available to the rural population in 1959, on the basis of actual grain output in that year.

Third, if, however, the absolute net grain procurement for 1959–60 (which is the record high of 47.57 million tonnes) is applied to the hypothetical output of 187 million tonnes in 1959, it can be shown from the relevant statistics in Table 11.3, that per capita grain availability for peasants would have been reduced from 281 to only 249 kilograms in 1959–60. This is, however, still considerably higher than the actual amount of 222 kilograms. Although this shows the extent to which the massive State grain-procurement in 1959–60 affected the peasants, their situation, without the misfounded reduction in sown area, was still relatively bearable in 1959–60.

Fourth, to complete our argument, we may further apply the same net procurement ratio (relating the actual amount of 47.57 million tonnes for 1959–60 to the hypothetical output of 187 million tonnes for 1959), to the lower hypothetical output for both 1960 (167 million tonnes) and 1961 (166 million tonnes) in order to estimate potential per capita grain availability and to see whether they could have survived the extraordinary downturn in weather conditions in these two years. The estimated figures are 219 and 214 kilograms respectively which are quite comparable with the realized ones. This implies that, even without the sown-area curtailment in 1959, the peasants could not have overcome the food crisis resulting from both bad weather and massive State extortions.

In short, the results of our simulations suggest that without the unwarranted sown-area curtailments and the massive State farm procurement, the peasants would have been able to survive, relatively comfortably, the unusually bad weather prevailing from 1959 through 1961. Without the sown-area reduction, they could also have survived the confiscatory State procurement quotas in 1959–60. This is not the case, however, in 1960–1 and 1961–2, when both weather and grain output deteriorated in tandem. The misplaced policy measures not only brought chaos to rural China and precipitated virtual total collapse in 1959–61, but, as our simulations indicate, their enormous impact continued to be felt throughout the first half of the 1960s. But before we examine the opportunity costs involved, we should first discuss the other important aspect of our simulation results, that is, the ability of Chinese peasants to have survived the disastrous policy miscalculations in favourable weather conditions.

Could the Peasants have Survived Policy Miscalculation, even if the Weather had been Good?

Assuming that the weather during 1959–61 had been as good as in the bumper harvest year of 1958, could the peasants have survived massive State exploitations and the policy miscalculation associated with the 'three-three-system'?

In an attempt to answer this, we may apply the weather-predicted yield of 1,457 kilograms per sown hectare for 1958 to the curtailed sown-area size of 1959 shown in Table 11.3. The resultant hypothetical grain output for 1959, net of the massive State procurement of 47.57 million tonnes, would have left the peasants with a per capita grain availability of 221 kilograms. This is nearly identical to the actual level of 222 kilograms which forced peasants in many areas to eat up their seed and feedgrain reserves for 1960.

Note, however, that the estimated hypothetical figure of 221 kilograms in 1959 is based on the actual population for that year. The population figures given in Table 11.3 reveal that growth in 1959 had already considerably slowed down compared with the earlier years 1952–7. This seems to reflect the stress resulting from the 1958 Leap, and the emerging grain-supply shortages of 1959. If instead of the actual population, the figure projected for 1959 (on the basis of the higher average population growth rate of 1952–7) is used, then the estimated hypothetical per capita grain availability falls to only 217 kilograms.

This discrepancy between the two hypothetical grain availability figures may be taken as a clear sign of the existence of a positive Malthusian check. That is to say, the population growth was reduced in 1959 by starvation or hunger, in order to provide the existing population with essential subsistence grain.

The same simulation for 1960 and 1961 yields similar conclusions about potential grain consumption in those years. The most critical variable in the simulation remains national sown area. As shown in Table 11.3, there were some increases in sown area in both 1960 (total 122 million hectares) and 1961 (121 million hectares) compared with 1959 (116 million hectares). But the total sown area in 1960 and 1961 was still not restored to the record level of 1957 (total 134 million hectares) prior to the consecutive curtailments in 1958 (128 million hectares) and 1959. Lack of seedgrain was clearly an important cause of the slow recovery of sown area. The seedgrain shortfall represents of course the cumulative impact of both the contraction in sown area (due to the 'three-three-system' and the irrigation campaign) in 1958–9, and excessive State grain procurements in 1958–9 and 1959–60.

What would the grain output and per capita availability in 1960 and 1961 have been had the weather been as good as in 1958? The favourable weather-predicted yield of 1,457 kilograms per sown hectare, as estimated for 1958, would generate, on the basis of the actual grain-sown area in 1960 and 1961, a potential grain output of 178 and 177 million tonnes for these two years. Could the peasants have survived the high amounts of State grain extraction at these two output levels? Two different assumptions may be made about possible State procurements. The first is to assume the net procurement ratio to be the same as in 1959, i.e. 28 per cent of total grain output. The second is to fix for 1960 and 1961 the total level of net State procurement at the all-time record of 1959, i.e. 47.57 million tonnes. With these assumptions we postulate that the government would have continued with its large-scale grain extortion from the peasants, given the buoyant mood of 1958.

However, the estimated potential per capita grain availability, net of State procurements on either of the two assumptions is very similar: 226 kilograms per head according to the first assumption, and 230 kilograms per head according to the second in 1960. For 1961, the respective figures are 221 and 224 kilograms. Taken together, these figures indicate hardly any improvement over the actual standard of 222 kilograms for 1959, and they all fall into the 'poor' category in terms of calorie equivalent. Moreover, the population base used for our estimates, i.e. the actual population, represented only those peasants who had survived the Malthusian check in 1960 and 1961, as discussed earlier. For the purpose of our simulations it is more appropriate to make use of the projected, higher 'unrestrained' population size. This would inevitably result in a considerably lower per capita grain availability for 1960 and 1961.

Note that the weather-predicted yield of 1,457 kilograms per sown hectare for 1958 was consistently used throughout our simulations. We may instead use the actual 1958 yield of 1,568 kilograms (the highest during 1952–8). The latter higher figure seems less appropriate as a basis for the measurement. As discussed earlier, the extra gains partly reflect the extraordinary sown-areas shifts from low-yielding to high-yielding crops in 1958. Nevertheless, for the sake of completing our simulations, we apply it to the curtailed sown-area base in 1960 and 1961 to estimate the respective potential output and net per capita grain availability during these two years. The estimated output is 192 and 190 million tonnes respectively for 1960 and 1961. By maintaining the two assumptions made about the possible size of State grain procurements (i.e. 28 per cent of total output or 47.57 million tonnes as realized in 1959), these output estimates generate a potential per capita availability of 243 and 254 kilograms for 1960, and 237 and 247 kilograms for 1961, based on the hypothetical population in the two years.

A similar simulation made for 1959 yields an even lower figure of 234 and 240 kilograms respectively, mainly due to the fact that the sown area was curtailed to a larger extent in 1959 than in 1960 and 1961. All these figures which are gross of seed and feedgrain allowance are consistently and, indeed, substantially lower than the amount available for any of the years prior to 1959. They imply barely sufficient grain consumption and probably widespread starvation in poor areas if we bear in mind the considerable degree of inter-regional inequality in per capita grain output that existed.

The foregoing analysis of the impact of the contraction in grain-sown area and the confiscatory grain procurements clearly adds force to the contention made by the late Kenneth Walker that 'even without natural disasters, the agricultural depression was inevitable'.[38]

Reckoning the Opportunity Costs through the mid-1960s

The hypothetical estimates given in Table 11.3 also shed some light on other important aspects of the consequences of the contraction in grain-sown area in 1959. Several points emerge.

First, looking at the total grain output trend, given the prevalent weather conditions which determined our weather-predicted yield per sown hectare, a national sown area equivalent to that of 1956–7 would already by 1962, have resulted in a potential grain output comparable with the actual output of 1956, and not much below the record high (for 1952–7) of 195 million tonnes achieved in 1957–7. Such a recovery was, however, not achieved until 1964–5.

The second point concerns the State grain procurements—again net of resales to the rural sector. If the State procurement ratio had been maintained at the 1956–7 level of 16.1 per cent of total grain output, the hypothetical grain output of 167 million tonnes would have generated a total State procurement of 26.72 million tonnes in 1960, which is 13.5 per cent lower than the actual figure of 30.90 million tonnes during that year. But, starting from 1961, the hypothetical State grain procurements, based on the 1956–7 procurement ratio, consistently yield an amount higher than the volume actually procured by the State in every year up to 1964.

The third point is that for the period 1959–66 taken as a whole, the total potential grain output of 1,542 million tonnes, shown in Table 11.3, is 11 per cent higher than the actual figure. State grain procurements would have totalled 247 million tonnes based on the average procurement ratio of 16 per cent for 1956–7. This is only 6 per cent lower than actual procurements of 263 million tonnes for 1959–66. Indeed, with potential grain output already restored by 1962–3 to the actual 1956–7 level, and consistently exceeding the all-time 1958 record of 200 million tonnes from 1964 onwards (Table 11.3), one might expect that the assumed procurement ratio of 16 per cent would be gradually adjusted upwards. This would certainly have generated a total potential procurement higher than the level actually attained during 1959–66 taken as a whole.

The fourth point concerns per capita grain availability for the peasants. The simulations in Table 11.3 add up to a potential total of 2,227 kilograms for 1959–66. This is 12 per cent higher than the actual value of 1,993 kilograms. It is unlikely that the extra gain of 29 kilograms per year between 1959 and 1966 would have been significantly curtailed, allowing for marginal upward adjustments in the procurement ratio from year to year in line with the projected increases in total grain output.

More importantly, the potential per capita availability was estimated on the basis of the higher hypothetical, rather than the lower actual, population figures. As mentioned earlier, the absolute decline in total population from 1960 to 1964 is a clear sign that a positive Malthusian check was at work. Thus, the higher estimated potential per capita grain availability could have been achieved without the enormous sacrifice of human life which occurred in the early 1960s.

Finally, consideration should be given to the implications for the urban-industrial sector. In the immediate aftermath of the crisis, investment resources originally earmarked for the State's top priority industrial sectors had to be reallocated to the countryside as an emergency measure. For example, the government were compelled to reduce farm deliveries starting in 1960 and to increase the supply of chemical fertilizers. Investment in the urban industrial sector was also scaled down in order to accommodate shortages in the supply of both wage-food and agricultural raw materials. In addition, increased foodgrain imports and reduced exports rapidly drained the country's foreign exchange earnings which had previously primarily been reserved for imports of scarce producer-goods such as steel and machinery.

Some Reflections

Some general reflections on the origin and implications of the 1959–61 food crisis may be made by making use of the frame of reference associated with Professor Amatyr Sen's 'food-exchange entitlement' theorem.[39] Sen concentrated on the great Bengal famine of 1943 and concluded that even without any shortfall in average foodgrain availability, external economic and political factors could cause prices, wages, and employment to change drastically and adversely for consumers, thus depriving them of the necessary grain-purchasing power (i.e. food-exchange entitlement) to provide for survival. In other words, a sudden sharp change in relative prices can precipitate famine. In the largely non-marketized and non-monetized Chinese context, which until recently bore all the economic planning characteristics of the Henselian 'Natural-wirtschaft',[40] such a process seemed unlikely to occur. This is because any abrupt disruption in grain supplies would be prevented through rationing and State price control, from finding its expression in spiralling grain prices.

Similarly, one would not have expected a natural disaster in the collectively controlled Chinese countryside to cause widespread unemployment, as in a market economy, whether directly by dislodging peasants from farming or indirectly by reducing employment opportunities in the farm-processing and related industries, and so further reducing purchasing power or the entitlement claims for foodgrain. It was, however, precisely Sen's 'food-exchange entitlement' mechanism which, as described by Elizabeth Oughton, operated in the case of the Maharashtra (India) droughts of 1970–3, in that an initial series of relatively mild shortfalls in foodgrain production eventually culminated in a serious famine, causing widespread starvation and death.[41] A situation like the Bengal famine of 1943 was less likely to develop in China.

There is none the less a certain Chinese analogy to Sen's mechanism if the 1959–61 food crisis is traced to the communication breakdown of 1958–9 which culminated in the huge miscalculation of adopting the 'three-three system' and implementing unrealistically high compulsory grain quotas. This miscalculation resembles the malfunctioning of the market-price signals. Thus, the 1959–61 tragedy does bear comparison with the great Bengal famine although to the extent that bad weather in

1959–61 also helped to reinforce the food crisis—it also has similarities with the Maharashtra case.[42]

NOTES

1. Cf. *RMRB* editorial, 22 Mar. 1958.
2. See Table 11.1.
3. This was one of the major complaints levelled by Peng Dehuai against Mao at the Lushan Plenum held in July/Aug. 1959. Peng, in his famous *Yizian Shu* (letter of opinion) to Mao maintained that a total of 70 million peasants had been mobilized for the campaign, and argued that the gains could not compensate for the losses (*Debu shanshi*). The text of Peng's letter is published in Union Research Institute, *The Case of Peng De-huai 1959–1968* (Hong Kong, 1968). For a provincial report about how the campaign led to supply shortage of farm labour, see *HNRB*, 19 Oct. 1958.
4. For a detailed discussion of the background to the 1957–8 irrigation campaign see Michel Oksenberg, 'Policy Formulation in China: The Case of the 1957–1958 Water Conservancy Campaign', Ph.D. Thesis, Dept. of Political Science, Columbia University, 1969. Cf. also Chao Kang, *Agricultural Production in Communist China*, 134 on the impact of the campaign on cultivated area. For an example of a local report from Guangxi province, see *GXRB*, 9 Oct. 1958.
5. See e.g. *SCRB*, 11 and 14 Nov. and 8 Dec. 1958, and 21 Jan. 1959. The winding-up of the campaign was also prominently noted in a Taiwan source; see *MGNJ 1959*, 741.
6. A notable exception is from Hubei province, in *HUBRB*, 26 Feb., 1 and 11 Mar., and 10 Apr. 1959, in particular. The provincial Party authority admitted that in 1958 close planting led to reduced output in some places, and that adjustments were being made, although it was still pursued in principle.
7. Cf. K. R. Walker, *Foodgain Procurement and Consumption in China*, 142–5.
8. *HBRB*, 15 Nov. 1959 complained in connection with the anti-Rightist campaign that in Hebei province peasants in many localities already took advantage of the rectification campaign in the spring and summer to practise *dingtian daohu* (fixed fields for the peasant households), and *baochan daohu* (contracting output quotas to the households). The grain ration fields were presumably also redistributed to individual families on a per capita basis. Similar radical readjustments were also made in Shandong, as reported in *DZRB*, 14 Mar. and 4 May 1959, i.e. well in advance of the Lushan confrontation.
9. Cf. Audrey Donnithorne, *China's Economic System* (London: Allen & Unwin, 1967), 82–8.
10. Byung-Joon Ahn, 'Adjustments in the Great Leap Forward and their ideological legacy, 1959–62', in Chalmers Johnsons (ed.), *Ideology and Politics in Communist China* (Seattle, Wash.: University of Washington Press, 1973), 260–1.
11. See *DZRB*, 24 May 1959 for an example of the adjustments made at the provincial level (speech by the Deputy Governor of Shandong province).

12. Byung-Joon Ahn, 262. Cf. also Walker, *Planning in Chinese Agriculture: Socialization and the Private Sector 1956–1962* (London), 16–17 for details about the changes made.

13. Cf. Victor D. Lippit, 'The Great Leap Forward reconsidered', in *Modern China*, 1/1 (1975), 92–3; and Walter Galenson and Ta-chung Liu, in *New York Times*, 17 June 1969 as quoted therein. Many other scholars also held the same view either explicitly or implicitly. See Alexander Eckstein, *Communist China's Economic Growth and Foreign Trade* (New York: McGraw-Hill, 1966), 29–40; and *Economic Development: The Interplay of Ideology and Scarcity*, 287–8; Chao Kang, *Agricultural Production in Communist China*, 129–30 and 256–9; and Thomas G. Rawski, 'Agricultural employment and technology', in R. Barker and R. Sinha, *The Chinese Agricultural Economy*, 122–3.

14. Table AB.2.

15. Ibid.

16. Chao Kang, *Agricultural Production in Communist China*, 315. cf. also Donnithorne, *China's Economic System*, 130; and Leslie T. C. Kuo, *The Technical Transformation of Agriculture in Communist China* (New York: Praeger, 1972), 85 and 91–2.

17. Chao Kang, *Agricultural Production in Communist China*, 135 and 315.

18. Ibid. 129.

19. Ibid. 132.

20. See sources of Table AB.16 for the three years 1959–61. At the time, the drought also received sensational press coverage both inside and outside China. Apart from the catastrophic Yangzi and Huai basin droughts in summer 1959 e.g. during prolonged droughts in the North China plains in 1960, several sections of the lower reaches of the Yellow River, notably in Jinan and Fanxian, Shandong province, virtually dried up for more than 40 days between March and June 1960, and could be waded across. This was not only prominently reported in *RMRB* (20 Dec. 1960), 1; but was also clearly noted in Hong Kong and Taiwan; see *MGNJ 1961*, 678 and *ZGZK* 23/5 (1960), 11–14; 35/12 (1961), 9–10; and 39/2 (1962), 9–11.

21. The seasonal process of salt accumulation is shown as: March–June: peak period of accumulation; July–August: freed or reduced during the rain season; September–November: slow reaccummulation; December–February: accumulation relatively stable. See Liu Chuntang, Wang Shenghou, and Zuo Hangqing, 'Yudong yanjian diqu xiaomai goubo baomiao fengchan jishude yanjiu' (A study of the techniques of protecting the seedling and reaping good harvests for duct-sown wheat in the saline-alkali area in Northern Henan) in *TRXB* 14/1 (1966), 32–8; and esp. *NYKX* 8 (1963), 40. For a similar study in Anhui, see *TRTB* 3 (1964), 45; and *TRTB* 5 (1964), 19–20; in Hebei, *Gailiang Yanjiandi Chuang Gaochan* (Improve saline-alkali land to reap high yield) (Shijiazhuang: Hebei renmin chubanshe, 1974), 6–7; and in Shandong, Shandong Provincial Research Institute for Water Conservancy Science, *Yanjiandi Gailiang* (Improving Saline-Alkali Soil) (Jinan: Renmin Chubanshe, 1974), 12–13. For a general and comprehensive study of the distribution and meteorological background of saline and alkaline soil in China, see *HQ* 15/16

(1962), 47–56. (The major problem in most of North China is salinization rather than alkalinization.) Thus it is of utmost importance that summer rains arrive in time every year to help leaching or diluting the accumulated salts; on this point see, e.g. *NYKX* 4 (1963), 41; and 8 (1965), 18; and *TRXB* 13/2 (1965), 172.

22. There was no lack of such scattered reports to this effect (as in *NYKX* 12 (1962), 20 as cited by Chao, 136), as well as appeals to suspend such inappropriate irrigation facilities (*RMRB*, 18 Dec. 1962, as cited by Chao, ibid.); but the majority of the irrigation projects built during the rigorous mass campaign in 1957–8 probably caught more dust rather than water in 1959–61. Note that the problem of salt being accumulated in irrigated areas without adequate drainage was not limited to the Great Leap Forward period; and there were indeed many reports from localities which had succeeded in desalinizing through improved drainage during the GLF period. See e.g. *TR* 9 (1959) (Salt from inadequate drainage since 1956, Shandong example), 5–6. *TR* 3 (1962) (Inadequate drainage ever since 1956, Shanxi example), 54–6; *TRTB* 5 (1962) (Salt successfully drained, north Jiangsu), 13; *TR* 3 (1962) (Several million *mu* of salinized fields in North China successfully tackled since GLF), 6; and *TR* 3 (1961) (Significant salt-leaching through a criss-cross system of river reservoirs in some localities in Hebei, Henan, and Shandong, 1959–60). Moreover, in some places, it was not necessarily a matter of inadequate drainage, but rather serious inundation from heavy downpours that resulted in increased salt sediments, esp. when flooding was followed by severe drought that helped rapidly to evaporate the flood water and left the salt behind. On this point see *TR* 3 (1962), 52–3, and *TRXB* 13/1 (1965), 98.

23. *HBRB*, 26 Sept. 1960 e.g. filed a current report at the time: as a result of serious shortfall of rain in Aug., alkaline could not be depressed into the subsoil. A similar report on Henan is available in *HNRB*, 11 Nov. 1960 and *TRTB* 4 (1962): the situation in 22 *xian* of Kaifeng prefecture (which traditionally suffered from mild salinization) was aggravated by prolonged droughts from 1959 (31–2).

24. Cf. *NYKX* 7 (1962), 29–32; and *TRTB* 3 (1966), 36, and 2 (1964), 29.

25. A good example is given in *TR* 1 (1962), 11 which refers to the dilemma encountered by peasants in the Shushui river basin in Shanxi province under severe drought pressures in Sept. 1959. They eventually yielded to the pressures. As a matter of fact, the entire controversy over the practice of *yinhuang guangai* (siphoning water from the Yellow River for irrigation) can be viewed in the same context. It is yield-raising, but unfortunately considered by many (though not all experts) as salt-conducive; see *TRYK* 5 (1959), 14; *NYKX* 4 (1961), 27–32; *TRTB* 12 (1963), 58–60, and 3 (1963), 19–20; *TR* 5 (1961), 12–16; and *ZGNB* 10 (1961), 17–19. Note also that during the extraordinary drought from July to September 1959, 6 bn. m^3 of water were siphoned from the Yellow River to irrigate a total of more than 2,133,000 ha. of arable land in Hebei, Henan, and Shandong provinces; see *HBRB*, 24 Oct. 1959.

26. Chao Kang, *Agricultural Production in Communist China*, 136.

27. The two cases cited by Chao seem to refer to the situation in 1958 and 1964 respectively.

28. It is not clear when exactly Liu Shaogi initially put forward the idea. But the system seems to have been first mentioned in *JXRB* 30 Sept. 1958. During his investigative tour to Jiangsu province, 19–28 Sept. 1958, Liu said that according to the experience of Hebei province, it should be better to plant less (*shaozhong*), achieve a high yield (*gaochan*), and reap more (*duoshou*), than to plant extensively (*guangzhong*) and have a thin yield (*boshou*). 'By doing so, in a few years, it would be possible to allocate a third of the arable land for grain cultivation, a third for tree-planting, and a third to lie fallow.' However, at the time, plans were already well under way in some provinces to adopt the system. In Guangdong, e.g. Chen Yu, the governor, proposed in his *Work Report* to the provincial People's Congress (*NFRB*, 26 Sept. 1958) to contract the arable area to 35 million *mu* in 1959 from the high of 48 million *mu* in 1957. Two months later (*NFRB*, 24 Nov. 1958), Zhao Ziyang, then the provincial party secretary, further reduced the proposed figure to 25 million *mu*, with the sown area for early paddy rice to be cut from 33 million *mu* in 1958 to 18 million *mu*. It is clear that the stipulated contraction was not fully implemented, as the first sign of a policy reversal (or readjustment) in this respect emerged at the National Agricultural Work Conference held in Jan. 1959; see *DZRB*, 19 Jan., 1 Feb., and 23 Apr. 1959; and *FJRB*, 30 Jan. 1959.

29. *HBRB*, 13 Aug. 1958. For mid-rice crop the comparable figure (as per *HNRB*, 1 Sept. 1958) was 401,508 kg./ha. This is respectively 51 and 73 times as much as the national average of rice yield in 1989. 'Wheat yield satellites' also began to proliferate from early June 1958; see e.g. *HNRB*, 6 and 12 June 1958, and *DZRB*, 10 and 12 June 1958.

30. Cf. Choh-ming Li, *The Statistical System of Communist China* (Berkeley, Calif.: University of California Press), 83–108.

31. *TJNJ 1990*, 373–4; and *HNRB*, 21 Oct. 1960. By autumn 1960 at the latest, the competitive claims for feed and seedgrain and basic human consumption had already become acute; see e.g. *DZRB*, 10 Nov. 1960. See also *DZRB*, 5 Nov. 1960 and *HBRB*, 10 and 14 Dec. 1960 for evidence of widespread searching by peasants for 'wild vegetables' (*yecai*) and other similar ingredients for surviving the famines (*duhuang*).

32. *TJNJ 1990*, 89. This refers to the year-end figure rather than the mid-year figure shown in Table 11.3. For a broader study of the extent of famine and mortality consequences in 1959–61 see Penny Kane, *Famine in China: 1959–61* (London: Macillan Press, 1988), chs. 5 and 6; and Nicholas Lardy, *Agriculture in China's Modern Economic Development*, 148–53.

33. Byung-joon Ahn, 'Adjustments in the Great Leap Forward and their ideological legacy, 1959–62', 265 and 271.

34. We follow Kenneth Walker in using the Chinese metaphor of *jieze eryu* (draining the pond to catch the fish) to describe the catastrophic effect of excessive State grain procurements. See his *Foodgrain Procurement and Consumption in China*, 146–64.

35. *TJNJ 1984*, 412.

36. *TJNJ 1984*, 397.

37. This is the method used by Walker, *Foodgrain Procurement and Consumption in China*, 146–9.

38. Ibid. 129.
39. A. Sen, 'Starvation and exchange entitlements: a general approach to its application to the Great Bengal Famine', *Cambridge Journal of Economics*, i (1977), 33–59; 'Ingredients of famine analysis: availability and entitlements', Oxford University and Cornell University Working Paper, No. 210 (Oxford, 1979); and *Poverty and Famines: An Essay on Entitlement and Deprivation* (Oxford: Oxford University Press, 1981); all as cited in E. Oughton below.
40. K. Paul Hensel, *Einführung in die Theorie der Zentralverwaltungswirtschaft* (Introduction into the Theory of the Centrally-Administered Economy) (Stuttgart: Gustav Fischer Verlag, 1979; 1st edn. 1959). This work represents a pioneer attempt made in the West to develop a Walrasian system of equilibrium for the conventional Soviet-type economy based exclusively on physical planning and control. For a discussion of Hensel's Model, see J. M. Montias, 'Planning with material balances in Soviet-type economies', in *American Economic Review*, 49 (1959).
41. E. Oughton, 'The Maharashtra droughts of 1970–1973: An analysis of scarcity', *Oxford Bulletin of Economics and Statistics*, 44/3 (1982), 169–97. For a global comparative evaluation of the causes of famines, see John W. Mellor and Sarah Gavian, *Famine: Causes, Prevention, and Relief* (International Food Policy Research Institute, Washington, DC), repr. No. 98 from *Science*, 235 (Jan. 1987).
42. Cf. Nicholas Lardy, *Agriculture in China's Modern Economic Development*, 152, for a similar interpretation.

V
Economic and Policy Implications

12

The 1985–1991 Perspective

The main body of this study traces agricultural instability in China up to 1984, the year in which the promulgation of *Document No. 1* (on 1 January 1984) effectively sealed the fate of rural collectives. Not only was collective farmland reparcelled to individual households (as had been the case increasingly since 1982), but peasants' leasehold rights were generally extended from three to fifteen years after 1984.[1] Moreover, starting in 1985, the compulsory State monopoly farm-purchase scheme (in force since 1953) was replaced by a system of contractual procurements. Coupled with successive farm-price deregulation measures since the early 1980s, the decollectivization and relaxation of physical and bureaucratic output control returned Chinese rural conditions to those which had emerged from the 1949–52 land reform, and even to a situation which increasingly resembled that of the pre-war era, save for the emergence of large-scale private land ownership.[2]

This is not the place to discuss what long-term solution the radical rural reforms may involve in the future, but what seems certain is that increased marketization and monetization of the rural context, albeit without the privatization of landed properties *de jure*, has and may increasingly become a potential source of instability in Chinese agriculture. It is against this background that this chapter seeks to examine the relative impact of the weather in Chinese agriculture under the changing institutional and policy framework in rural China in the recent past.

We begin in the first section with an examination of the overall trend and sources of long-run agricultural instability towards the 1990s, viewed against the pattern of changes in the earlier periods. The second section gives a year-by-year analysis of the relationship between weather conditions and agricultural fluctuations. Special reference is made to the controversial year 1985, in which the dramatic decline in grain production was widely interpreted as a reflection of mistakes in farm-pricing policy (committed in conjunction with the abolition of the forced procurement system) rather than of adverse weather. An attempt is made in the third section to isolate the relative weather influence on farm yields in the late 1980s, and to examine the impact of peasants' responses to hedge against income losses on yield stability. Based on such microeconomic analysis, the last section evaluates the sources of overall grain-output fluctuations by breaking down total losses into those caused by contraction of sown area and reduction in yield per sown hectare. The exercise focuses especially on 1985 and the 'great deluge' of 1991, in comparison with other earlier episodes of great instability. This will help to shed light on the changing relative influence of the weather and policy factors in Chinese agriculture.

Trends and Sources of Long-Run Instability

Fig. 12.1 shows changing instability indexes for the various technological and policy periods from 1952 to 1990 in relation to weather instability. The three physical measures (i.e. grain-sown area, output, and yield), and the associated value measure of GVAO are identical to those which have been applied in the foregoing chapters. Clearly farm production and related economic activities (encompassed by GVAO) had stabilized quite substantially by the end of the 1980s. This is true not only for the comparison of the broad periods between 1952–66, 1970–84, and the reform decade of 1978–90, but also for the transition from the subperiods 1952–8 up to 1970–7 and the post-decollectivization years 1985–90. The sole exception is the remarkable but temporary reversal in 1978–84.[3] Improved agricultural stability was achieved against a climatic background which in the late 1980s remained as volatile as in the 1950s (see Fig. 12.1). Evidently the improvement reflects long-term technological progress in Chinese agriculture had reduced the risks of floods and drought.

The reversal in 1978–84 of the stabilizing trends for both grain-sown area and yield (and hence increased output and GVAO instability) from the Cultural Revolution period of 1970–7 is easily understandable. It underlines the dramatic institutional and policy changes (towards rural diversification) in the initial post-Mao years from the strong pro-grain strategy of the Cultural Revolution (when rural resources were massively concentrated on grain production to ward off fluctuations). Moreover, it occurred at a time when the computed weather instability index reached an all-time low of 16 per cent for the period (1978–84), compared with 36 per cent for 1952–8, 28 per cent for 1970–7, and 38 per cent for 1985–90 (see Fig. 12.1).

Once the dust of the institutional upheaval had settled, however, all the measurement standards continued to show a stabilizing trend during 1985–90 (see Fig. 12.1). Reduced agricultural instability does not, however, imply that the weather has become irrelevant. Any *residual* year-to-year fluctuations of, say, grain yields, from the long-term technological trend-line, may still be due to weather volatility.

In our previous regression analysis, in Chapter 7, we came to the conclusion that by the early post-Mao years, 1978–84, the influence of

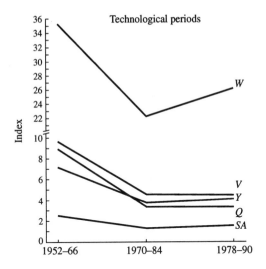

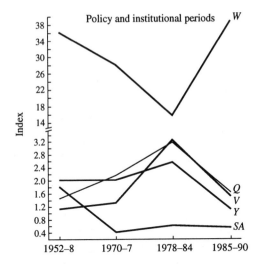

FIG. 12.1. A comparison of the instability indexes for grain-sown area (*SA*), output (*Q*), and yield (*Y*), and gross value of agricultural output (*V*), with that of the weather (*W*) between the different technological and policy periods, 1952–90

Notes: GVAO instability indexes for 1978–90, 1978–84, and 1985–90 are based on 1980 prices; those for the earlier periods are all in 1952 prices. The weather instability index is taken as the average of the yearly percentage fluctuations (ignoring the plus and minus signs) of the weighted *shouzai* area from the 1927–90 means; see Appendix chapter for detailed explanation.

Sources: Tables AB.3(*b*), AB.3(*c*), AB.12(*a*), AB.12(*b*), and AB.15.

the weather on Chinese agriculture in relation to policy had become increasingly blurred. This is because with the magnitude of farm fluctuations reduced on account of an improved technological hedge, peasants' responses to post-Mao readjustments in farm-prices and income policy (and the concomitant organizational changes) tended to dilute the relative weather influence. We also concluded that despite the policy changes, the effects of major weather events were nevertheless still clearly visible in terms of fluctuations in grain production.

We now extend the analysis to the post-decollectivization years, 1985–90, by focusing again on the weather and grain-yield relationship. As noted earlier, the choice of the yield variable, instead of grain output or sown area, may help to isolate the possible impact of policy-induced sown-area shifts and abstract from price influence. However, we also make use of the GVAO measure to examine how disruption of the farm sector proper can spill over to the farm-related rural output. Table 12.1 shows the estimated regression equations for both the broad post-Mao reform period 1978–90 and the two subperiods 1978–84 and 1985–90. The estimated results of the other periods prior to 1978 are also set out in order to lend an historical perspective against which to interpret the implications of the post-Mao changes.

A preliminary methodological remark is in order. It concerns the weather index which is used in the regression estimates. As defined earlier, the index is no more than the yearly percentage fluctuations of the weighted *shouzai* area from the long-term mean. However, unlike the previous indexes which were based on the 1952–84 mean, those used here, including those for all the periods prior to the bench-mark year 1984, are all based on the 1952–90 means. Despite this computational change, the regression equations based on the new weather indexes (as shown in Table 12.1) differ only very slightly from and in a manner consistent with the variations in the earlier estimates between the different periods (cf. Table 7.2). This makes it possible to interpret the latest regressions against the earlier findings. Several points may be made.

First, comparing the short policy-periods, the statistical significance of weather instability as an explanatory variable of grain-yield (and especially GVAO) variations has consistently declined from 1970–7 and 1978–84 and on to 1985–90. This trend underlines yet again the improved technological hedge, and it particularly highlights the point made earlier that any residual fluctuations in both physical yield and GVAO were becoming increasingly sensitive to farm-policy changes relative to the weather factor. This is likely to continue through the 1990s, given increased rural commercialization, improved income and consumption standards of peasants, and enhanced employment alternatives. All these factors have made peasants less vulnerable to policy demands than in earlier periods, especially the Cultural Revolution when draconian

TABLE 12.1. The relative scale of weather influence on agricultural stability in China, 1952–1990 (estimated regression equations)

	Intercept	Regression coefficient	t-value	r^2	t-significance
Technological periods					
1952–66 (Y)	−1.26992	−0.16728 W	3.68449	0.54672	0.003
(Va)	−0.39022	−0.13431 W	2.38565	0.32481	0.034
1970–84 (Y)	0.00023	−0.10856 W	3.58907	0.60291	0.004
(Va)	0.03438	−0.08789 W	2.18094	0.64473	0.050
1978–90 (Y)	2.84168	−0.14936 W	3.81018	0.63220	0.003
(Va)	−0.75767	0.01790 W	0.29245	0.01110	0.776
(Vb)	1.66404	−0.07689 W	1.94172	0.33330	0.081
(Vc)	1.65386	−0.07447 W	1.84336	0.31460	0.095
Policy/institutional periods					
1952–8 (Y)	−4.07200	−0.12293 W	3.62627	0.76944	0.022
(Va)	−0.26286	−0.00689 W	0.18433	0.01183	0.863
1970–7 (Y)	−1.48086	−0.06923 W	1.89661	0.4752	0.117
(Va)	−1.27465	−0.05527 W	2.80563	0.6589	0.038
1978–84 (Y)	0.01873	−0.10407 W	1.67395	0.37902	0.169
(Va)	0.01373	−0.04643 W	0.53484	0.14683	0.621
(Vb)	2.60303	−0.11038 W	1.63773	0.41250	0.177
(Vc)	2.48042	−0.10631 W	1.52748	0.38210	0.201
1985–90 (Y)	2.77093	−0.08697 W	1.19299	0.36150	0.318
(Va)	−0.96895	0.05420 W	0.38579	0.06510	0.725
(Vb)	2.27047	−0.04596 W	0.54843	0.12400	0.622
(Vc)	2.26450	−0.04792 W	0.57878	0.13160	0.603
The entire post-war context					
1952–90 (Y)	0.01102	−0.15943 W	6.98547	0.84727	<0.0001
(Va)	0.32193	−0.11465 W	1.98018	0.84385	0.056

Notes: See Table 7.2 for the definition of grain yield (Y) and GVAO (V) instability, and text for the explanatory weather variable (W). (Va) stands for GVAO gross of *cuban* (hamlet-operated) industries at 1952 prices (for periods involving years before 1978) or 1980 prices (for periods starting from 1978); and (Vb) and (Vc) GVAO net of *cuban* industries at 1980 and 'comparable' prices respectively. All regressions are corrected for first-order autocorrelation by the maximum likelihood method.

Sources: Tables AB.2, AB.12, and AB.15.

controls on occupational and rural migration left Chinese peasants wholly exposed to the government's *diktat* for maximizing grain output.

Second, the influence of weather none the less remained traceable throughout the 1980s. This is especially so if the longer reform period,

1978–90, is taken as a whole. The figures in Table 12.1 reveal that the weather and grain-yield relationship was indeed as pronounced in the post-Mao years 1978–90 as in 1970–84, when conditions were dominated by the highly stabilizing pro-grain strategy of the Cultural Revolution. Admittedly the two periods overlap for several important years, 1978–84. But the finding still seems to contradict the conclusion drawn from the comparative regressions for the short policy-periods, 1970–7, 1978–84, and 1985–90, which taken together indicate a long-run reduction in weather influence. Perhaps this is a matter of chance biases associated with detrending problems. We really cannot be sure. However, it seems appropriate to resort to the estimated regressions for the three shorter periods in interpreting the long-run changes in weather influence relative to possible policy factors. Note that the three policy periods concerned represent relatively well-defined entities, compared with the other two longer heterogeneous periods.

The third point concerns the behaviour of GVAO. It is understandable that its relationship with the weather has not been as pronounced as that of the physical grain-yield series throughout the different periods under observation. However, the relationship changes from period to period in a manner strikingly consistent with the variations in respect to grain-yield instability. It should also be noted that GVAO net of *cuban* (hamlet-operated) industries tended to vary even more closely with weather conditions compared with that gross of *cuban* industries. The contrast may not appear to be clear enough to justify any definitive judgement in this respect. One possible explanation, however, is that the bulk of GVAO proper (gross of *cuban* industries) comprises farm-related output, while *cuban* undertakings really represent small-scale family handicrafts which may not at all be farm-attached and are hence less weather-sensitive.

The Weather versus Policy Factors in the Post-Collectivization Years

Fig. 12.2 extends our analysis of the yearly weather and grain-yield relationship for 1970–84 (shown in Fig. 8.2) to the reform period of 1978–90. Note that the absolute scale for the abscissa and ordinate in the diagram is the same as that used in Fig. 8.2. This facilitates a direct visual comparison between the relative loci of the co-ordinates for the various years. During 1978–84, the weather and yield relationship in the present diagram is almost identical to those located in the previous diagram. We can therefore focus our attention on the years since 1985.

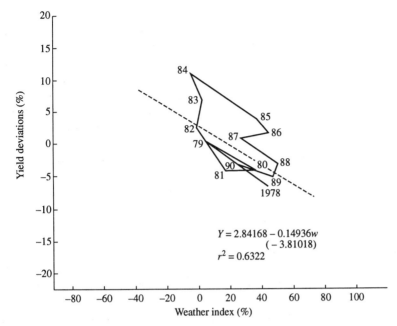

Fɪɢ. 12.2. The instability of annual average grain yield per sown hectare (expressed as percentage deviations from the long-linear trend value) in relation to the size of *shouzai* area (expressed as percentage deviations from the 1952–90 mean) for China, 1978–90

Source: Table 12.1.

The year 1985 merits detailed consideration. Fig. 12.2 shows a sharp downturn in grain yield per sown hectare, relative to the trend value, from 1984 (10.93 per cent above trend) to 1985 (only 3.57 per cent above trend). This was associated with a remarkable deterioration of weather conditions, in terms of *shouzai* area, from minus 4.8 per cent to plus 37.87 per cent from the 1952–90 mean. The sharp decline in 1985 in grain yield and output has attracted much attention in both China and the West, the prevalent view being to attribute the losses to policy rather than weather factors. The former include, in particular, the so-called *daosanqi* farm-pricing policy which was introduced in early 1985.

The new pricing policy stipulated that following the abolition of two-tier grain quotas and above-quotas price system, any grain sold to State procurement agencies under the new contractual system would fetch a single price which was a weighted average of the standard quota and the higher above-quota prices, respectively embracing 30 per cent (*san*) and 70 per cent (*qi*) of total purchases. This new arrangement thereby reversed (*dao*) the previous practice whereby the bulk of quota deliveries was priced at the standard price and any above-quota sales attracted an attractive price premium.[4]

The new 'proportionate price' formula on the surface suggested increased price and income benefits for peasants. But in reality it resulted in lowered average procurement prices for the country as a whole in 1985 compared with earlier years.[5] The main reason is that the State no longer undertook to buy up all surplus output; and with reductions made in State procurements in 1985 (in response to the mounting pressure on grain-storage facilities resulting from excessive procurements in the wake of the extraordinary harvest of 1984), many grain-rich peasants, who had earlier reaped substantial price premiums for above-quota sales, now saw their *average* price reduced under the new scheme.[6] This threatened to reduce peasant income and impair input incentives.

The situation in 1985 was also aggravated by successive increases since 1984 in the prices of such essential farm inputs as chemical fertilizers, pesticides, farm implements, and machine and diesel oil.[7] Moreover, the agricultural tax reform, also introduced in 1985, to substitute cash levies for grain tax in kind based on the *daosanqi* price, served to increase the peasants' burden and in many places wholly to offset the benefits obtained from increases in grain-procurement prices between 1980 and 1984.[8] Thus, 'under the double pressures of direct and indirect agricultural taxes, peasants were excessively burdened and not willing to grow grain'.[9]

Meanwhile, State procurement prices for cash crops remained higher than for grain, and the liberalization of, and hence increases in farm subsidiary output prices in 1985, also helped to reinforce peasants' inclination to move away from grain to non-grain, and from farm to non-farm engagements.[10]

More generally, it has been argued that accelerating increases in the general price level during 1985 (as a result of the dramatic monetary expansion by the end of 1984) also rendered the new grain contract prices ineffective as a 'support' or 'reserved' price for peasants.[11] All these policy factors clearly point towards reduced peasant incentives as the source of the remarkable national losses in grain production in 1985.

There is yet another, probably more important factor. This concerns inter-provincial grain transfers. The *daosanqi* price formula worked strongly against the interests of grain-surplus provinces in favour of deficit provinces, in the same manner as it discriminated against grain-rich peasants. For Hunan province, for example, any transfer of grain to Guangdong (a persistently grain-deficit province) under the new pro-portionate price implied continuous and enhanced 'subsidies'. Thus, while Guangdong could rely on cheap grain supplies and thereby free resources for employment in lucrative foreign trade and investment business, land-locked Hunan had no opportunity to share its neighbour's precious foreign-exchange earnings. This resulted in renewed attempts by the provincial authorities in Hunan to blockade grain transfers, and so reduce output. This is likely to have had a much greater impact on national grain output than similar sporadic action taken by individual peasant households.[12]

Chinese observers themselves have also cited weather conditions in 1985 as an important factor accounting for the remarkable downturn in grain production. In general, however, the weather factor has been considered no more than a supplementary contributory factor, compared with the negative impact of policy and institutional changes.[13]

The complexities surrounding the dramatic decline in grain output in 1985 really warrant a separate study. In relation to our present analysis, however, it is important that a distinction be made between the possible influence of the policy factors in terms of the total planned sown area on the one hand, and the application of current inputs to the effectively sown area on the other. By the time of sowing the 1985 crops the terms and conditions of State procurements under the new contractual system were already clear. Peasants should have known how to adjust the planting acreage in accordance with their benefit and cost calculations. There is no doubt that the various negative policy measures caused widespread reductions in grain-sown area and hence overall output in 1985 (see below). It is less clear, however, whether similar factors affected current inputs to sown areas and hence grain yield per sown hectare. None the less, from Fig. 12.2 it seems undeniable that the weather was deeply implicated in the remarkable downturn in national average grain yield in

1985. As two specialists from the research office of the Party Secretary-General's headquarters put it:

Starting from late spring (in 1985), both North and South (China) were covered by low temperature and drizzling climate, with various types of plant disease widespread, such as wheat rust and rice blast. Especially after late June, drought prevailed over the entire South region, with summer drought covering a total of more than 200 million *mu* (13.33 million ha.). Thus, early rice crops were overwhelmed by high temperature and late rice crops were also affected. (On the other hand), Liaoning, Jilin, and Heilongjiang provinces (in the North), and part of Shandong, Zhejiang, and Shanghai suffered from continuous rainfall. The area affected by floods and rainstorms was more than 85 million *mu* (5.67 million ha.). Among these the three North-eastern provinces accounted for more than 66 million *mu* (4.4 million ha.). Grain production suffered quite seriously.[14]

This description should not of course be taken as conclusive evidence of the weather impact. But barring any other plausible interpretation the weather was probably more important than anything else in causing the dramatic reduction in grain yield in 1985.

The same holds true for the year 1986, for which Fig. 12.2 reveals a parallel deterioration in weather conditions and grain yield per sown hectare.[15] Interestingly, despite continuous increases in State grain-procurement prices in absolute terms, and relative to farm input prices as well, in 1987 and at least through 1988[16] yield per sown hectare did not show any signs of recovery. This is again basically consistent with known weather conditions, however (Fig. 12.2). Only in 1990 did a powerful recovery occur.[17] In the words of Li Peng, the Chinese premier, the bumper harvest of 1990 (which significantly surpassed the all-time record of 1984 in terms of both total grain output and yield per sown hectare) was 'to a very large extent due to "harmonious wind and smooth rain" (*fengtiao yushun*) rather than anything else'.[18]

Relative Weather Influence:
A Microeconomic Viewpoint

Fig. 12.3 shows to what extent grain yield losses (or gains) in relation to the trend value for each year from 1978 to 1990 may be accounted for by the weather. For 1980, for example, 64 per cent of the average loss of 118 kilograms per sown hectare for the country as a whole was due to adverse weather. By contrast, however, in 1985, while according to the weather conditions there should have been an average loss of 95 kilograms per sown hectare, actual yield was nevertheless 120 kilograms above the trend.

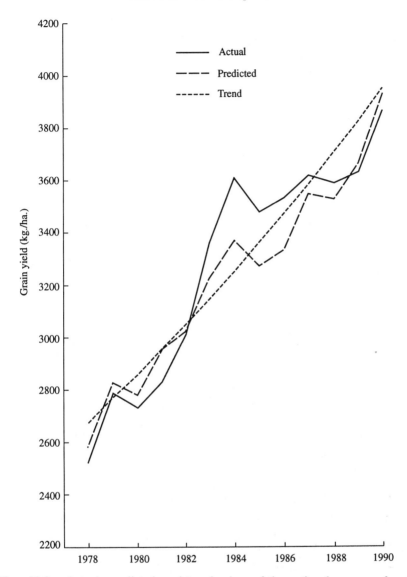

FIG. 12.3. Actual, predicted, and trend values of the national average of grain yield in China, 1978–90 (in kg./ha.)

Source: Table 12.1.

The estimates are all based on the regression equation shown in Table 12.1 for 1978–90. The comparable regression estimates for the shorter policy periods, 1978–84 and 1985–90, are not used, but are incorporated into the longer trend, 1978–90, so that the comparative differences in peasant incentives during the two periods may more readily emerge in relation to the weather influence. None the less, the usual reservations associated with such highly aggregative measurements should be borne in mind for purposes of interpretation.

In general terms, the changing relationships between the actual and weather-predicted yields, relative to the trend-line, as shown in Fig. 12.3, seem generally to fit quite well with what one would expect from the likely impact of the radical rural institutional and policy changes occurring in 1978–90. Several interesting points emerge.

Let us first compare the two bumper harvest years of 1979 and 1984. In 1979 the actual yield gain (by 19 kilograms per sown hectare above the trend) was nearly three and a half times lower than the weather-predicted gain of 63 kilograms per sown hectare. Exactly the reverse was true, however, in 1984 when the respective figures were 356 and 116 kilograms per sown hectare. This seems to suggest that in the early post-Mao years (when the collective-distributive structure was still intact) peasants' incentives did not have any marked effect on output, despite the substantial price benefits offered in 1979.[19] However, effective decollectivization starting early in 1984 prompted them to take advantage of good weather conditions in that year. Indeed, and this was happening in a year which was the worst in terms of relative price incentives offered by the government to the peasants.[20]

The same holds for the year 1983, but curiously not for 1982, although in terms of both price incentives (i.e. the lack of such incentives) and weather conditions, both years were comparable with 1984. The likely explanation is that decollectivization had already started to gain momentum in 1983, but the official pronouncement of the demise of the people's commune system was not made until the promulgation of the new PRC Constitution in December 1982.

Turning to the post-decollectivization years, it is interesting that as in 1984, 1985 saw grain yield per sown hectare still in excess of the weather-predicted yield, despite the remarkable downturn. The same goes for 1986, 1987, and 1988, although these years were admittedly also associated with increased price incentives which seem to have been deliberately designed to reverse the declining output trend since 1985.[21]

Granted that the offered price benefits from the mid-1980s were comparable to those of the early 1980s, that the actual grain yield per sown hectare should have fared better than the weather-predicted yield in the post-decollectivization years is probably best explained in terms of peasants' incentives. That is, a positive incentive impact on input utilization—and so sown area yields—could more easily be realized in a decentralized, independent household farming regime, rather than through improvements in farm output price in an unaltered collective-distributive framework.

Interestingly, in both 1989 and 1990, when the weather showed definite signs of improvement (see Fig. 12.2), especially in the latter year, the actual grain yield per sown hectare fared worse than the weather-predicted one (see Fig. 12.3). This occurred side-by-side with a deterioration in relative price benefits offered to peasants,[22] coupled with renewed centralized physical control of farm output and input allocation after 1989, which tended to return basic rural organizational parameters to those of the pre-decollectivization era.[23]

None the less, the parallel movement between actual and weather-predicted grain yield per sown hectare, as shown in Fig. 12.3, clearly shows that the weather has remained a factor to be reckoned with. The weather impact cannot be more expressively brought to light than by the way it bears on peasants' income. In a separate study we have shown that peasants' net income (i.e. gross sales receipts for farm produce and non-farm output minus costs of material inputs) during the period 1978–88, was more closely correlated with weather conditions (as defined in this study), than with State farm procurement and farm-input prices.[24]

This finding is both interesting and important in relation to our analysis of the economic role of the weather in Chinese agriculture in the post-collectivization era. Several points emerge.

First, given the overriding concern of peasants to maximize income, it is likely that, with input and output prices assumed to be fixed, they would attempt where possible to increase the output volume, by expanding sown acreage or raising yield per sown hectare, subject to the familiar law of diminishing returns, i.e. increasing costs, in order to equalize marginal cost with marginal revenue. Once sown area is determined, it may be supposed that income will vary with physical yield per sown hectare, leaving the weather as the only possible culprit for income losses in such circumstances. This simple model helps to capture the basic elements of the observed weather and income relationship.

Second, if the familiar 'abnormal agricultural supply function' (i.e. expanded output compensating for falling output prices in order to maximize or maintain a desired level of farm income) holds true, then it can be argued that with a deterioration in weather conditions more intensive efforts will be made by peasants to hedge against possible yield losses. If this is the case, then the role of the weather would seem to be even more powerful in accounting for residual variations in peasant income.

Third, in the wake of adverse changes in prices, i.e. increases for farm inputs, reductions for output, similar compensatory manipulations are likely to be followed to maximize income by way of expanding sown acreage, raising yield per sown hectare by increasing inputs, or economizing on input use in order to reduce costs. This is similar to the built-in income incentive to hedge against the weather impact. In these circumstances, any observed variations in farm yield, and hence peasant income, may be wholly attributable to the weather.

Fourth, adverse changes in farm input and output prices for, say, grain crops, will inevitably reduce, *ceteris paribus*, their relative profitability as against that of cash crops (or non-farm activities). This is very relevant to the policy of grain price (*daosanqi*) reform of 1985, which presumably resulted in rural resources being diverted from the less lucrative grain sector. There is no doubt that with increased rural economic diversification providing an increase in competitive employment opportunities, consequent price changes were increasingly intertwined with the weather factor as an important source of instability after the mid-1980s. However, apart from the radical possibility that the price-induced input curtailment could lead to a total abandonment of the area sown, such a price effect seems to have been comparatively limited. Moreover, the peasants' need for income may also have helped to lessen or balance potential yield losses arising from marginal input readjustments.

Fifth, a particularly important factor is reduction in total sown area prior to scheduled sowing, because hectare for hectare this will have significantly greater effect on overall output than any marginal curtailment in yield per hectare sown due to current input reductions. This seems to have accounted for a considerable part of the grain output decline in 1985 (see next section). However, any sown-area reallocation in favour of more profitable economic crops may well help to balance or overcompensate for the resultant losses. Viewed in this way, fluctuations in peasants' income (which are essentially a function of farm and farm-related output including the flourishing rural industries) may again be explained by the weather more than anything else, as our regression estimate for 1978–88 strongly suggests (see footnote 24).

Finally, a word or two should be said about the regression result that peasants' net income is positively correlated with increases in farm-input prices, but negatively with farm-procurement prices. This unexpected relationship clearly helps to underline the point just made, for any potential gains (losses) in output and income via procurement price incentives, or disincentives in the case of increased input prices may indeed be outweighed by the weather effect in certain circumstances. For example, a positive output response in the face of higher procurement prices in a particular year may be overturned by the exigencies of climate.

1985 and the 'Great Deluge' of 1991 Compared

We have hitherto attempted to isolate the possible impact of peasant behaviour in relation to policy changes from the weather factor in terms of fluctuations in grain yield per sown hectare or peasant income. We now examine how, in aggregate terms, annual fluctuations in total grain output may be broken down into changes in yield per sown hectare (subject to peasant behaviour *vis-à-vis* weather conditions) and pre-sowing sown-area readjustments (made largely as a response to price changes). The break-down may help more exactly to highlight the weather impact on overall Chinese agricultural performance relative to policy influence. We concentrate again on the controversial year 1985 but special reference is also made to 1991, when catastrophic summer floods in East and Central China caught the attention of the world media.

Table 12.2 shows that in 1985 out of an overall reduction of 6.93 per cent in grain output for the country as a whole, virtually half may be attributed to losses in yield per sown hectare. The other half appears to have been brought about by losses from sown-area curtailment. The reduction in sown area in 1985 was indeed the largest, in both absolute and relative terms, since the losses incurred in 1959 in conjunction with the 'three-three system'. It also amounted to three and a half times the absolute hectare losses of 1991 (Table 12.2). While the sown-area losses in 1991 appear to have been mainly brought about by the disastrous floods (which occurred during the cropping transition from the summer harvests to sowing for the autumn crops, rendering wide stretches of farmland unsowable for the rest of the year), in 1985 there seems to have been no particular weather event which could have had a similar impact on the sown-area size.[25]

Rather it seems that the drastic sown-area reductions in 1985 were wholly attributable to the *daosanqi* price formula. This reflects the difficulties which State procurement agencies encountered late in 1984 in purchasing from the peasants and stockpiling the exceptional harvests due to limited warehouse and financial capabilities. At any rate, the difficulties the peasants encountered in selling grain (*mailiangynan*) certainly impaired their incentives for 1985 sowing.

None the less, even without the sown-area contraction, the reduction in yield per sown hectare alone by 128 kilograms (3.55 per cent) from the 1984 level would have reduced the 1985 output by 3.4 per cent. It is not possible to make an exact estimate of the relative contributions of weather and policy factors to the overall reduction, but there is no doubt that a considerable part of the loss was caused by adverse weather (see

TABLE 12.2.   Losses in grain output due to reductions in sown area and yield per sown hectare in China in years of great instability, 1931–1991

	Losses/reductions in						Shares in output losses by reductions in	
	Output		Sown area		Yield			
	(million tonnes)	(%)	(million hectares)	(%)	(kg./ha.)	(%)	Sown area (%)	Yield (%)
1934	9.14	8.45	0.14	0.21	129	8.25	2.56	97.46
1937	13.15	11.84	2.88	4.12	128	8.06	34.80	65.28
1959	30.00	15.00	11.59	9.08	105	6.70	60.58	40.61
1960	26.50	15.59	(6.41)	(5.52)	293	20.03	(35.40)	135.36
1972	9.66	3.86	(0.36)	(0.30)	86	4.06	(7.71)	107.90
1980	11.56	3.48	2.03	1.70	51	1.83	48.91	51.73
1985	28.20	6.93	4.04	3.58	128	3.55	51.65	49.40
1991	10.95	2.45	1.15	1.02	57	1.45	41.40	58.60

Notes: Absolute losses (or gains in the case of the bracketed figures) are measured against the levels of the preceding year, rather than the 'trend variables'. The reason for this is twofold. First, almost all the 'preceding years' concerned can be considered as normal or good-weather years. This may help to capture the real magnitude of year-to-year fluctuations. The year 1959 appears to be problematic as a basis for measuring the losses in 1960. But the grain yield (1,463 kg.) per sown hectare in 1959 was nevertheless the same as in 1957, which is generally considered to have been a normal year. Second, subject to the periods chosen, the actual yield or output for some bad years (see e.g. Fig. 12.3 for 1985) may turn out to be higher than the estimated trend output or yield, although admittedly when compared with good years in the same trend (cf. e.g. 1984), the discrepancy may be substantially narrowed. For ease of illustration, we resort to the method used here.

The percentage shares of reductions in sown area (*As*) and yield (*Ys*) in total output loss are derived as per the formulae:

$$As = Y_{t-1}(A_{t-1} - A_t)/(Q_{t-1} - Q_t) \times 100, \text{ and}$$

$$Ys = (Y_{t-1} - Y_t)A_t/(Q_{t-1} - Q_t) \times 100,$$

where *A*, *Y*, and *Q* stand respectively for sown area, yield per sown hectare, and output, and, the year concerned.

Sources: Table AB.2 and *TJNJ 1992*, 352, 358 and 365.

Fig. 12.2). However, judging by the extremely high absolute yield loss of 128 kilograms per sown hectare in 1985 (compared with 57 kilograms in 1991, 51 in 1980, and 86 in 1972—all very bad weather years), it seems certain that the various negative policy measures described earlier also led the peasants to curtail farm inputs *after* sowing, thereby seriously aggravating the yield losses.

As a rough measure, if we assume that the weather and policy factors had an equal share in this respect, then the extent of weather-caused yield loss (i.e. about 1.8 per cent) in 1985 is basically in line with that of, say, 1980 (1.83 per cent) and 1991 (1.45 per cent) (see Table 12.2). Note that unlike 1985, both 1980 and 1991 saw no major change in agricultural policy that could have upset the balance of the influence of the weather.

If this analysis is correct, then in 1985 *only* one-quarter of the large output decline of 6.93 per cent (from the 1984 record) can be attributable to the weather. This is a significant finding. It implies that the bench-mark changes in rural institutional arrangements in 1985 (abolition of compulsory delivery quotas, coupled with the *daosanqi* price-reforms) brought about for the first time major grain-output losses (i.e. 28 million tonnes) comparable with those of 1959 (30 million tonnes), or 1960 (26.5 million tonnes) in absolute terms; or nearly half (6.93 per cent) of the percentage losses for 1959 (15.00 per cent from 1958), or 1960 (15.59 per cent from 1959). This was by any standard a disastrous loss compared with those of 1991, 1980, or for that matter 1972 (see Table 12.2).

In other words, misconceived agricultural policy within an increasingly decentralized rural context characterized by centralized price signals can readily precipitate a disaster (as in the case of 1985) comparable in nature (though not necessarily in scale) to that occurring under a highly centralized system with physical-bureaucratic control.

Setting aside institutional and policy influences, the role of weather conditions has emerged as a clear determinant of long-run agricultural instability in China. From the vantage-point of the early 1990s, the relative influence of the climate was reduced quite remarkably in the 1980s compared with earlier decades, although in absolute tonnage, output losses remained substantial. As shown in Table 12.2, output losses in the three disastrous years, 1980, 1985 (discounting possible man-made factors), and 1991 were of the order of 10 million tonnes, with the percentage losses of around 3.5 per cent in 1980, 2.0 per cent in 1985 (discounting again the non-weather factor) and 2.5 per cent in 1991. This spectrum of changes seems to be perfectly consistent with the higher losses in 1972 (shown in Table 12.2) in yield (by 4.06 per cent) and output (by 3.9 per cent or even higher if compensation by sown-area increases is discounted).

The events of 1991 merit further comment. In terms of our weighted *shouzai* area, it ranks next to the catastrophic years of 1960 and 1961.[26] Yet the losses in 1991 in terms of grain output (by 2.45 per cent from 1990) and yield (by 1.45 per cent) proved to be far less than the disastrous figures for 1960 (15.59 and 20.03 per cent respectively) (see Table 12.2). Note especially that if the compensatory output gains from sown-area expansion (or rather recovery by 35.40 per cent) are discounted, the output loss rate for 1960 is even higher. Special account has of course to be taken of the peculiar situation in 1960, when with part of the seedgrain being eaten up due to dramatic shortfall in food supplies, sowing density and hence yield per sown hectare were reduced. None the less, even discounting this, the 'net yield losses' which may be attributable to the weather would still have proved to be almost insurmountable in 1960.

A comparison with the year 1959 is equally revealing in this respect. Grain output loss in 1959 was 15 per cent, and yield loss per sown hectare 6.7 per cent from the 1958 level. Yield losses alone (which accounted for 40 per cent of total output loss) would have reduced the 1959 grain output by 6 per cent (see Table 12.2). This is markedly higher than the comparable loss rates for 1991. Note too that the implied weather-induced yield loss in 1959 probably represents a minimum, by virtue of the fact that so much effort was made by peasants to hedge against the negative weather impact as a matter of self-survival in the wake of the unwarranted sown-area contraction in that year. Note also that the same enforced area contraction accounted for 60 per cent of total output loss, which is higher than during any other episode of major instability (Table 12.2).

A similar comparison can be made between 1991 and the pre-war years 1934 and 1937. Grain-yield losses in these two years of bad weather were consistently higher than in later years (1960 excepted for the reason just explained). The losses would indeed be higher were it not for the pre-war statistical practice of discounting sown areas which suffered a full loss in its harvest statistics.

Some General Observations

There is no doubt that as a result of improvement in flood and drought control technology during the past several decades, agricultural production was increasingly and substantially stabilized from the 1930s to the 1980s. Against this background, the 'great deluge' of 1991 represents a touchstone of the way in which the weather can impinge upon present-day Chinese agriculture. What is significant is that the relative losses of both grain yield and output were the lowest ever when compared with earlier episodes of great instability.

None the less, the case of 1985 clearly shows that policy miscalculations can still cause substantial upsets, especially through sown area and hence output fluctuations. Perhaps the 1985 episode should be regarded as a rare exception, since it marked the radical transition from a collective to a market-orientated system in Chinese agriculture. Inexperience in manipulation of relative prices under the new system seems to have been the main reason for the large-scale grain-sown area displacement and unexpected output reduction in that particular year. With improved experience, the scale of such possible readjustments may be expected to decline and become marginal in the future. This seems to have been increasingly the case since 1985.

NOTES

1. For an elaboration of the implications of Document No. 1 see Y. Y. Kueh, 'The economics of the "Second Land Reform" in China', *CQ* 101 (Mar. 1985), 122–31.
2. See Robert Ash, 'Agricultural policy under the impact of reform', and Terry Sicular, 'Progress and setbacks in agricultural planning and pricing', both in Y. Y. Kueh and Robert F. Ash (eds.), *Economic Trends in Chinese Agriculture* (Oxford: Clarendon Press, 1993), for a detailed examination of the changes made since 1985.
3. Fig. 12.1 reveals that the instability indexes for both grain yield and output in 1952–8 were relatively small compared with that of the other later periods. This does not imply, however, comparative agricultural stability in the 1950s, but is rather the result of the *volatile* 'yield-for-area' and 'area-for-yield' substitution processes, explained in Ch. 3.
4. Cf. Sicular, 'Progress and setbacks in agricultural planning and pricing', 68.
5. See Li Bingkun, 'Liangshi duice: xunqiu he zhu ru duozhong ciji yinsu' (Policies on grain production: searching for and injecting various incentive factors), in *NCJJ* 5 (1986), 3.
6. For a practical numerical example from Hunan province see Li Zhiqiang, 'Guanyu woguo liang mian shengchan pai hui buqian de zhengce yinsu fenxi' (An analysis of the policy factors accounting for the stagnation in grain and cotton production in China), in *NCJJ* 4 (1988), 9. See also Liu Yunqian, 'Tantan fazhan liangshi shengchan he liutong wenti—dui fangkai liangshi jiagede yixie kanfa' (Issues on promoting grain production and circulation: some views on freeing grain prices), in *NYJJWT* 6 (1986), 23 for a similar discussion.
7. This was widely discussed with practical examples, see e.g. Zhao Changfu and Ming Haijun, 'Liangshi shengchan yu jiazhi guilu' (Grain production and the law of value), in *NYJSJJ* 4 (1986), 39; *NYJJWT* 2 (1986), 5; and *NCJJ* 5 (1986), 3.
8. Peng Chuanbin, 'Jiage biandong dui liangshi shengchan de ying xiang yu duice' (The effect of price changes on grain production and our counter policy measures), in *NCJJ* 6 (1986), 21 gives a practical example from Hubei province.
9. Ibid.
10. Liu Yunqian, 'Issues on promoting grain production', 23.
11. Chen Jian, 'Lijie he bawo muqian zhongguo nongcun xingshi de suoshi shi shenme' (What is the key to understanding the current rural situation in China?), in *NYJJWT* 8 (1987), 23.

12. Many Chinese authors have dealt with the problems of inter-provincial transfer, if only implicitly. See e.g. He Kang (then Minister of Agriculture), 'Jianchi shenru gaige, cujin nongye wending xietiao fazhan' (Carrying on the reform to promote agricultural development in a steady and coordinated way), in *NYJSJJ* 2 (1986), 2; Zhongguo nongye kexueyuan liangjing zu, '1986 nian woguo nongye shengchan cuishi fenxi' (An analysis of the trends in agricultural production in China in 1986), in *NYJSJJ* 3 (1986), 11; and Duan Yingbi, 'Liangshi liutong tizhi bixu da gaige' (The grain circulation system must be vigorously reformed), in *NYJJWT* 11 (1986), 39.

13. See e.g. *NYJJWT* 1 (1986), 17, and 2 (1986), 5; and *NYJSJJ* 3 (1986), 9.

14. Wu Xiang and Lu Wenqiang, 'Liangshi wenti mianmian guan' (An all-round look at the grain problem), in *NYJJWT* 1 (1986), 17.

15. Total *shouzai* area in 1986 was indeed larger than in 1985; see Table AB.15 and *NYJSJJ* 8 (1986), 3 for an elaboration about the major areas affected.

16. See Kueh, 'Food consumption and peasant incomes', in Kueh and Ash (eds.), *Economic Trends in Chinese Agriculture*, 237–9.

17. Grain output and yield per sown hectare in 1990 increased respectively by 6.7(9.5) and 5.6(8.3)%. The bracketed figures are revised SSB statistics which were not published until mid-1992, and could not therefore be incorporated in time into our regression analysis and Fig. 12.2

18. *DGBHK*, 2 Dec. 1990.

19. For an elaboration on this point, see Kueh, 'China's new agricultural-policy programme: major economic consequences, 1979–1983', in *Journal of Comparative Economics*, 8/4 (1984), 358.

20. State retail prices of farm inputs in 1984 increased by 8.9%, but State farm procurement prices by only 4.1%; see Kueh, 'Food consumption and peasant incomes', 237.

21. Ibid.

22. Average farm procurement prices increased by 15% in 1989, but farm input prices by 19%. With the respective changes in 1990 being a negative 2.6% and a positive 5.5%, the terms of trade have further moved against the peasants.

23. Kueh, 'The state of the economy and economic reform', in Kuan Hsin-chi and Maurice Brosseau (eds.), *China Review* (Hong Kong: Chinese University Press, 1991), 10.20. It should of course be noted that by virtue of the estimates, the residuals formed by the differences between the actual and the predicted values, which may be positive or negative, sum to zero in theory. So there will necessarily be 'above-prediction' years and 'below-prediction' years. Given that understanding, the exercise is still meaningful, in that explanations are sought for circumstances which led to more or less yield than that predicted by weather alone. The findings for 1989 and 1990 when, even with relatively good weather the actual yield was below the predicted one, should be viewed from this perspective.

24. The estimated regression equation is

$$Y = 4.7375 - 0.5306P_1 + 0.2331P_2 - 0.1507W$$
$$(1.991) \quad (-2.467) \quad (0.2755) \quad (-1.682)$$
$$r^2 = 0.73 \text{ (bracketed figures are } t\text{-values)},$$

where *Y* stands for the detrended yearly percentage fluctuations in peasant net income per head; P_1 and P_2 the yearly percentage changes in farm-procurement prices and retail prices of farm inputs respectively; and *W* the weather index (similarly defined as in this study). See Kueh, 'Food consumption and peasant incomes', 269 for details.

25. For a detailed, though preliminary account of the 1991 flood, see Cheng Yuk-shing, 'The economic impact of the summer floods', in Kuan Hsin-chi and Maurice Brosseau, *China Review 1992* (Hong Kong: Chinese University Press, 1992).
26. Cf. Table AB.15 and *TJNJ 1992*, 385.

13

Conclusion

This volume is a study of long-run agricultural instability in China from the 1930s to the early 1990s. Centring on grain production (which is the most important sector of Chinese agriculture), a rigorous attempt has been made to examine the extent of weather influence in relation to the impact of technological and institutional changes over the past six decades.

The annual extent of sown area affected by various types of natural calamity is taken as a proxy of climatic disturbance for purposes of analysis. This is supplemented by detailed precipitation statistics for selected periods, notably the 1930s and the Great Leap Forward of 1959–61. The derived 'weather index' is related to fluctuations in grain production (net of the technological trend) in order to determine the possible degree of correlation.

The analysis is carried out separately for periods of different length which are demarcated and defined in terms of technological and institutional parameters, as well as patterns of policy variations, as these may have affected Chinese agricultural stability. A comparison between the various periods shows how technological progress in Chinese agriculture has helped to lessen the weather impact and highlights the way in which variations in rural institutional and policy arrangements from period to period have influenced peasants' behaviour relative to the weather influence.

In addition to inter-period comparisons, designed to display the *long-run* trends and sources of agricultural instability, an attempt has also been made to reveal the pattern and sources of year-to-year variations in grain production in the different periods under study. This helps to determine more exactly the relative scale of weather influence under given technological and institutional settings. By contrasting, for example, good weather with bad policy, or bad weather with good policy (in terms of price and income incentives for peasants to hedge against weather exigencies), it is possible to reveal the relative strength or weakness of weather influence compared with man-made factors.

Our weather analysis is also extended beyond grain production to embrace the gross value of agricultural output (GVAO). This includes not only the cropping sector (grain and cash crops), animal husbandry, forestry, and fisheries, but more importantly rural industry which relies largely on the farm sector for input supplies. As such, the GVAO measure may help to show how climatic disturbance in the farm sector spills over into the entire rural economy.

The major findings of the study may be summarized as follows:

1. From the period 1930–7 to 1952–8, grain production stabilized remarkably in terms of the average magnitude of yearly fluctuations. Thus, contrary to popular perceptions among Western scholars, agricultural collectivization in the 1950s helped to stabilize rather than destabilize farm output, in the absence of any major technological progress. Such stabilization affected both sown area and yield per sown hectare, and hence total output. This reflects the fact that collectivization, coupled with increased restrictions on peasant migration and occupational mobility left no alternatives for Chinese peasants other than to hedge against potential adverse weather conditions as a matter of self-survival. The situation in the 1950s was therefore very different from that of the 1930s when in the wake of large-scale floods or drought, peasants often abandoned their fields, sown or yet to be sown, and joined the bandwagon of rural refugees.

2. Grain production, especially in terms of sown-area input, was further stabilized from 1952–8 to 1970–7. This reflected the combined impact of technological improvements in flood and drought control, and the enhanced institutional hedge provided by the Cultural Revolution, when collectivist controls and mass rural labour mobilization were carried to extremes in the fight against natural disasters, notably the great North China drought of 1972. Moreover, the extremely prograin strategy of the time (epitomized in the slogan 'take grain as the key link'), as well as the drive for regional grain self-sufficiency, demanded massive concentration of rural manpower and material resources to ensure stability.

3. The post-Mao policy adjustment of the prograin bias in the early 1980s resulted in dramatic cropping shifts in favour of cash crops, and thus helped reverse in 1978–84 the declining instability in the grain-sown area, and so overall output. The destabilizing trend was underlined by rural decontrol which enabled peasants increasingly to take advantage of available price benefits by expanding non-grain acreage. It was also complicated by the disastrous Yangzi floods in 1980 and Sichuan floods in 1981. Both floods seriously affected rice output, but were followed by dramatic improvements in the weather in subsequent years, until the all-time harvest record of 1984.

4. The reversal in the stability trend during 1978–84 was, however, swiftly arrested in 1985–90. The latter period was characterized by the effective decollectivization of rural China, but it saw none the less continuous stabilization in Chinese agriculture by virtually all the measure-

ment standards adopted, i.e. grain-sown area, yield, and output, as well as GVAO. More remarkably, this process of stabilization took place amidst increased weather instability, compared with the preceding period 1978–84. This suggests that agricultural development in China had already reached the stage whereby flood and drought control technology tended increasingly to replace the institutional hedge in limiting the magnitude of agricultural fluctuations and losses.

5. The improved technological hedge as an agricultural stabilizer is even more evident in the comparative degress of agricultural instability for the broader technological periods 1952–66, 1970–84, and 1978–90. Set against the background of 1931–7, these periods present a consistent spectrum of reduced instability. The 'great deluge' of 1991 is a real touchstone in this respect. Grain-sown area and yield per sown hectare were reduced by only one and 1.5 per cent in 1991, resulting in a marginal loss of 2.5 per cent in total grain output. This is a small fraction of weather-induced losses during earlier episodes of great instability, such as in 1959–60, 1934 or 1937.

6. The long-run decrease in agricultural instability implies a reduced scale of weather influence. This is confirmed by regression estimates of the weather and the grain-yield relationship for the various periods under study. The yield series is particularly apt for such analysis as it allows abstraction from price influence and policy-induced sown-area fluctuations. However, regressions on the broader GVAO measure also consistently show a similar, though a less coherent relationship, reflecting the less weather-sensitive nature of the non-farm sectors. Thus, both grain production and the rural economic growth in general have been subject to diminishing weather influence over recent decades.

7. The *residual* fluctuations in grain yield (i.e. left over from the institutional and technological hedge) can none the less still be identified with known weather events on a yearly basis. This holds true not only for the Cultural Revolution period 1970–7 (when the highly stabilizing pro-grain strategy, coupled with bureaucratic and physical target control, helped to minimize any possible policy interference), but also for the 1980s (when the radical reorientation of agricultural policy, together with rural decollectivization, strongly influenced peasants' input and output decisions to 'dilute' the weather influence). For example the Yangzi (1980) and Sichuan (1981) floods obviously overwhelmed the peasants' growing income incentives (as offered by successive increases in farm procurement prices from 1979 onward) to result in a drastic downturn in grain yield per sown hectare during the two years concerned. Similarly for 1985, which, despite possible negative policy implications (see below), displayed clear parallels in a substantial decline in grain yield and a drastic deterioration in the weather conditions. Likewise, in the absence

of any major recovery in peasant incentives, yield in 1990 (again in terms of deviation from the trend value) recovered strongly, as did weather conditions, from the record low of 1989.

8. Major policy readjustments, rare as they have been, at times caused large-scale disturbance in Chinese agriculture, independent of the weather impact. Thus in 1985 the abolition of compulsory quotas and acreage targets, coupled with the misconceived 'proportionate price' (*daosanqi*) reform, was mainly accountable for the 3.6 per cent loss in grain-sown area. This alone made up more than half (i.e. 51.7 per cent) of the 6.9 per cent decline in total grain output, an all-time record loss since the early 1960s. Adverse changes in price incentives in 1985 also led to substantial diversion of input resources to non-farm economic activities, aggravating the adverse weather impact on grain yield per sown hectare (loss by 3.6 per cent). On balance, therefore, the large output losses in 1985 were much more a matter of policy misconception than of weather exigencies.

9. The only other episode similar to that of 1985 in the post-war history of Chinese agricultural development is the 1959–60 saga. Grain-sown area was curtailed by nearly 10 per cent in 1959, wholly as a result of the 'three-three system'. This accounted for 61 per cent of the total 15 per cent loss in grain output. Grain yield per sown hectare was reduced as a result of bad weather by 6.7 per cent to make up the 39 per cent balance of losses. Unlike 1985, however, the per hectare loss in 1959 probably represents a minimum due exclusively to the weather rather than to policy. This is because, in the wake of the then imminent food crisis due to an unwarranted sown-area contraction (in the 1958–9 winter and 1959 spring sowing), peasant initiative was strongly directed towards protecting the fields sown from the adverse weather, notably the summer drought. The self-survival urge came on top of the institutional hedge, and certainly helped to minimize the per hectare loss rate in 1959. Note that the 6.7 per cent loss was still lower than that of 8.3 per cent sustained in 1934, when, as in 1959, virtually the whole of East and Central China was covered by a summer drought of similar scale and intensity.

10. The dramatic shortfall in food supply (from the 1959 sown-area curtailment reinforced by excessive State procurement), coupled with inevitable weather-caused losses, compelled the peasants to eat into their seed and feed reserves (for livestock) for 1960. This initially resulted in a lower sowing density. But with the 1959 summer drought persisting through the 1959–60 winter sowing, and spreading into North China to affect most of the 1960 spring (summer-harvested) crops, the average per hectare grain yield in 1960 was further reduced by an alarming 20 per cent. Despite a sharp recovery in sown area by 5.5 per cent, total grain output fell unprecedentedly by a further 15.6 per cent in 1960, on top of

the major losses of 15 per cent incurred in 1959. This is the policy and agronomic background to the extraordinary human tragedy which unfolded in full in 1960–1 to engulf tens of millions of innocent Chinese peasants.

11. In reconstructing the story of 1959–60, it is therefore important to make a distinction between grain-output losses due to sown-area contraction and yield declines per sown hectare. Regression estimates based on grain and weather statistics for 1952–66 indicate that in per hectare terms, the losses in 1959 (and 1961 as well) could all be attributed to the weather. In 1960, however, only about 70 per cent of per hectare losses could be accounted for in this way, leaving the balance (30 per cent) to be explained by the man-made factors. These included reduced sowing density due to depletion of seed reserves, forced slaughter and starvation of livestock (pigs essential for supply of organic fertilizers and draught animals for ploughing and other farmwork) due to feed shortage, and of course the physical effects on farm-workers due to malnutrition. The 70 per cent weather-induced yield losses for 1960 seem to suggest that the weather was also implicated as a conspirator in the human tragedy. But our hypothetical estimates based on available population and grain-supply data show that, had it not been for the unwarranted sown-area curtailment and excessive grain-procurements in 1959, the majority of peasants who perished would have been able to survive the natural catastrophe of 1959–61.

12. To an extent, the 1985 incident resembles that of 1959–60, in that policy-induced sown-area fluctuations outweighed the weather factor (as it impinges on yield per sown hectare) as a source of major agricultural disruption. The comparison also points to the fact that within the decollectivized rural context (as in 1985), the effect of misguided policy management of prices in controlling the rural output mix might have been comparable with that of a highly centralized system with physical and bureaucratic control (as in 1959), in which any policy error could make itself felt at grassroots levels of decision-making.

13. There is none the less a decisive difference between the rural environment in 1985 and that of 1959–60. With a substantially higher level of food security, Chinese peasants in the 1980s were less subject to State extortion than in the early 1960s. This security allowed them a certain, albeit limited, degree of occupational and migration mobility. This explains the observed diversion of resources away from the grain sector, in pursuit of income maximization in 1985, and the resultant increase in grain production instability. By contrast, in the absence of such flexibility, peasants in 1959–61 simply found themselves locked into a hopeless situation. Indeed, mass starvation without the possibility of

rural exodus during the 1959–61 catastrophe made the situation even starker than the 'grand agonies' of the past.

14. Finally, a word about the widely held view that, because of its vast geographical size, good and bad harvests occurring in different regions of the country in a particular year normally cancel out each other. This would suggest that national output should not fluctuate dramatically. The comment has often been made, perhaps justifiably, with respect to the 1959–61 disasters. There is no doubt that the unwarranted sown-area contraction in 1959 occurred throughout the country and resulted in across-the-board reductions in grain output. Likewise, the desperate food crisis in the worst-affected areas also demanded massive inter-provincial grain transfers, and thereby resulted in a pervasive shortfall in seed grain and indiscriminate reductions in sowing density and output in some provinces. However, the major regional centres of weather disturbance and output losses in 1959–61 which had a substantial influence on the national aggregates were readily identifiable. The same holds true with other episodes of major instability. Our regional analysis reveals that in most, if not all cases, fluctuations in the national total output could easily be traced to disturbances in one or two major grain regions, notably, the East, Central, or North China.

Looking ahead, the implications that may be drawn from the major findings of this study are many. The fundamental question is whether the relative significance of weather influence will continue to diminish as suggested by the long-run trend in recent decades. The answer may appear to be negative, judging from the fact that agriculture in many advanced countries, notably, the USA, Europe, and Russia, is still by no means weather-proof. Indeed, fluctuations in grain and cereal production in these countries have often turned out to be even more substantial than in present-day China.

However, the relative stability of Chinese agriculture should be interpreted against a peculiar Chinese background which is likely to remain intact for many years to come. The background is that, compared with the advanced Western countries, per capita grain availability in China (376 kilograms in 1991 or 390 kilograms in 1990) remains substantially lower; the margin of grain surplus is still meagre; the overwhelming majority of the rural labour force (431 million in 1991) is still engaged in grain-cropping; and virtually every province is still heavily committed to grain production.

The widespread geographical distribution of cropping activities will, by virtue of the 'spatial risk-dispersal function', ensure basic stability in overall output. More importantly, with the income and livelihood of around 850 million rural residents directly or indirectly linked to grain production, there clearly exists an enormous 'subsistence urge' by the peasant masses to ward off any untoward eventuality. Such built-in human factors will certainly continue to be available to reinforce the improved 'technological hedge' and help minimize the effect of farm fluctuations in China in the foreseeable future.

Improved peasant income and consumption standards during the past decade, coupled with increased rural decontrol, may have left Chinese peasants still susceptible to government policy manipulations as a source of instability, as the case of the *daosanqi* price reform in 1985 has revealed. However, the present situation clearly is still far removed from the fully marketized and monetized rural context of, say, the USA, where the immense 'wheat plains' or 'cornbelts' are visually dotted more with tractors and harvesters than with scores of farm labourers. Specifically, in the labour-scarce US context, where the 'abnormal agricultural supply function' certainly does not hold, major weather adversities are likely to help raise the real opportunity cost to a prohibitive level for many farm inputs and investment. For basically self-sufficient Chinese peasants (many indeed displaying zero marginal productivity), however, the chances of any significant employment alternatives remain a distant luxury.

Thus, the case of 1985 probably represents a one-off historical incident, in which modest peasant responses to adverse price and income incentives resulting from inexperienced government manipulation with relative prices, generated considerable disorder. With trial and error, however, the scale and frequency of such policy mishaps are likely to be curtailed. In fact, the system of contractual purchases introduced in 1985 in conjunction with the *daosanqi* proportionate price reform, and intended to replace compulsory delivery quotas, in many ways reflected continued monopsonistic extortion by the State. This involves in particular areas with a greater degree of production instability. Thus, the established 'institutional hedge' is likely to remain operative for many years to come.

Moreover, to the extent that industrial growth in China is still constrained by the supply of food and cash crops as inputs (either directly or via export earnings for financing import-related capital prospects), it does not seem likely that State control of the terms of trade for agriculture will give way completely to uncertain market forces. Equally important is the fact that State procurements designed to balance regional grain deficits probably cannot be completely abolished, despite increased autonomous inter-provincial transfers.

Finally, the national food balance is likely to continue to be strained by increases in population. The figures in Table 13.1 show that, apart from the years 1986 and 1985 (which benefited from the spectacular 1984 harvests), China has persistently remained a net grain importer. This

TABLE 13.1. China's per capita grain availability (A) and grain exports (X) and imports (M)

	1984	1985	1986	1987	1988	1989	1990	1991
A kg./head	394	363	368	324	360	367	390	376
X								
Million tonnes	319	932	942	737	717	656	583	1,086
$US100	7.4	13.6	13.1	10.1	11.9	11.9	10.2	15.8
M								
Million tonnes	1,041	600	773	1,628	1,553	1,658	1,372	1,345
$US100	8.3	10.0	10.8	17.5	19.0	29.9	23.5	16.4
X − M								
Million tonnes	−722	332	169	−891	−816	−1,002	−789	−259
$US100	−10.9	3.6	2.3	−7.4	−7.1	−18.0	−13.3	−0.6

Notes and sources: *TJNT*, various issues. The dollar figures for 1985–91 are given. For 1984 they are converted from the original Renminbi account by using the exchange rate of one $US1 to 2.2 yuan as implied in the comparative official $US and RMB accounts of total trade volume (respectively 49.77 bn. and 109.71 bn.) for 1984.

threatens to cut into precious foreign exchange holdings for priority imports for major capital projects, and should certainly enhance the built-in imperative to stabilize grain production.

Indeed, the State has taken further measures to counter emerging trends towards disinvestment in rural social overhead capital associated with decollectivization, notably in irrigation and flood control capacity. Such factors make Chinese agriculture very different from its US or European counterpart. Thus, ironically, grain production will probably remain more stable in China than in advanced Western countries in the foreseeable future.

Appendices

Appendix A

Weather Patterns and Climatic Disturbances in China, 1921–1984

Introduction

The main purpose of this appendix is to formulate a 'weather index' for China as a whole which can serve as a basis for measuring average annual changes in weather conditions. Such a weather index is needed to fulfil the major objective of this study to analyse the relationship between observed year-to-year changes in aggregate grain output and yields on the one hand, and climatic variations on the other hand. Our approach is, basically, to make use of the SSB (State Statistical Bureau) statistics about the size of the *shouzai* area (i.e. farm areas covered by various types of natural disasters, notably floods and drought),[1] as an indicator of weather disturbance. The methodological problems involved in this respect are diverse and complicated. They warrant detailed separate treatment, hence this appendix.

Moreover, the *shouzai* area is merely a 'weather proxy', not a measure of the weather *per se*. This raises the important question of why we do not make use of the monumental *500-Year Weather Record, 1470–1979*, made available by the State Meteorological Bureau (SMB).[2] The SMB study is a direct meteorological measurement of the weather in which precipitation data available for the modern period have been compiled to determine the degree of floods and drought for different localities in China. However, the SMB study focuses essentially on summer rainfall, omitting the other two important cropping seasons of spring and autumn.

Nevertheless, it is important to make use of the SMB study, incomplete as it may be, to verify the reliability of the *shouzai* area records. Thus, before we embark on the formulation of our own 'weather index', a systematic reconciliation of the SMB's meteorological records with the SSB's *shouzai* area statistics is carried out in the second part of this chapter. This will underline the pro-summer bias of the SMB study and its unsuitability as a measurement of the annual average of weather changes in China.

The analysis is preceded by a brief discussion of the basic ecological relationship between the weather and agriculture, with special reference to China. This provides a broad perspective for looking at how the weather may impinge on Chinese agriculture, and, in particular, on grain production which is the main concern of this study. The discussion also helps to shed some light on the major practical difficulties involved in any attempt to establish a precise agricultural weather index for China. Bearing in mind the major problems involved it will help in interpreting and evaluating the usefulness and limitations of our *shouzai*-area-based weather index.

This study concludes with a brief comparison of the formulated weather index with a number of the so-called 'categorical weather indexes' compiled by a number of Western scholars, which have been widely used for gauging the major direction of weather changes in China. The categorical weather index classifies different weather years into broad categories of good, bad, and average, with possible subdivisions for each of the three categories. We will show that, in contrast to our weather index, which is a strictly differentiated quantitative weather scheme, the categorical weather index possesses little operational significance for our analytical purpose of examining the impact of weather disturbance on Chinese agriculture.

Finally, a word about the time-span of this study. While the SMB record only covers the years down to 1979, our *shouzai* area analysis is extended to 1984 by using the SSB statistics. The initial year of the period under study is, however, 1921 (not the benchmark year of 1931), which is generally used in the main body of our analysis, by resorting to the SMB's records of drought and floods. This gives a longer historical perspective to an examination of the long-term patterns of weather changes and climatic disturbance.

Basic Ecological Weather – Yield Relationships

Climatic conditions affect plant growth and crop output mainly by way of solar radiation, precipitation, and temperature changes. The entire ecological process has been well studied in the West and in China. In general terms, solar radiation facilitates the process of photosynthesis, by which plants, with the aid of the cholorophyll pigment, utilize the energy of solar radiation to produce carbohydrates (i.e. starch, sugars, cellulose, etc.) out of water and carbon dioxide. The photosynthetic rates vary with temperature. For most plants in temperate and tropical regions the optimum temperature exceeds 25 °C. The photosynthetic rates reach a maximum between 30 and 37 °C and then drop sharply at higher temperatures.[3]

Water shortages in plants result in dehydrated protoplasm and thus a reduced photosynthesis capacity. In addition, once the leaf loses its turgidity, the stomata guard-cells close, thus preventing any further intake of CO for photosynthesis. In general, the rate of photosynthesis declines noticeably after a reduction of approximately 30 per cent in the water content of the leaves. When 60 per cent of the leaf moisture is lost, photosynthesis usually ceases.[4] The effect of a water deficit in the plant on yield can be conveniently gauged by comparing variations in yield with evapotranspiration, which is the same as the 'consumption use' of water by the plants, apart from the negligible amount used in the metabolic activities. Fig. AA.1, which is based on experimental and field data obtained in North America, shows the generalized relationship between the two variables, whereby the 'potential values' refer to a situation with unlimited application of water. It is clear that yield not only varies closely with the adequacy of water supplies, but in the case of drought much also depends on the stage of plant growth at which it occurs. A water deficit which develops at the moisture-sensitive period of a plant will do much more damage to yield than if it occurs at other periods.

Of the three fundamental ecological elements of plant growth, solar radiation is a rather stable variable, although its absolute magnitude varies from area to area. Generally speaking, its annual total decreases

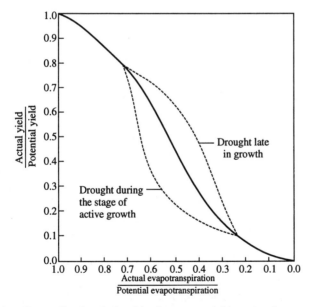

FIG. AA.1. Generalized relationship between yield and adequacy of water application

Source: adopted from Chang Jen-hu, *Climate and Agriculture: An Ecological Survey* (Chicago: Aldine Publishing Co., 1968), 214.

with increases in latitude. However, countries far to the north or south of the equator, such as Spain, Italy, Australia, and Japan, normally have substantially higher rice yields than tropical countries such as India, the Philippines, and Thailand. Similarly, rice yields in China measured in kilograms per sown hectare for such northern provinces as Ningxia (7,815 kg.), Liaoning (6,255 kg.), and Shanxi (6,188 kg.), compare very favourably with those in such southern provinces as Guangdong (4, 358 kg.), Guangxi (4,238 kg.), and Fujian (4,440 kg.).[5] This is contrary to the popular view which assumes rice to be a tropical crop, unsuitable for growing in the temperate zone. The most decisive climatic factor is that rice is a summer crop and in the summer months the northern region enjoys much stronger solar radiation than the rainy southern areas.

In contrast to solar radiation, temperature and precipitation may, however, deviate significantly from the seasonal averages for any part-icular area. The relative importance of these two climatic elements differs of course from place to place. Japan, for example, is similar to the British Isles, rarely lacking rainfall for summer crops (rice, millet, maize, and soya beans), while per hectare yields for such winter crops as wheat and barley have consistently proved to be negatively correlated with precipitation. Extensive regression analysis based on county data in Japan has revealed that rice yield and temperature display a close positive correlation, with deviations in summer temperature accounting for no less than 65 per cent of the standard deviation in rice yield and output for the country as a whole in the two decades from the 1920s to 1940s.[6]

In China, however, the reverse tends to be true. The Chinese land mass is dominated by the continental type of climate, with summer temperatures consistently higher than in Japan for areas of similar latitude. Nevertheless, the annual interchange of the Siberian and Pacific monsoons causes annual rainfall in China to be concentrated in the summer months. High summer temperature and moisture have together made China the world's largest agricultural country. Yet the arrival of the summer monsoon does not adhere to any fixed calendar schedule. This results in relatively large precipitation variability. As shown in Map AA.1, in virtually all regions, the annual variability is greater than the maximum of 16 per cent recorded for any locality in the British Isles. In New York State the average is only 9 per cent.

The smaller the annual precipitation, the greater the variability tends to be. Thus, within the Yellow River basin, average annual variability consistently exceeds 20 per cent compared with only 15 per cent for the Yangzi River basin and South China. More important, the monthly rainfall variability is even greater. In the Yellow River basin, for example, the record is 40–75 per cent for April, May, and June, and 25–60 per cent for July and August.[7] This has profound implications especially for

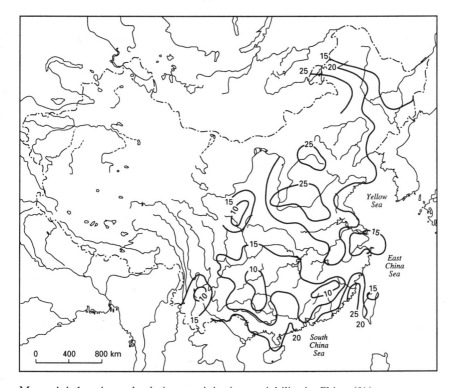

MAP. AA.1.   Annual relative precipitation variability in China (%)

Source: adopted from Zhu Kezhen, *Zhu Kezhen Wenji* (Collected Essays of Zhu Kezhen) (Beijing: Kexue Chubanshe, 1979), 455.

the wheat crop in the North China plain, i.e. the 'Wheat Belt' of China, comprising mainly Hebei, Shandong, and Henan provinces. Note that the critical moisture-sensitive periods for wheat growth are the stages of jointing, earing, and flowering, and these occur in April and May each year.

The erratic arrival of the summer rains often causes disastrous flooding and waterlogging in China. Moreover, summer rainstorms are frequently accompanied or rather precipitated by meteorologically different, but allied, tropical cyclones or typhoons. Japan is also frequently visited by harmful summer typhoons although, for topographical reasons, China is in a less favourable position than Japan. Immense river basins and low coastal deltas make China prone to widespread and prolonged inundation, whereas in the hilly and mountainous Japanese islands, flood-waters can be more easily drained into the rivers and oceans.[8]

Temperature is also a significant yield determinant in China. Very often the withdrawing winter monsoon suddenly reverses its course in the late spring or early summer, causing what is customarily known among the Chinese peasants as a 'Daochunhan', i.e. a 'reversed spring chill'. The result is widespread frost killing, or 'chill injury' to, sprouting or maturing plants in North and South China.[9] In addition, sowing and transplanting normally require temperature to rise to a certain level, as well as demanding sufficient soil moisture and sunshine. However, 'forced' sowing or transplanting in the absence of adequate temperature often has to take place, either to meet the extremely tight multiple-cropping schedule, or to prevent the seedlings from becoming over-mature.[10] In either case, the final output is bound to be reduced.

Problems of Aggregate Weather Measurements

Granted that rainfall is the single most important climatic determinant of Chinese agriculture, the question is whether it is possible to establish a 'comprehensive' national agricultural weather index for the whole of China, by compiling, first of all, a national precipitation index. This might then serve as a basis on which to incorporate other important meteorological variables in order to obtain a properly weighted system of weather measurements.

The crux of the problem involved here does not lie primarily in the lack of hard meteorological data. The number of meteorological observatories and weather stations in China totalled 317 in 1952, and 1,647 in 1957. Thereafter it gradually increased to reach a total of 2,568 in 1984, covering virtually every *xian* in China.[11] The network was much less developed during the 1930s. Nevertheless, the forty-odd observatories and smaller weather stations then in existence were fairly widely dispersed, at least within China proper. Though the temperature and precipitation data generated at that time were scanty, Buck made use of them for his pioneering demarcation of China's agricultural areas.[12]

To formulate a comprehensive agricultural weather index, it is necessary to establish, first, a system of well-classified and suitably disaggregated agricultural regions or areas. This should take into account regional variations, not only in terms of basic agro-climatic parameters, but also of yields and output, which depend, in turn, on differences in the cropping pattern, the cropping index, and the level of agricultural technology.

Systematic efforts seem to have been made in the early 1960s by the Chinese Academy of Science to modify and reclassify Buck's classic contribution into a four-tier system, comprising four basic agricultural regions and 129 subareas.[13] Unfortunately very little is known about the underlying criteria and one wonders whether these were specific enough to be suitable for the formulation of the required system of regional weights. In recent years, similar studies have been extensively carried out in China, albeit at a more confined regional and local level.[14] Nevertheless, the currently available, most authoritative Chinese classification of agricultural areas, published by the Chinese Academy of Science, gives only a general framework with broad regional demarcations similar to those of Buck.[15]

Another practical difficulty is that any demarcation of agricultural areas is likely not to coincide with provincial borders. Local statistics, probably disaggregated down to the *xian* level, would be required to make such a demarcation. Presently available agricultural statistics are, however, aggregated at the provincial level. Availability of statistics apart, the sheer amount of data involved would render the task almost insurmountable.

The Record of the State Meteorological Bureau

Setting aside the complicated problems concerning the regional differences in agricultural response to weather changes, if we look at the weather *per se*, the most comprehensive and systematic survey available of long-term climatic patterns in China is the monumental contribution first published in 1981 by the State Meteorological Bureau (SMB) in Beijing. The contribution, which was cited in the introduction to this appendix as the *500-year Weather Record* (hereafter the SMB study), not only maps on a year-to-year basis (from 1470 to 1979) the regional distribution of drought and floods, but also gives detailed moisture gradings for the 120 stations (localities) surveyed. Each station represents one or two *diqu* (prefectures), and the sample is large enough to cover the whole of China. It is therefore worth while to conduct a detailed examination of the SMB study in relation to our effort to formulate a national weather index.

Nature and coverage of the SMB study

A simple, ordinal index system of 1, 2, 3, 4, and 5 has been adopted by the SMB to designate, respectively, 'great flood' (*dalao*), 'flood' or 'mild flood' (*bianlao*), 'normal' (*zhengchang*), 'drought' or 'mild drought' (*bianhan*), and 'great drought' (*dahan*). For the earlier periods, the gradings were mainly based on verbal records from local *xian* gazetteers. For the more modern periods, when precipitation records became available, however, the following standardized formulae were applied:

Grades 1: $Ri > (R + 1.17\delta)$
2: $(R + 0.33\delta) < Ri < (R + 1.17\delta)$
3: $(R - 0.33\delta) < Ri < (R + 0.33\delta)$
4: $(R - 1.17\delta) < Ri < (R - 0.33\delta)$
5: $Ri < (R - 1.17\delta)$,

where R stands for the long-term mean of precipitation for the period from May to September, Ri the corresponding precipitation for the year concerned, and δ the standard deviation. For North-east and North China, the June to September precipitation value was used instead of the May–September figure. The specific parameters shown for the five-tier classification were based on the frequency distribution of floods and drought obtained from the historical records. This excludes drought and floods during spring and autumn, but includes those occurring in the summer, whenever such adverse weather prevailed over two or more seasons within the same year.

Thus, the entire SMB study essentially represents a summer weather record. The omission of the precipitation value for May and April for the North China plain is especially unfortunate, since it is precisely these two months that represent the most critical moisture-sensitive period for winter wheat. Nearly the entire cropping season for winter wheat is excluded from the SMB survey. Likewise, the weather for early rice crops in many areas of the south is not covered by the SMB study. The crop weather for spring wheat, gaoliang, millet, maize, mid-rice, etc. is only partially covered. Nevertheless, most of the major weather events do normally take place during the summer monsoon period. It is therefore worth while to make use of the SMB study to examine the possible patterns of climatic change in China.

The derivation of the flood and drought indices

A long-term index of flood and drought may be compiled by adopting the following formulae:

$$FI = \frac{2(N1 + N2)}{N} \qquad DI = \frac{2(N4 + N5)}{N}$$

where FI and DI stand for the flood and drought indices respectively, $N1$, $N2$, $N3$, and $N4$ the number of localities (stations) graded as 'great flood' (1), 'flood' (2), 'drought' (4), 'great drought' (5), and N the total number of stations including those graded as 'normal' (3).

The computed value of FI or DI varies between 0 and 2, indicating the relative geographic extent of flood or drought in general. A DI value of 1.20, for example, implies that 60 per cent (i.e. $1.2 \div 2 \times 100$), of the stations for which gradings are made are under drought of different degrees of intensity. If *DI* falls below 0.60, there may, in turn, have been widespread flood, since less than 30 per cent (i.e. 0.6 ÷ 2.0 × 100) of the area as represented by the stations are under drought or great drought. A senior SMB analyst, from whom the above formulae are borrowed, classified those years falling within the range of *DI* 0.60 to 1.20 as 'drought-dominated' years, and those with a *DI* above 1.20 as the 'most severe drought years'.[16] The classification may of course also be applied to the flood index. And the computed drought and flood indices may be simply added together to become a single national weather index based exclusively on a moisture standard.

To obtain a more coherent index of the weather conditions in China proper I have excluded the twenty stations west of the 100° E longtitude and the two stations on Taiwan. This leaves a total of ninety-eight stations covering virtually the whole of the Chinese agricultural system, including the North-east. The results of the computation for the years 1921–79 are presented in Fig. AA.2, and show the flood index separately, as well as the combined flood and drought index. These indices probably represent the best national aggregate measurement of long-term weather changes that can presently be obtained, being directly derived from hard meteorological data, and are systematically weighted by numerical values relating to weather events at the local level.

Many of the years known to be affected by severe floods or drought can be identified in the graphical indices. Take flood, for example: the years 1931, 1949, 1954, 1956, and 1964 have all been well recorded. The disastrous Yangzi River floods of 1931 and 1954 have become classic phenomena in modern Chinese history. It was the great 1964 flood in the North China plain which prompted Mao to make his famous national call to tame the Hai River. Records for more recent years have not yet been included in the SMB study. But if the catastrophic floods of both 1980 (in the middle reaches of the Yangzi) and 1981 (in the Sichuan basin) were included, there would be a total of six peak-flood years during the past six decades, i.e. an average of one every 10 years. And, as revealed in Fig. AA.2, the peaks seem to be relatively well dispersed along the time-path.

This cyclical weather pattern has led many Chinese meteorologists to resort to the classical theory of black sunspots for an explanation.[17] Interestingly, long-term regression analyses based on the SMB's statistics for the past 500 years have strongly borne out the correlation between the changing accumulations of sunspots and the larger-scale cyclical oscillations of floods and drought.[18] The drought cycles for the past six decades are nevertheless neither as pronounced nor as regular as the flood cycles, although most of the dramatic increases in the drought index shown in Fig. AA.2 can also be accounted for by known exceptional

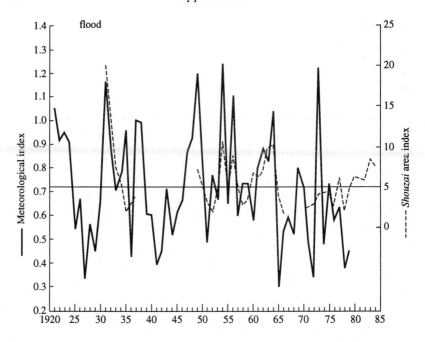

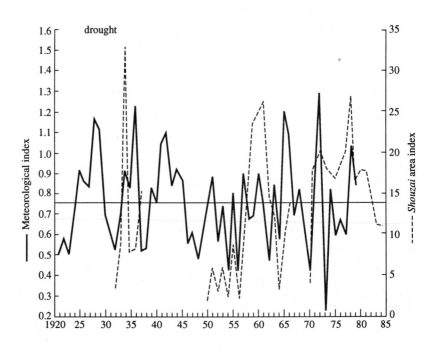

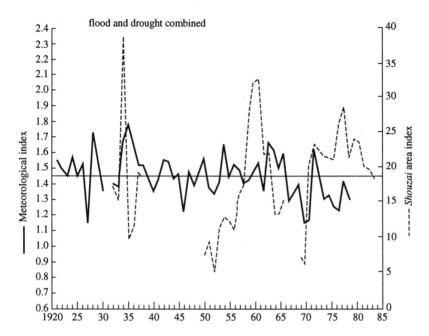

Fɪɢ. AA.2 Comparison of the meterorological (precipitation-based) flood drought index and the *shouzai* area index in China, 1921–1984

Notes: See text for the formulae for deriving the meteorological flood and drought indices. The flood and drought *shouzai* area indices are given as a proportion (%) of total sown area. They are not weighted by the relative severity (*chengzai chengzai*) of losses. The horizontal lines in the graphs denotes the long-term average of the meteorological (rather than the *shouzai* area) indices.

Source: Table AB.19.

droughts, notably the 1928-9 drought of the North-west and the great North China drought of 1972.[19]

Surprisingly, however, the drought index for the three years 1959-61 clusters along the long-term average line in Fig. AA.2. This is much below the zone classified as 'most severe droughts'. Roughly the same relative position is maintained on the combined flood and drought index for the three years 1959-61 (also in Fig. AA.2). Taken together, the derived drought and flood indices imply that the weather in any of these three years was better than in 1963, 1964, and 1966. Even more curiously, the year 1960 is portrayed here as being a better weather year than either 1956 and 1957; and 1959 as better than the good weather year of 1955, comparable with 1958, which has been widely considered as one of the best weather years of the last three decades.

A similar problem is discernible for the 1930s. For example, probably no one familiar with the Republican period would agree that the drought in 1934 was substantially less severe than that of 1936 as is suggested by the computed drought index in Fig. AA.2.

Verification of the SSB's shouzai *area statistics*

The issues relating to the discrepancy between the computed SMB's drought and flood indices and the known weather events discussed above must be resolved before any weather index can be used for present analytical purposes. The most logical approach is to examine how the meteorological indices derived from the SMB records relate to the SSB statistics in terms of the size of *shouzai* area (i.e. sown area covered by floods and drought). The *shouzai* area statistics are also plotted in Fig. AA.2, alongside the derived SMB's indices, as a proportion of the yearly national total sown area. This facilitates a direct comparison with the SMB's flood and drought indices, which also represent the percentage of the national total of stations (i.e. areas) affected.

Note, first of all, that in the formulae used for computing the flood and drought indices, 'great flood' (weather grade 1) and 'flood' (grade 2) are treated identically, as are 'great drought' (grade 5) and 'drought' (grade 4). For purposes of consistency, no distinction is made between the two categories of *shouzai* areas as adopted in the SSB statistics between areas sustaining a crop loss of below and above 30 per cent on account of natural disaster, notably floods and drought.[20] Moreover, the numerical scales for the three *shouzai* area diagrams are made identical in Fig. AA.2. This enables a direct visual comparison of the possible differences in the impact of flood and drought in terms of sown areas affected.

Taken together, the three *shouzai* area diagrams in Fig. AA.2 convey the impression that the influence of floods was much more limited than that of drought. This is conceivable, given that the scale of flooding and waterlogging very often varies with local topographical conditions, while invisible drought normally displays no such clear boundaries. But does this mean that the drought area statistics are necessarily less accurate and that the SMB's meteorological index for drought, or for that matter, for floods, should therefore be preferred as an indicator of real weather changes? To answer this question, a closer examination of the relationship between the two sets of indices (i.e. the precipitation-based indices versus the *shouzai* area indices) is needed.

Simple linear regressions were run on the *shouzai* area indices against the three different meteorological indices for flood, drought, and both combined. The results are summarized in Table AA.1. They show that the flood series represents, statistically speaking, a much better 'fit' than the drought series (except perhaps for the 1970–9 subperiod to which we

TABLE AA.1. Correlation between the meteorological indices (X) of flood and drought and the *shouzai* area indices (Y) in China, 1931–1979

	Regression equation	Correlation coefficient
Flood series		
1931–7	$Y = -4.457647 + 14.856287X$	0.5424
1949–61	$Y = -0.814224 + 7.843953X$	0.7363
1949–66	$Y = -0.548992 + 8.158293X$	0.7135
1970–9	$Y = 2.692197 + 1.559333X$	0.3009
1949–79	$Y = 0.390887 + 6.488175X$	0.6509
1931–79	$Y = -0.608644 + 8.442817X$	0.5750
Drought series		
1932–7	$Y = 6.469846 + 7.885059X$	0.2007
1950–61	$Y = -6.292348 + 24.496818X$	0.4421
1950–66	$Y = 2.625486 + 11.071340X$	0.2957
1970–9	$Y = 10.89365 + 9.324510X$	0.5047
1950–79	$Y = 6.969152 + 9.484551X$	0.3071
1932–79	$Y = 7.032577 + 8.960889X$	0.2753
Flood & drought combined		
1932–7	$Y = 11.460977 + 4.594489X$	0.0703
1950–61	$Y = -30.47776 + 32.709512X$	0.3177
1950–66	$Y = -3.463506 + 14.002364X$	0.1825
1970–9	$Y = -6.482538 + 21.107963X$	0.5106
1950–79	$Y = 13.63409 + 3.654284X$	0.0723
1932–79	$Y = 14.099063 + 3.216128X$	0.0643

Source: Table AB.19.

will turn shortly). That is, the relationship between the metereological index and the *shouzai* area index is much more consistent for floods than for drought. This is not surprising, because floods normally occur in China during the summer months which are the main focus of the SMB study, whereas drought may prevail during the spring and autumn as well, thereby falling outside the orbit of the SMB analysts.

A few words about the exceptional period of 1970–9 are in order. The year 1973 probably holds the key to the lack of correlation in 1970–9 between the flood index and the *shouzai* area statistics. As revealed in Fig. AA.3, the exceedingly large flood index for 1973 (1.22) stands in a totally disproportionate relation to its below-average *shouzai* area index of 4.2 per cent (of total sown area). Curiously enough, the historical record does not show 1973 to have been a serious flood year, but its computed flood index is nevertheless virtually identical to that of the catastrophic flood years of 1931 and 1954.

A plausible explanation for this is that the 1973 flood index was strongly dominated by grade 2 (flood) rather than grade 1 (great flood) stations. Grade 2 stations make up nearly 60 per cent of the total number of grade 1 and 2 stations for 1973, compared with, say, 1931 and 1954, for which the respective shares amount to only 31 per cent and 41 per cent. The point to be made here is that many of the stations graded as 2 in 1973 may not have fallen in areas sufficiently flooded or inundated as to warrant inclusion in the SSB's *shouzai* area statistics. Note that a grade-2 station enjoys a relatively large range under the SMB's classification formula shown earlier.

More important perhaps, most of the grade-2 floods in 1973 were concentrated in areas south of 30°N latitude, i.e. south of the Yangzi River basin. This is clearly reflected in Fig. AA.3, which shows 1973 to have been one of the few exceptional cases throughout the entire period (1921–79) under study that had such a regional bias in flood incidence. These areas, i.e. in South China, enjoy relative abundance of rainfall and in particular stability in its temporal distribution, so that the transition from 'normal' (grade 3) to 'flood' (grade 2) may not be as clear-cut as farther north. In contrast, the 1931 and 1954 floods, apart from being dominated by grade-1 stations, were both concentrated in the Yangzi River basin (25–35°N latitude). These areas certainly have a greater degree of precipitation variability, both annual and monthly, compared with South China.

The implications should be clear, especially if we also consider that the SMB's gradings are based on mean precipitation for the period from May to September. The case of 1973 probably represents a fairly even distribution of excess rainfall over the five-month period, whereas both 1931 and 1954 saw highly concentrated downpours within relatively short

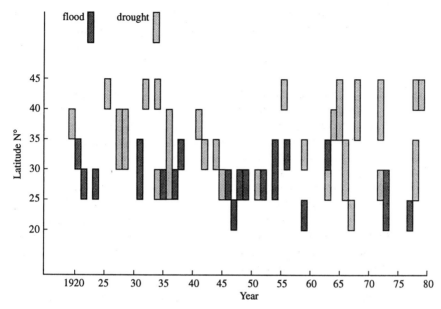

F IG. AA.3 Temporal and spatial distribution of 'severe' droughts and floods in China, 1920–1979

Notes and sources: 'Severe' drought and floods are determined according to the formula:

$$\overline{DF} = \frac{1}{n}\sum DF,$$

where *DF* stands for the values of the SMB's ordinal drought or flood indices (i.e. 1, 2, 3, 4, and 5 as explained in the text), *n* the number of the monitoring weather stations, and $\overline{DF}$ the averaged drought–flood index for the latitude band concerned. 'Severe' drought is taken to be $\overline{DF} \geqslant 3.8$, and 'severe' flood $\overline{DF} \leqslant 2.2$. The gradings as well as the entire schematic presentation for 1920–77 are adopted from Xu Ruizhen and Wang Lei, *Woguo Jin Wubainian Hanlao de Chubu Fenxi* (A Preliminary Analysis of Droughts and Floods in China in the Past Five Hundred Years), in SMB (Weather and Climatic Institute, Research School of Meteorological Science) (ed.), *Quanguo Qihou Bianhua Xueshu Taolunhui Wenju* (Proceedings of the National Academic Symposium on Climatic changes) (Beijing: Kexui Chubanshe, 1978), 54–5. I have made use of the comparable SMB data given in *Zhongguo Jin Wubainian Hanlao Fenbu Tuji* to extend the analysis to 1979. It seems likely that the data used by Xu and Wang are from the same sources. See also Feng Peizhi, Li Cuijin, and Li Xiaoquan, *Zhongguo Zhuyao Qixiang Zaihai Fengxi* (An Analysis of the Major Meteorological Calamities in China) (Beijing: Qixiang Chubanshe, 1985) for a more detailed year-to-year descriptive account of the distribution of floods and drought (and other major climatic disturbances).

282 *Appendix A*

intervals. In addition, the entire Yangzi River basin east of Sichuan represents a much less effective drainage system than the hilly and mountainous South China. All these points suggest why the exceedingly high meteorological flood index was accompanied by a low *shouzai* area index in 1973.

Turning to the drought indices, it is clear that the pro-summer bias of the SMB study singularly accounts for the low degree of correlation between the meteorological index and the *shouzai* area index. As a matter of fact, the estimated lack of correlation between the two indices involves not only the post-war period (1950–66) or for that matter 1950–61, which covers the three anomalous years 1959–61, but also, worse still, the pre-war period 1932–7, as is shown in Table AA.1.

Moreover, there seems to be no consistent statistical bias to show that the SSB's *shouzai* area statistics are disproportionately large in relation to the meteorological drought index for any particular periods or year. This can be shown by relating the drought area index (AD) to the meteorological drought index (DI) for those years which show a greater value in the former relative to the latter. The years in question can be detected from Fig. AA.2. The computed 'discrepancy ratios' (AD/DI), as given in Table AA.2, reveal that the ratios for the three years 1959 (i.e. 34.3), 1960 (28.1), and 1961 (36.2) are comparable to those of the pre-war years of great drought, i.e. 1934 (ratio 35.9) and 1937 (29.2); or for that matter, say, 1975 (28.1) and 1977 (33.9). There are also years in which the computed 'discrepancy ratio', is either substantially lower or higher than those cited here. These all suggest that the discrepancy between the

TABLE AA.2. The 'discrepancy ratio' between the drought area index (AD) and the meteorological drought index (ID) in China for selected years, 1934–1979

Year	AD/ID	Year	AD/ID
1934	35.9	1971	24.9
1937	29.2	1973	91.5
1950	24.5	1974	21.0
1958	21.9	1975	28.1
1959	34.3	1976	27.5
1960	28.1	1977	33.9
1961	36.2	1978	25.8
1962	31.5	1979	19.8

Source: Table AB.19.

meteorological drought and *shouzai* area indices probably has its origins in drought which occurred outside of the SMB's summer focus, but which entered into the SSB's *shouzai* area statistics. This can be illustrated in the case of three major drought years 1959–61.

The case of 1959–1961

Consider the critical years 1960 and 1961. In 1966, Lu Wu, the SMB's Deputy Director, gave the following *post mortem* analysis:

In the whole wheat area in North China, the shortfall (in precipitation) lasted for ten months consecutively from September 1959 through June 1960. In April, when the winter wheat crops were entering the stage of jointing to earing, the precipitation was 70 to 90 per cent lower than normal. The dry spell reoccurred in 1961 during the critical period of crop growth in the same winter wheat areas, resulting in yet another year of great losses.[21] (This statement can be verified by our independently compiled precipitation data for April of both 1960 and 1961; see Chapter 10 of this study.)

Note that the SMB's moisture indices cover only the months June to September for North China and the North-east regions. Thus, the extreme shortfall in the spring moisture in both 1960 and 1961 is left out of the SMB's deliberation.

In addition, being based on the mean precipitation for May to September (or June to September in the case of North China and North-east), the SMB's moisture gradings tend to conceal the sharp monthly rainfall fluctuations in the summer and their impact on cropping activities. The standard drought-resistant targets adopted in the National Programme of Agricultural Development for 1956–67 for irrigation projects represent a capacity to withstand a drought of not more than 30 to 50 days, or 50 to 70 days in the double-cropped areas.[22] Needless to say, many of the irrigation works completed by the late 1950s did not meeet these norms.[23] Even for such relatively water-rich areas as Shanghai, a drought may emerge in the summer if adequate rainfall is not available for 15 to 20 days. It becomes serious if the soil is not soaked for more than 30 consecutive days.[24]

This is precisely what happened in the summer of 1961 and in 1959. As is revealed in Map AA.2, the July precipitation in 1961 throughout virtually the entire Yangzi rice and wheat area and the Sichuan basin was over 50 per cent lower than the long-term mean. The situation in summer 1959 was very similar in terms of the dominating cyclonic patterns identified by the Research Institute of Atmospheric Physics of the Chinese Academy of Sciences.[25] Indeed, the negative precipitation deviations in 1959 are geographically much more consistent and wide-

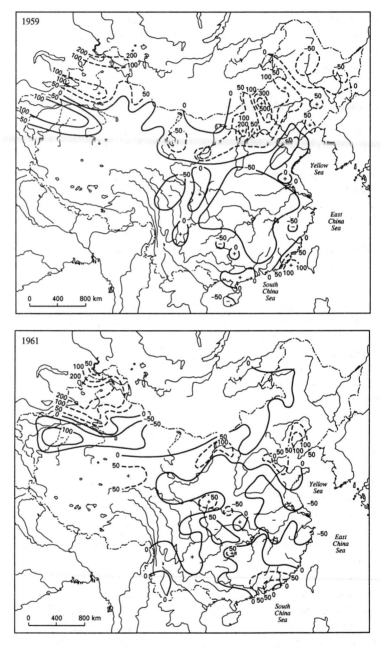

MAP AA.2. Regional distribution of radical precipitation shortfall in China in summer 1959 and 1961 compared (percentage deviations of July precipitation from the long-term mean)

Source: Research Institute of Atmospheric Physics, Chinese Academy of Science, *Zaihaixing Tianqide Yice he Yufang* (Forecast and Prevention of Calamitous Weather), (Beijing: Kexue Chubanshe, 1981), 9–10.

spread than in 1961. Worse still, if precipitation for 16 June to 15 July is taken as the measurement, then in both 1959 and 1961 the departures from the long-term mean are even more drastic.[26]

It should now become clear why in the schematic presentation of the temporal and spatial distribution of drought and floods in Fig. AA.3, both 1960 and 1961 are so blank. The pro-summer bias of the SMB study totally excluded the extraordinary spring droughts of the North China Plain in both years. And the adopted precipitation mean for moisture gradings covered too extensive a period (May to September) to capture the relatively short-term, but nevertheless disastrous, summer drought.

One or two remarks are in order with respect to the other major drought years, notably 1934 and 1972. The 1934 drought was also a summer drought, meteorologically and geographically comparable to that of 1959. As in 1959, 1960, and 1961, 1934 displays an extremely large *shouzai* area index in relation to the meteorological drought index shown in Fig. AA.2 (and Table AA.2). This is partly accounted for by the fact that the national sown-area statistics for the 1930s are heavily weighted by the more accessible provinces, i.e. precisely where the droughts prevailed. This explains why the *shouzai* area index, defined as a proportion of total sown area, is much higher than that for the years 1960 and 1961.

By contrast, the great North China drought of 1972 was an all-time high, in terms of the meteorological index, in the entire period (1921–84) under study. By the SMB's meteorological standard, the 1972 drought ranks among the five top droughts ever to occur in the past 500 years.[27] Yet the 1972 *shouzai* area index, as shown in Fig. AA.2, is considerably lower than that for 1960 and 1961, or indeed 1934. This can be explained by the fact that the North and North-west, both host to the disastrous 1972 drought, have a smaller multiple-cropping index, and hence a comparatively smaller total sown area when compared with the South. Against the background of a smaller sown-area base (without the possible benefits from the risk-aversion function of the multiple-cropping practice), it is understandable that the enormous meteorological impact was bound to bring about a catastrophic loss in 1972. This is why the 1972 drought, confined though it was in terms of regional coverage (note especially the lower per hectare yield in the North), helped greatly to depress the national grain output.

Conclusion

The elaborate effort made to reconcile the *shouzai* area indices with the SMB's meteorological indices suggests that there are no gross inconsistencies on the part of either of the two series. The observed discrepancies between the two sets of indices reflect differences in both the coverage and the methodology applied. However, it should now be clear that the *shouzai* area indices more readily capture the diverse weather changes from the viewpoint of agricultural meteorology. They are, of course, not weather indices *per se*, but they seem to be a good weather proxy. We may now turn to look at the problems involved in using the *shouzai* area to formulate a national weather index.

Shouzai *Area as a Weather Proxy*

A number of statistical and methodological problems are involved in any attempt to use the *shouzai* area statistics (i.e. sown area covered by natural disasters) in order to formulate a comprehensive weather index and analyse the causal relationship between weather variability and agricultural instability in China. For example, it is important that areas which sustain crop losses on account of factors other than weather are not given undue emphasis. We begin by examining the nature of the *shouzai* area statistics and their reliability.

Reliability of the shouzai *area statistics*

The area data used in this study are official statistics published by the State Statistical Bureau (SSB) for the period from 1949 to the present and those collected by the National Agricultural Research Bureau (NARB) for 1931–7.[28] In addition to drought and floods, both series cover windstorms, frosts, and hail. The NARB series also takes account of insect pests and plant diseases. The latter may not be totally weather-related. Drought and flood are consistently the most important factors, accounting for overwhelming shares in the *shouzai* area total. Drought and flood may of course partly result from deforestation, soil erosion, and inappropriate cultivation, as opposed to the weather. But these are long-term factors which cannot easily account for short-run weather variability on a relatively large scale.

The main problems associated with both the SSB and NARB series arise from the methods used to collect and compile the statistics, and the possible motivation of the reporting agencies or the responsible national authorities to falsify the extent of *shouzai* area. A careful scrutiny of available statistics reveals, however, no gross inconsistencies in either the post-war or pre-war periods.

By far the most unreliable period would seem to be the three crisis years 1959–61, in view of the statistical fiasco of 1958–9.[29] Surprisingly, however, recent SSB figures showing the sown areas affected in 1960 and

1961 by natural disasters amount to 65.46 and 61.75 million hectares respectively. Both are much higher than the rounded-up figure of 60 million hectares published in the early 1960s for both years, which was regarded by many Western scholars as an exaggeration to cover up policy and human errors associated with the Great Leap Forward. Moreover, the new series reveals that 1960 and 1961 remain the hardest-hit years, while the severity of disasters in 1959 (totalling 44.63 million hectares of *shouzai* area) was only surpassed in 1977–8.

There seems to be no reason why the present SSB should seek to exaggerate 1959–61 *shouzai* area statistics, considering the fact that current policy is to repudiate the 'leftist' errors of the Great Leap Forward. It is now clear that earlier official data for 1959–61 were exclusively given in terms of the more comprehensive terms of *shouzai* area rather than the more narrowly defined *chengzai* area. The latter category refers to an area which is, literally speaking, 'calamitously affected' (*chengzai*) by natural disasters and is statistically defined as having an average crop yield loss per hectare of more than 30 per cent (below the 'normal' yield) on account of the disasters. The former category, however, includes the non-*chengzai* proportion as well, i.e. those areas which sustained a yield loss of less than 30 per cent. It thus refers to areas which are 'covered' (*shouzai*) by natural disasters in general. In our subsequent discussion we shall consistently use the two original Chinese terms of *shouzai* and *chengzai* (or non-*chengzai* as the balance between the two different categories) rather than the more ambiguous official SSB translations of 'covered' and 'affected' areas.

The 1959–61 *shouzai* figures cited above have nevertheless been frequently misinterpreted in the West as the *chengzai* area, to be related to the 1949–57 *chengzai* series (presented in Table AA.3). This inevitably gives the impression of an abrupt increase in the areas affected by natural disasters in 1959–61.[30] Worse still, both the *shouzai* and *chengzai* figures are often taken to refer to the arable area, whereas they refer, in fact, to the sown area.[31] This of course tends to exaggerate the size of *shouzai* area as a proportion of the country's total cultivated area.

At the local level, it is conceivable that, within the framework of collectivization, the Chinese peasants and local cadres were motivated to boost the size of *shouzai* areas, since relief-grain dispatches and the amount of compulsory deliveries demanded were related to the claimed losses. However, in view of the meagre grain margin available and the government's urge to appropriate the agricultural surplus, the counter-vailing power of the authorities can be assumed to have been at least as strong, or even stronger than the ability of the peasants to cheat in this way. In fact, it is the Ministry of Civil Affairs (rather than the Ministry of Agriculture, or Ministry of Water Conservancy), which is responsible for

TABLE AA.3. Comparison of the old and new series of sown area affected (*chengzai*) or covered (*shouzai*) by natural disasters, 1949–1961 (million ha.)

	Total *chengzai* area				Flood			Drought		Total flood and drought	
	Old			New	Old		New	Old	New	Old	New
	Liao	Zhou	Xu	TJNJ	Chen	Xu	TJNJ	Xu	TJNJ	Xu	TJNJ
1949	—	—	8.52	8.53	8.52	8.52	8.53	—	—	8.52	8.53
1950	—	—	5.12	5.12	4.71	4.71	4.71	0.41	0.41	5.12	5.12
1951	—	—	3.78	3.78	1.48	1.48	1.48	2.30	2.30	3.78	3.78
1952	—	—	4.43	4.43	1.84	1.84	1.84	2.60	2.59	4.44	4.43
1953	5.33–6.00	6.33	3.87	7.08	3.20	3.20	3.20	0.67	0.68	3.87	3.88
1954	10.67	12.09	11.84	12.59	11.31	11.52	11.31	0.33	0.26	11.85	11.57
1955	7.33	7.61	7.21	7.87	3.07	3.07	3.07	4.13	4.14	7.20	7.21
1956	15.33	15.33	—	15.23	10.99	—	10.99	—	2.06	—	13.05
1957	14.67	—	—	14.98	5.67	—	6.03	—	7.40	—	13.43
			RMRB			*RMRB*		*RMRB*		*RMRB*	
1958			6.67 (26.67)	7.82 (30.96)		1.20 (3.87)	1.44 (4.28)	6.00 (31.33)	5.03 (22.36)	7.20 (35.20)	6.47 (26.64)
1959			— (43.33)	13.73 (44.63)		—	1.82 (4.81)	— (33.33)	11.17 (33.81)	—	12.99 (38.62)
1960			20–26.67 (60.00)	24.98 (65.46)		—	4.98 (10.16)	— (40.00)	16.18 (38.13)	—	21.16 (48.29)
1961			— (60.00)	28.83 (61.75)		—	5.40 (8.87)	—	18.65 (37.85)	—	24.05 (46.72)

Notes: *Chengzai* area refers to farmland which was covered by natural disasters within the overall context of *shouzai* area, but which sustained crop losses of 30% and above the normal output. Figures in parenthesis are *shouzai* area.

Sources: Liao: From Liao Luyan, Ministry of Agriculture, in *XHBYK* 5 (1958), 129. The figures were presumably jointly verified by the Ministries of Interior Affairs, Irrigation, and Agriculture.

Zhou: Zhou Boping, 'Liangshi Tonggou Tongxiao shi fan buliao de' (It is impossible to oppose the unified purchases and sales of grain), *LSYK* 7 (1957), 4.

Xu: Xu Kai, 'Zhongguo de guangai shiye' (China's irrigation construction), *ZGSL* 10 (1956), 9. (Xu was Director of the School of Surveying and Design of the Ministry of Irrigation.)

Chen: Chen Lian, 'Ba chulao baijai nongye jianshi de zhongyao diwei' (Prioritize the elimination of waterlogging in agricultural construction work), *JHJJ* 1 (Jan. 1958), 15.

TJNJ: Table AB.15.

RMRB: 1958: Overall area: *RMRB* 3 June 1959: *Shouzai* area *more* than 400 million *mu* and *chengzai* area *more* than 100 million *mu*. We take the two rounded figures (note the conversion ratio of one hectare to 15 *mu*) and they are consistently lower, but already quite close to the new *TJNJ* figures. Flood area: *RMRB* 14 Oct. 1958 (communiqué of the Ministry of Agriculture on achievements in farm irrigation construction): *shoulao* (waterlogged) area 58 million *mu* and *chengzai* area 18 million *mu*. Presumably 'according to the amount and pattern of rainfalls, *shouzai* areas should have amounted to a total of around 100 million *mu*', but the size in 1958 was reduced on account of 'success in water conservancy work'. Note that the two figures are very close to, in fact even lower than the new *TJNJ* figures. Drought area: *RMRB*, ibid. The two (*shouzai* and *chengzai*) figures are substantially higher (by 19 and 40% respectively) than the new *TJNJ* figures, but they are specifically referred to, in the same Ministry's communiqué, as 'cumulative' total. This implies overlappings in areas covered by drought within the same cropping seasons. See notes to Table AB.16 for detailed explanation.

1959: Overall area: *RMRB* 23 Jan. 1960, a national total of 650 million *mu* which 'is equal to 30 per cent of total sown area'. This clearly implies that the official *shouzai* or *chengzai* area should be taken as 'sown area' rather than 'cultivated area'. See notes to Table AB.16 for further explanation. The *shouzai* area figure as given is virtually identical with the new *TJNJ* figure. See also *RMRB* 23 Dec. 1959 for a similar figure (rounded down to 600 million *mu*) billed as having been covered by droughts, floods, winds, and insects, etc. No *chengzai* figures are available. Drought area: *RMRB* 23 Dec. 1959: a total of 500 million *mu* in 20 provinces/municipalities were covered by droughts (*shouhan*: this is a summation of areas covered in the spring, summer, and autumn. *RMRB* 23 Jan. 1960 also reported that nearly 400 million *mu* (26.67 million ha.) in 8 provinces were seriously (*yanzhong*) covered by drought. It is difficult, however, to bill this figure as *chengzai* area, for the *chengzai* criterion as adopted in *TJNJ* (above 30% crop loss) seems to be more stringent than the one used in the 1950s. See notes to Table AB.16 for explanation.

1960: Overall area: *RMRB* 29 Dec. 1960: a total of '900 million *mu* of farm area were covered by natural disasters at varying degrees; . . . among which 300 to 400 million *mu* particularly seriously (Zhongzai)'. (See identical figures in *RMSC 1961*, 274.) We take the two latter figures to be *chengzai* area.) *RMRB* 1 Oct. 1960: out of the total of 900 million *mu* 300 million *mu* was *chengzai*. However, in the year-end report, the 900 million *mu* figure is related to the 'cultivated' (instead of 'sown') area so that over half of the total cultivated area was affected. This is different from the 1959 report (*RMRB* 23 Jan. 1960) quoted above. I cannot explain the changes made. At any rate, if the 900 million *mu* figure is related to total sown area, the percentage share would be considerably lower.

1961: Fang Chung, 'All-round achievement in China's economy', in *Beijing Review* (23 Aug. 1963), 8–10.

conducting the *shouzai* and *chengzai* area surveys and compiling the relevant area statistics.[32] This specialization is clearly construed as a safeguard against possible falsification by the other ministries: whether in an effort to unload their portfolio responsibility for any mismanagement, or to claim for inappropriate relief aid or fund appropriations for reconstruction of devastated areas. Note that the Ministry of Civil Affairs was throughout charged with planning and organizing the dispatch of relief aid. It is interesting that even during 1959–61, the *chengzai* area, which normally entitled the peasants to relief grain, was on average no more than 40 per cent of the total *shouzai* area. This *chengzai* ratio is virtually identical to the 1949–84 mean.

The NARB statistics are less susceptible to similar scrutiny. Where charity relief-dispatches were made, however, the pre-war peasants naturally responded in the same way, although their attempts to fool sophisticated statistical investigators were rarely successful. Take the 1931 flood, for example, which, under the direction of J. L. Buck at the University of Nanking, was comprehensively surveyed, with the authorization of the National Flood Relief Commission, by means of the same team of investigators as that used in his *Land Utilization in China*. Buck said:

> There is a general trend toward underreporting the family's available stock on foodgrain, fuel, and hay, and overreporting requirements of foodgrain, seeds, etc. . . . Fortunately such inaccurate answers are generally not tenable, for the questionnaires provide many cross-checkings. For example, the reported requirements should not exceed the amount of losses. Seedgrain must be compatible with the usual seed rate per *mu*, and foodgrain not more than what the family normally can consume.[33]

In short, there seems to be no particular built-in bias in either the SSB or the NARB series of *shouzai* area statistics. In fact, a comparison of the two series made in Table AA.4 shows that the means and standard deviations for the overall *shouzai* ratios (i.e. *shouzai* area as a percentage of the total sown area) are comparable for all three periods 1931–7, 1952–66, and 1970–84. The comparability between the two earlier periods (with means of 18.9 vs. 17.6; standard deviation 8.9 vs. 7.8) is especially remarkable, considering the fact that the institutional settings of the pre-war and post-war years were totally different. However, physical, rural conditions were still generally comparable, in the absence of any major agricultural technological progress. For 1970–84, the somewhat lower standard deviation (4.9) reflects the stabilizing effect of improved flood control and drainage capacity. If this is so, it raises the legitimate question of whether the *shouzai* area can still be properly used

TABLE AA.4. Sown area covered (*shouzai*) by floods and droughts as a proportion of total sown area in China, 1931–1937 and 1952–1984 compared (%)

1931–7		1952–66		1970–84	
1931	20.1	1952	5.0	1970	6.2
1932	17.5	1953	11.1	1971	19.9
1933	15.4	1954	12.9	1972	23.5
1934	38.9	1955	12.4	1973	22.5
1935	9.8	1956	11.0	1974	21.5
1936	11.3	1957	16.1	1975	21.2
1937	19.3	1958	17.5	1976	21.2
		1959	27.1	1977	26.1
		1960	32.1	1978	28.7
		1961	32.6	1979	21.2
		1962	21.8	1980	24.1
		1963	22.1	1981	23.6
		1964	13.3	1982	20.1
		1965	13.4	1983	19.6
		1966	15.3	1984	18.3
Mean	18.9		17.6		21.2
Standard deviation	8.9		7.7		4.8

Notes: The figure for 1931 refers to flood area only, and as in the case of the other years 1931–7 that for 1932 does not include the two North-east provinces of Heilongjiang and Jilin for which grain output, sown area, and yield data are not available for a consistent comparison with other provinces. Since the grain-sown area, rather than *total* sown area, is used as base for 1931–7 to derive the *shouzai* area proportion, the estimated mean tends to be biased upwards in relation to 1952–7.

Sources: Tables AB.2, AB.14, and AB.15.

as a measure of changes in the weather over a long period. We shall turn to this problem in the next section.

Another way to look at the statistical problem is to make an inter-temporal comparison of the spatial patterns of drought and flood distribution. Tables AA.5 and AA.6 show the provincial distribution of drought and floods as a percentage share of total sown area covered by natural disasters for 1931–7 and 1952–61 for which a more complete set of data is available. A close scrutiny of the figures in both tables reveals

Appendix A

TABLE AA.5. Provincial percentage share of grain-sown area covered (*shouzai*) by natural disasters in 1931–1937, and the Hirschman–Gini concentration index (FD = flood and drought, AD = all natural disasters)

	1931	1932		1933		1934		1935		1936		1937	
	FD	FD	AD	FD	AD	FD	AD	FD	AD	FD	AD	FD	AD
Hebei	—	5.91	5.14	4.11	4.86	15.84	15.76	31.55	19.91	6.28	4.40	12.25	12.25
Shanxi	—	0.96	0.93	0.76	0.88	3.95	3.86	19.99	13.13	6.68	7.59	5.04	5.04
Chahaer	—	—	—	—	—	0.05	0.06	0.02	0.02	0.90	0.61	—	—
Suiyuan	—	—	—	1.02	0.92	1.20	1.18	1.46	0.99	1.33	0.98	0.22	0.22
Liaoning	—	—	—	—	—	—	—	—	—	—	—	—	—
Jilin	—	35.91	34.49	—	—	—	—	—	—	—	—	—	—
Heilongjiang	—	28.15	27.03	—	—	—	—	—	—	—	—	—	—
Jiangsu	30.65	—	0.01	8.52	7.73	13.78	13.54	1.31	3.58	9.91	8.99	4.10	4.10
Zhejiang	3.99	—	—	4.38	4.25	5.41	5.29	2.49	10.05	7.46	6.00	1.18	1.18
Anhui	15.54	9.58	9.29	14.15	13.31	9.43	9.30	9.11	6.01	3.65	2.93	2.89	2.89
Fujian	—	—	—	3.06	2.77	1.21	1.18	0.18	0.95	1.78	1.35	0.02	0.02
Jiangxi	4.67	1.33	1.30	7.75	7.02	6.78	6.62	1.57	1.94	4.47	3.13	0.26	0.26
Shandong	6.98	1.59	2.52	3.65	4.39	11.64	11.55	9.95	9.21	3.52	18.18	7.82	7.82
Henan	18.37	12.55	14.48	17.08	20.41	12.38	13.40	3.39	2.13	13.50	9.10	27.20	27.20
Hubei	11.77	—	—	6.73	6.10	6.02	5.87	8.30	13.97	6.27	6.73	5.84	5.84

Hunan	8.04	1.50	1.44	13.12	11.89	7.33	6.87	2.18	5.65	2.45	2.65	1.27	1.27
Guangdong	—	0.05	0.05	3.08	3.04	0.37	0.49	3.05	3.21	5.01	3.49	1.24	1.24
Guangxi	—	—	—	0.93	1.39	0.38	0.37	—	—	—	—	1.75	1.75
Sichuan	—	—	—	0.47	0.46	0.14	0.15	1.82	3.47	21.26	15.84	18.47	18.47
Guizhou	—	—	—	0.23	0.22	0.03	0.04	0.79	0.66	0.90	0.90	1.15	1.15
Yunnan	—	—	—	0.09	0.09	—	—	—	0.08	1.63	1.51	2.69	2.69
Xizang	—	—	—	—	—	—	—	—	—	—	—	—	—
Shaanxi	—	2.22	2.28	10.46	9.63	3.86	3.84	2.28	1.75	1.79	3.79	5.42	5.42
Gansu	—	0.18	0.36	0.37	0.58	0.15	0.24	—	2.90	0.99	1.64	1.11	1.11
Qinghai	—	0.05	0.07	0.02	0.03	—	0.00	—	—	0.24	0.21	0.09	0.09
Ningxia	—	—	—	0.03	0.03	0.05	0.05	0.53	0.38	0.00	0.001	—	—
Xinjiang	—	—	—	—	—	—	—	—	—	—	—	—	—
TOTAL	100.00	100.00	100.00	100.00	100.00	100.00	100.00	100.00	100.00	100.00	100.00	100.00	100.00
Concentration index	42.52	48.77	47.60	31.98	32.50	31.90	31.85	41.15	32.79	31.65	31.05	37.73	37.90

Source: Table AB.14.

TABLE AA.6. Provincial percentage share of total sown area covered by floods and droughts in 1952–1961 and the Hirschman–Gini concentration index

	1952	1953	1954	1955	1956	1957	1958	1959	1960	1961
Hebei	1.15	20.14	14.77	9.00	16.33	n.a.	8.35	5.16	18.66	15.06
Shanxi	—	1.62	3.16	10.17	3.23	7.86	5.57	3.59	10.06	7.32
Inner Mongolia	n.a.	—	—	n.a.	n.a.	n.a.	7.66	0.18	n.a.	—
Liaoning	n.a.	—	—	n.a.	0.52	5.37	1.32	1.86	0.23	3.99
Jilin	1.01	3.16	2.07	1.35	5.22	2.78	7.95	0.71	2.00	0.06
Heilongjiang	—	—	n.a.	—	5.01	n.a.	2.64	0.37	n.a.	0.20
Jiangsu	5.75	n.a.	11.74	n.a.	11.63	3.41	n.a.	5.65	1.68	5.72
Zhejiang	3.21	4.35	2.83	—	4.36	1.64	1.78	0.77	1.43	1.65
Anhui	21.43	8.68	14.49	3.45	9.88	2.49	24.63	10.05	3.88	5.98
Fujian	0.10	2.43	n.a.	3.57	2.01	1.10	0.05	0.45	1.21	0.96
Jiangxi	n.a.	5.18	5.85	0.60	n.a.	n.a.	4.24	2.28	0.14	0.53
Shandong	25.12	35.53	14.91	44.14	8.20	30.62	11.40	13.52	16.00	12.97
Henan	19.51	0.48	14.89	n.a.	17.95	26.71	6.51	20.51	18.95	7.99
Hubei	10.69	4.05	7.50	n.a.	n.a.	6.29	1.78	9.98	—	5.57
Hunan	1.60	0.12	0.87	0.18	3.59	4.19	1.88	3.58	3.54	6.62
Guangdong	0.21	8.50	4.43	19.13	2.84	2.96	2.02	2.59	7.70	2.93

Guangxi	3.14	1.27	6.39	6.10	1.22	5.84	1.12	1.17	7.07
Sichuan	0.08	n.a.	n.a.	1.14	0.92	4.68	11.06	9.89	10.21
Guizhou	7.00	1.08	1.90	1.62	1.16	1.59	2.06	2.35	1.55
Yunnan	n.a.	0.14	n.a.	—	0.96	0.09	0.75	0.27	—
Xizang	—	—	—	—	—	—	—	—	—
Shaanxi	—	—	n.a.	—	n.a.	—	3.74	0.84	2.89
Gansu	—	—	n.a.	—	—	n.a.	—	n.a.	n.a.
Qinghai	1.21	—	—	0.36	0.15	—	—	—	0.72
Ningxia	—	—	—	—	—	—	—	—	—
Xinjiang	n.a.	—	—	—	0.06	—	—	—	—
TOTAL	100.00	100.00	100.00	100.00	100.00	100.00	100.00	100.00	100.00
Concentration index	41.14	33.85	50.70	32.21	42.92	33.19	32.40	35.60	29.15

Notes and Sources: The concentration index as estimated for 1955 and 1957 covers a percentage share of 0.12 each for Shanghai (1955) and Beijing (1957). Table AB.16.

that the regional centre of the impact of natural disasters shifted from time to time. Compare, for example, the two provinces of Jiangsu in the east and Sichuan in the west during the 1930s. Their percentage share of the *shouzai* area (droughts and floods combined) in the national total changed from 8.52 and 0.47 in 1933, to 13.78 and 0.15 in 1934, 1.31 and 1.82 in 1935, 9.91 and 21.26 in 1936, and finally 4.10 and 18.47 in 1937. Similar changes occurred between Guangdong in the south and Hebei in the north in the period 1952–61. The discernible geographic shifts in the relative weights of the provincial *shouzai/chengzai* areas are basically consistent with the changing temporal distribution of known floods and droughts.

The Hirschman–Gini concentration indices computed for the provincial *shouzai/chengzai* area data in Tables AA.5 and AA.6 are also quite revealing. It is understandable that the index should be comparatively high for exceptionally good weather years such as 1932 (48.8), 1952 (41.1), and 1955 (50.7). In such good years, the bad weather was confined to a relatively small number of provinces. The indices for 1959–61 are incomplete as the data may be, generally speaking, neither smaller nor greater than those for the other years, including the 1930s. And the two meteorologically comparable years of 1934 and 1959 have virtually an identical index.

Formulation of the weather index

Granted that the *shouzai* area statistics are broadly acceptable, we now turn to the methodological problems involved in using the national aggregates as a weather proxy for formulating a comprehensive weather index. When considering the post-war period, for which statistics for both the *shouzai* and *chengzai* areas are available, different weights must clearly be furnished for each set of area statistics to take into account the varying severity of weather anomalies. This is because a year which had a greater *shouzai* area (e.g. 1974) might actually have been less 'affected' than another year which had a smaller *shouzai* area but a larger *chengzai* proportion (e.g. 1975). In the latter case the absolute *chengzai* area may turn out to be larger than in the former. Since the degree of crop losses incurred for the *chengzai* area is higher (over 30 per cent), compared with that for the non-*chengzai* area (below 30 per cent), the combined losses for both the *chengzai* and non-*chengzai* areas may be larger for the year with a smaller *shouzai* area total.

To facilitate intertemporal comparisons, the official *shouzai* area series for 1952–84, inclusive of natural disasters other than floods and drought, may thus be converted into a weighted index by using the following formula:

$$W = An\sqrt{\frac{Ln}{Lc}}Ac\sqrt{\frac{Lc}{Ln}} \qquad (1)$$

where W stands for the weighted *shouzai* area, Ac the yearly '*chengzai*' area, An the '*shouzai*' area net of the *chengzai* proportion, and Lc and Ln the respective percentages of output losses sustained thereby. A crucial problem is how to determine the loss percentages. For Ln, 15 per cent seems to be appropriate which is simply the mean of the narrow range of 0 per cent to 30 per cent loss used by the SSB to define the non-*chengzai* area. As for the *chengzai* area, the empirical evidence available for the 1950s generally suggests that Lc falls within the range of 55 to 65 per cent as shown in Table AA.7.[34] A 60 per cent loss has been adopted as the weight. Before the weighted *shouzai* area index is presented, a number of points still need to be made.

First, through rearrangement, equation (1) is equal to

$$W = \frac{1}{\sqrt{(Lc \times Ln)}}(AcLc + AnLn) \qquad (2)$$

This implies that the derived weighted *shouzai* area is roughly proportional to the total output loss in all of these areas. That is to say, if both Lc and Ln are taken to be the effective loss rates for the years concerned, then the estimated weighted *shouzai* area represents no more than a proxy for the output or yield changes for that particular year. This could of course defeat the entire purpose of constructing an independent weather index, as one would come close to using the dependent variable as proxy for the explanatory variable. To avoid being tautological, Lc and Ln must therefore be held constant as a set of 'weather weights' to be consistently applied to all the years under study.[35] In other words, we are using the two standard loss rates as independent factors to qualify the changing weather conditions as captured by the non-weighted weather series.[36]

Note especially that the 60 per cent loss rate for the *chengzai* area assumes great weight in the total *shouzai* area series. This standard rate is derived from the effective loss percentages for the 1950s and may thus be taken to represent the 'real' weather impact, for in the absence of modern technology farm production in China in the 1950s was greatly affected by the weather. With the accelerated use of modern technology starting in the mid-1960s, however, the effective loss sustained was bound to decline. Viewed in this way the two *constant* weather weights help to capture the comparable real scale of weather disturbance for the later years.

Secondly, the two *constant* weather weights, as used here, may understate or overstate the relative severity of weather adversity inasmuch

TABLE AA.7. Percentage of grain crops destroyed by natural disasters causing over 30 per cent loss, 1949–1957

	Grain losses		'Normal' grain yield (kg./ha.)		Loss percentage	
	Total (million kg.)	(kg./ha.)	Actual value	5-year moving ave. of (3)	(2)/(3)	(2)/(4)
	(1)	(2)	(3)	(4)	(5)	(6)
Series (1)						
1949	5,700	668	1,029	1,209	64.9	55.3
1950	2,600	508	1,155	1,209	44.0	42.0
1951	3,150	833	1,220	1,209	68.3	68.9
1952	2,922	660	1,322	1,266	49.9	52.1
1953	7,500	1,067	1,317	1,318	81.0	81.0
1954	8,850	703	1,314	1,357	53.5	51.8
1955	6,400	813	1,417	1,384	57.4	58.7
1956	12,200	801	1,414	1,435	56.6	55.8
1957	17,550	1,168	1,460	1,464	80.0	79.8
1949–57	66,872	840	1,294	1,317	64.9	63.8
Series (2)						
1953–6	37,500	734	1,366	1,374	53.7	53.4

Notes and Sources: (1) For 1949–57 (except 1952) in series (1) see *RMRB*, 22 Dec. 1957, 2 and *XHBYK* 5 (1958), 129; the 1952 figure was derived from a linear regression equation fitted with the absolute *chengzai* area data (sources of Table AB.15) and the grain loss data for 1949–51 and 1953–7. The equation is $Y = -1599.71 + 1020.8X$ $(r^2 = 0.82)$, where $Y =$ grain loss and $X = $ *chengzai* area. The *chengzai* area for 1952 is from Table AB.15. For series (2) see Xiao Yu, 'How to allocate agricultural investments', *JHJJ* 9 (1957), 5–7.
(2) Column (1)/*chengzai* area.
(3) *NYNJ 1980*, 35.
(4) For 1949–50 the average yield of 1949–53 was used. For each of the other years the mean covers alternately the two preceding and the two following years and the year concerned. See ibid. for 1958 and 1959 yields.

as the effective losses for periods with comparable farm-technologies (say, the 1950s) may be greater or smaller than the assumed standard loss rates. It is impossible to correct for such biases. Theoretically, an 'over-stated' scale of natural disasters may help to cover up losses incidentally caused by man-made errors. However, such coincidence seems unlikely in view of the random nature of agricultural policy changes in China and the indiscriminate application of our *constant* weather weights.

Thirdly, the *chengzai* area as a proportion of total *shouzai* area declined steadily over the years. This trend is particularly marked in the case of the flood area, with average *chengzai* proportion declining from 63 per cent for 1952–66 to 45 per cent for 1970–82. This is mainly due to the enormous efforts made since the 1950s to tame the major rivers and to build up the country's drainage and irrigation capabilities. In view of this, the indiscriminate procedure used to convert the *shouzai* area series inevitably tends to underestimate the scale of weather disturbance in the later years, or to overstate the correlation between weather and crop yield inasmuch as agricultural instability was reduced over time as a result of technological advance. The distortions may not be too serious, for the *chengzai* proportion declined only very gradually over the years. What is clear, however, is that the declining *chengzai* ratio may help to dampen the amplitude of fluctuations in the weighted *shouzai* area index, thus lessening over time its special properties as a workable weather proxy. But this is of little relevance for the time-span covered in this study.

With these qualifications we may now look at the weighted *shouzai* area index, which is presented here as yearly percentage deviations from the 1952–84 mean, although it can also be given in terms of absolute hectare size. From Table AA.8, 1961 remains the hardest-hit year, while both 1952 and 1970 enjoyed exceptionally good weather. In addition to the relatively large annual weather variability, there is a clear pattern of long-term movement which fits the notion of a decadal weather cycle. Thus, nearly a decade of favourable weather in the 1950s culminated in large-scale natural disasters in the early 1960s. This was followed by a

TABLE AA.8. The computed weather (*shouzai* area) indices for China, 1931–1937 and 1952–1984

1931	20.06	1952	−71.60	1963	21.91	1974	−23.01
1932	6.27	1953	−40.99	1964	−21.26	1975	−12.62
1933	15.35	1954	−21.71	1965	−28.01	1976	1.55
1934	38.88	1955	−42.34	1966	−29.30	1977	28.89
1935	9.82	1956	−10.29	1967	—	1978	53.61
1936	11.33	1957	−2.04	1968	—	1979	12.02
1937	19.26	1958	−29.17	1969	—	1980	47.40
		1959	13.50	1970	−73.72	1981	26.91
		1960	85.97	1971	−29.41	1982	7.74
		1961	95.99	1972	21.67	1983	10.17
		1962	15.27	1973	−21.53	1984	2.69

Notes: The 1931–7 series is shown as a proportion of total sown area covered by floods and droughts with the exception of 1931, for which only flood area data are available. The 1952–84 series represents, however, yearly percentage deviations of the weighted *shouzai* area from the 1952–84 mean, covering all natural disasters.

Sources: Tables AB.5 and AB.14 for 1931–7, and Table AB.15 for 1952–84.

gradual improvement until the climax of 1970, when the weather again entered another decade of deterioration closely resembling the pattern of the 1950s. If this decadal cyclical weather pattern is a good indicator, the exceptionally good weather of 1982, 1983, and 1984 may have signalled the advent of another decade of favourable weather in China. It can be expected, however, that any good or bad decades are bound to be interspersed with some bad or good years, since nature is never totally predictable.

Another noteworthy point is that, in the weighted *shouzai* area index, in only three years (1956, 1957, and 1976) are the percentage deviations within 10 per cent of the 29-year mean. This situation is somewhat similar to the conclusion drawn by James McQuigg from his crop weather studies for the US Corn Belt states. Thus, he advises against reference to 'average weather', but favours instead a focus on 'large scale, long-term variability of weather events from year-to-year or from decade to decade'.[38]

Can similar weather cycles be established for other countries or regions of the world? In the post-war period, at least, the world has seen many parallel large-scale meteorological disturbances. The 1954 Yangzi floods, for example, were matched in the USA by severe flooding in Chicago, an extremely rare occurrence. 1954 has also been identified (meteorologically) as a wet year on a global scale. The weather disturbance in China during 1959–61 should be seen in a global perspective. The British journal *Nature* pointed out that after a decade (1948–58) of remarkable increases in crop yields throughout the world, which promoted a wave of optimism about the world's food problem, the rate of growth of food production unexpectedly slowed down, because of 'generally poor levels of agricultural yield in less developed areas'.[39] Unfavourable weather was cited by *Nature* as one of the main causes.

Likewise, the exceptionally bad weather years of 1974–5 in the USA were preceded by 'about 15 years of remarkably favourable crop season weather'.[40] Note that the early and mid-1970s were also characterized by the great Sahelian drought in Africa, the Maharashtra drought of India, 1973–4, and by the North China drought of 1972–3. Similar examples could be taken from Australia and Europe, including the Soviet Union.

Turning to the pre-war years 1931–7, a similar weighted *shouzai* area index cannot be established, becaused the *shouzai* area statistics do not allow for a distinction to be made between the *chengzai* and non-*chengzai* areas. Granting that the 1931–7 *shouzai* area statistics are not susceptible to serious bias, we simply relate them to the total sown area, as shown in Table AA.8, to generate a *shouzai* area index in terms of percentage of sown area covered by natural disasters. This is used in the main body of this study, alongside the 1952–84 *shouzai* area index, in order to analyse the weather and yield relationship.

The 'categorical' weather index compared

Before applying the weighted *shouzai* area index as a weather proxy for analysing long-term agricultural instability in China, mention should be made of the 'categorical weather index' compiled by a number of scholars (including two Chinese scholars) for Chinese agriculture in the post-1949 period. As shown in Table AA.9, each of the years since 1952 is placed in different qualitative weather categories, distinguishing basically between good, average, or poor years. In most cases, such a 'categorical weather index' is used to describe general weather conditions in a particular year, or rather, the changing overall harvest results from year to year. But it has also been fitted into an analytical scheme for the purpose of separating the relative influence of agricultural policy and weather variability.[41]

It is not entirely clear, however, how these 'categorical' weather indices were constructed. None grew out of any quantitative simulation, including any from which feedback from the US meteorological satellite intercept programme might be available. Thus, the broadly defined weather categories represent at best some highly generalized weather perceptions. To the extent that some of the categorizations are actually based on the realized harvest (rather than meteorological) conditions,

TABLE AA.9. Categorical weather indices compiled by Western and Chinese scholars for China, 1952–1980

	Jones	Petrov and Molodtsova	US Dept. of State	Tang	Feng	Jao
1952	G		G	G	G	G
1953	A		A	A	A	A
1954	P		P	P	P	P
1955	G		G	G	A	G
1956	P		P	P	A	A
1957	A	A	A	A	A	A
1958	G	G	G	G	G	A
1959	A	A	A	A	P	P
1960	P	P	P	P	P	P
1961	P	P	P	P	A	A
1962	G	G	G	G	P	
1963	A	A	A	A	P	
1964	G	G	G	G	A	
1965	A	A	A	A	A	
1966		P	A	A	A	
1967		G	G	G	G	

Appendix A

TABLE AA.9. *cont.*

	Jones	Petrov and Molodtsova	US Dept. of State	Tang	Feng	Jao
1968		A	A	A	A	
1969		A	A	A	A	
1970		G	G	G	G	
1971		A	A	A	P	
1972		P		P	P	
1973		A		A	A	
1974				G	G	
1975				A	A	
1976				P	A	
1977				P	A	
1978				A	P	
1979				A	G	
1980				P	P	

Notes: Different terms are used for different weather categories by the authors; but they have all been grouped under the three divisions 'good' (G), 'average' (A), and 'poor' (P). These three standard terms are uniformly used by Jones, US Dept. of State, and Tang, Petrov and Molodtsova substitute 'bad' for 'poor'. The original Chinese terminology is, for Feng, 'comparatively good' (*jiaohao*), 'general' (*yiban*), and 'comparatively poor' (*jiaocha*); and in the case of Jao, 'basically normal' (*jiben zhengchang*), 'not sufficiently normal' (*bugou zhengchang*), and 'abnormal' (*fanchang*).

Sources: Jones, US Dept. of State, and Tang: Anthony M. Tang, *An Analytical and Empirical Investigation of Agriculture in Mainland China, 1952–80* (Taipei: Chung Hua Institution for Economic Research, 1984), 95–8. Tang's original notes read as follows: 'Crop weather classifications for 1952–65 are taken from E. F. Jones, 'The Emerging Patterns of China's Economic Evolution,' *An Economic Profile of Mainland China*, Joint Economic Report, US Congress, i (Feb. 1967), 93. For 1966–71, the classification is adopted from US Dept. of State, *People's Republic of China*, and from *Issues in US Foreign Policy*, 4 (Oct. 1972), 23. There is agreement between these two sources on the earlier overlapping years. For 1972–80, the estimates are ours as summarized from weather and crop conditions reported in various Chinese and outside sources, including the weekly world weather reports in USDA's *News*, the department's annual situation reports on mainland China, the *Far Eastern Economic Review*, *Yearbook of Chinese Communism* (*Chung Kung Nien Pao*) (Taipei), and relevant CIA reports. Mainland China Project Office, International Economic Division, USDA express essential agreement with our weather classifications for the entire period, except for the suggested plus and minus signs as further qualifications of the weather type for five of the years.'

Feng: Feng Peizhi, Li Cuijin, and Li Xiaoquan, *Zhongguo Zhuyao Qixiang Zaihai Fengxi* (An Analysis of the Major Meteorological Calamities in China) (Beijing: Qixiang chubanshe, 1985), 12.

Jao: Jao Xing, 'Qixiang gongzuo wei nongye jishu gaige fuwude chubu yijian' (Preliminary views on meteorological work to serve agri-technical reform) in *ZGNB* 7 (1963), 12.

one wonders whether it is meaningful at all to make use of such categorical weather indices for any analytical purpose.

Fig. AA.4 compares the weighted *shouzai* area index with the more complete, categorical weather index compiled by Anthony Tang, by grouping the individual years from 1952 to 1980 under his three weather categories and by plotting these against our completed *shouzai* area index. The three means shown in the Fig. AA.4 differ from one another in a manner that is consistent with Tang's three weather categories. The likely implication of this is twofold: either the categorical index has

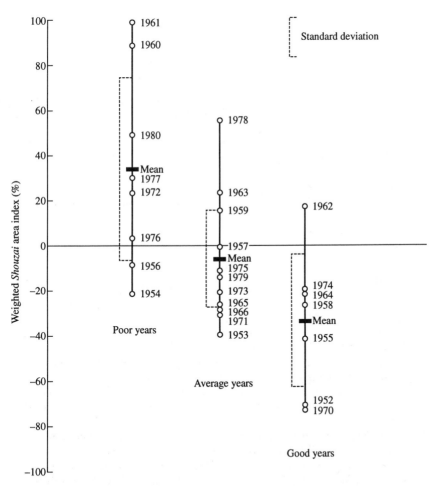

FIG. AA.4. The weighted *shouzai* area index measured against the three weather categories constructed by A. M. Tang for China, 1952–1980

correctly captured the changing *overall* weather for China as a whole, or it is tautological to the realized harvest standards. We do not know which is the case.

From an analytical point of view, the quantitative *shouzai* area index seems to constitute a more useful approach. Note that in Fig. AA.4, 13 different years out of the comparable total of 27 years overlap across the 'poor' and 'average' weather categories. Exactly the same number of years may be categorized both as 'average' and 'good' years. Worse still, 10 different years may be lumped into either of the two extreme weather categories of 'good' and 'poor'. This leaves the usefulness of the categorical weather index in considerable doubt.

NOTES

1. Table AB.15.
2. State Meteorological Bureau (Institute of Meteorological Science), *Zhongguo Jin Wubainian Hanlao Fenbu Tuji (Yearly Charts of Dryness/Wetness in China for the Last 500-Year Period)* (Beijing: Ditu chubanshe, 1981).
3. Jen-hu Chang, *Climate and Agriculture: An Ecological Survey* (Chicago: Aldine Publishing Co.), 23–8.
4. Ibid. 114.
5. *TJNJ 1983*, 37.
6. Zhu kezhen, 'Lun Wugou qihoude jige tedian jiqi yi liangshi zuowude guanxi' (On the several characteristics of the climate in our country and their relations to grain crops production), in *DLXB* 30/1 (1964), 5.
7. Ibid. 7.
8. Ibid. 6.
9. Cf. Ch. 9 n. 9.
10. Kenneth Walker, 'Organization of agricultural production', 406–9.
11. *TJNJ 1984*, 189.
12. J. L. Buck, *Land Utilization in China*, 36.
13. Keith Buchanan, *The Transformation of the Chinese Earth* (London: G. Bell & Sons, 1970), 177–85.
14. See Cao Heguang and Luan Binghuan, *Nongye Quhua* (Agricultural Regional Demarcation) (Jinan: Shandong Kexue Jishu chubanshe, 1983).
15. Chinese Academy of Science, *Zhongguo Nongye Dili Zonglun* (A General Treatise on Chinese Agricultural Geography) (Beijing: Kexue Chubanshe, 1981).
16. Zhang Xiangong, 'An analysis of the index of droughts in the eastern half of China in the last 500 years' (as cited in Ch. 8 n. 6), 46–54.
17. Ibid. 50 and other contributors to the *Proceedings of National Symposium on Climatic Changes* (cited in Ch. 8 n. 6) including notably, Xu Ruizhen and Wang Lei, 'Woguo jin wubainian hanlao de chubu fengxi' (A preliminary analysis on droughts and floods in our country in the last 500 years), 59–60.

Cf. also Institute of Atmospheric Physics, *Forecast and Prevention of Calamitous Weather*, 14.

18. Xu Ruizhen and Wang Lei, 'Wogou jin wubainian hanlao de chubu fengxi'.
19. See Ch. 1 n. 7 for details about the relative scale of the 1972 drought.
20. Cf. Tables AB.19 with AB.15.
21. Lu Wu, 'Our country's climate and calamities of droughts and floods', *RMRB*, 6 Jan. 1966.
22. Leslie T. C. Kuo, *The Technical Transformation of Agriculture in Communist China* (New Praeger, 1972), 79, citing the *National Programme for Agricultural Development 1956–1967* (Art. 5).
23. *RMRB*, 4 Oct. 1958.
24. *Qixiang Zhishe*, 191.
25. Institute of Atmospheric Physics, *Forecast and Prevention of Calamitious Weather*, 8–13.
26. Ibid. 8.
27. Zhang Xiangong, as cited in Ch. 8 n. 6, 47.
28. Tables AB.14 and AB.15.
29. See Chao Kang, *Agricultural Production in Communist China*, 129 and 257; and the discussion in Ch. 11 of this study.
30. Kuo, *The Technical Transformation of Agricuture in Communist China*, 88–9; Michael Freeberne, 'The Role of natural calamities in Communist China', in *Current Scene: Developments in Mainland China*, 2/25 (1963), 1–13 and 'Natural Calamities in China, 1949–1961: An examination of the reports originating from the mainland', in *Pacific Viewpoint* 3/2 (1962), 33–72, and many other Western authors. See also Chao Kang, *Agricultural Production in Communist China*, 129 and 240.
31. A personal communication with the State Statistical Bureau (SSB) confirmed that the standard official *shouzai* and *chengzai* areas do refer to sown area rather than cultivated area, distinguishing basically between the summer and autumn crops. For a detailed explanation see notes to Table AB.16.
32. It has also been confirmed by the SSB by way of personal communication that while the *shouzai* and *chengzai* data are consolidated for publication by the Dept. of Agricultural Statistics of the SSB it is indeed the Ministry of Civil Affairs (Minzhengbu) from which the data originally derive.
33. Nanjing (Jinling) University, 'The 1931 floods in China: An economic survey', 203–4.
34. The table is adopted from Kueh, 'A weather index for analysing grain yield instability in China, 1952–1981', 73.
35. Notwithstanding the point made here, a complete time-series of loss rates as shown for the 1950s (Table AA.7) is not available for the subsequent years.
36. In discussing Kueh's weather index, Bruce Stone ('The composition of changes in foodcrop production variability in China, 1931–1985: A discussion of weather, policy, technology, and markets') seems to have had this point in mind, when he noted that 'the regression of Kueh's weather index on aggregate yield is also somewhat *circular*' (p. 14). It should however, be noted in this context that contrary to Stone's suggestion, '*Chengzai* area and probably even *shouzai* area are defined in terms of *actual* yield loss.' (ibid.), a

personal communication with the SSB reveals that the area statistics as compiled by the Ministry of Civil Affairs are obtained during the crop-growing seasons rather than after the close of the harvests. The procedure is therefore the same as that adopted by NARB and official authorities in charge of relief aid in the 1930s. Without being aware of the background, many readers of Kueh's index would probably share Stone's impression and thus regard the *shouzai* and *chengzai* areas as simply being a retrospective projection of harvest loss, 'from whatever factors, both weather- and non-weather-related' (ibid.).

37. This fits in perfectly well with generally perceived weather condition in these two years.
38. James D. McQuigg, 'Climatic constraints on food grain production', 387.
39. *Nature*, 4880, 11 May 1963. Interestingly, *vide RMRB*, 23 Jan. 1960 (p. 2), the natural calamities in the USA in 1959 were not as serious as in China, although total cereals output declined from 190 to 178 million tonnes, and wheat output was reduced by 23% from the 1958 level.
40. McQuigg, 'Climatic constraints on food grain production', 390; cf. also his 'Effective use of weather information in projections of global grain production', 313.
41. Tang, 'Trend, policy cycle, and weather disturbance in Chinese agriculture, 1952–1978', 339–48.
42. The diagram is reproduced from Y. Y. Kueh 'A weather index for analysing grain yield instability in China, 1952–81'. The given *Shouzai* area index is based on the 1952–81 time series, but it is only very slightly different from the one based on the 1952–84 series shown in Table AA.8, p. 229.

Appendix B
Statistical Tables and Sources

TABLE AB.1. Index of food and cereal production in China, USA, USSR, India, and the world as a whole, 1952–1985

	China	USA	USSR	India	World
1952–6 = 100					
1952	94	98	90	90	94
1953	95	97	95	101	98
1954	97	99	96	101	99
1955	105	102	104	102	102
1956	110	104	116	107	107
1957	112	102	119	106	108
1958	140	110	130	111	114
1959	97	110	133	115	116
1960	82	111	134	119	119
1961–5 = 100					
1961	86	97	103	100	93
1962	93	96	108	100	99
1963	99	103	81	103	99
1964	109	95	115	107	105
1965	113	109	93	91	104
1966	124	109	134	91	113
1967	127	123	115	109	118
1968	122	120	132	117	121
1969	123	121	126	121	122
1970	140	111	146	130	125
1971	146	141	142	128	135
1972	140	135	131	122	131
1973	154	141	174	135	142
1979–81 = 100					
1974	81	68	109	77	83
1975	85	83	76	93	86
1976	87	86	127	87	92
1977	84	88	111	100	92
1978	95	91	138	104	100
1979	102	99	102	91	97
1980	98	90	108	103	99

TABLE AB.1.   *cont.*

	China	USA	USSR	India	World
1981	100	111	89	107	104
1982	111	111	101	99	108
1983	122	71	110	121	105
1984	130	105	95	120	115
1985	121	114	107	121	117

Notes: The three time-series are not completely comparable. A distinction is first made between food (1950–60) and cereals (1961–73 and 1974–85) production as defined by the Food and Agriculture Organization (FAO). For 1952–60, the index for USSR also includes East Europe, while that for China refers to grain output by the conventional Chinese definition. For 1961–73 the world total (from FAO) is adjusted by replacing FAO's own estimates for China with the official Chinese figures for grain output.

Sources: China, 1952–60 and 1961–73: Table AB.2(*b*) and AB.2(*c*), USA, USSR, India, and world total; 1952–60: FAO, *The State of Food and Agriculture 1962*, 163–4; 1961–73: FAO, *Production Yearbook 1973*, 43, 46–7, and 50; and ibid. *1974*, 41–2; 1974–85: FAO, *Production Yearbook 1985*, 85–6.

TABLE AB.2.   Grain-sown area, output, and yield in China, 1931–1990
AB.2(*a*):   1931–1937

	Sown area		Output		Yield	
	Million ha. (1)	Index (2)	Million tonnes (3)	Index (4)	Kg./ha. (5)	Index (6)
1931	61.391	100.0	95.46	100.0	1,555	100.0
1932	62.973	102.6	105.43	110.4	1,674	107.7
1933	69.198	112.7	108.16	113.3	1,563	100.5
1934	69.056	112.5	99.02	103.7	1,434	92.2
1935	71.016	115.7	110.43	115.7	1,555	100.0
1936	69.938	113.9	111.10	116.4	1,589	102.2
1937	67.060	109.2	97.95	102.6	1,461	94.0

Notes: Definition and coverage: Grain as defined for the 1931–7 series includes rice, glutinous rice, wheat, maize, barley, oats, gaoliang, millet, prosomillet, sweet potatoes, field peas, and broad beans. The coverage is not exactly the same as that of the post-war series. The latter includes soya bean as well, which was classified as an oil crop in pre-war years. The post-war grain series, as shown in the subsequent tables, are normally given with breakdowns for major grain categories only, namely rice, wheat, maize, soya bean, and potato. The residual category includes gaoliang and millet, but not peas and beans.

TABLE AB.2. *cont.*

Comparative grain yield levels: Average grain yields per sown hectare were consistently higher in 1931–7 than in 1952–7 (Table AB.2(*b*)). This includes the years 1931 and 1934 which were respectively affected by the catastrophic Yangzi-Huai Rivers floods and Yangzi-Huai basins droughts. Note that despite this the average grain yield for 1934 was still higher than that in any of the years 1952–6, and was only slightly lower than that in 1957. Note also that the discrepancy in wheat yields between 1931–7 and 1952–7 was significantly greater than that in rice yields (compare Tables AB.3(*a*) and AB.3(*b*)). The possible reasons for the discrepancy have been discussed by a number of specialists including notably Ta-chung Liu and Kung-chia Yeh, *The Economy of the Chinese Mainland: National Income and Economic Development 1933–1959* (Princeton, NJ: Princeton University Press, 1965), 277–91; Chao Kang, *Agricultural Production in Communist China* (Madison, Wis.: University of Wisconsin Press, 1970), 214–25; Dwight Perkins, *Agricultural Development in China, 1368–1968* (Chicago: Aldine Publishing Company, 1969), 266–70, and most systematically, Thomas B. Wiens, 'Agricultural Statistics in the People's Republic of China', in Alexander Eckstein (ed.), *Quantitative Measures of China's Economic Output* (Ann Arbor, Mich.: University of Michigan Press, 1980), 44–107, and its appendix C, 297–306.

Implications on instability study: One major source of the observed grain yield discrepancy between 1931–7 and 1952–7 is the fact that crop statistics for the 1930s were generated mainly from the more advanced localities which were situated near major waterways, and hence were more easily accessible for statistical reporting. Note that the total sown area for the 1930s (Table AB.2(*a*)) was substantially smaller than that for the 1950s (Table AB.2(*b*)).

However, the absolute differences of grain yield between the two periods should not have any significant bearing in relation to the major objectives of this study. We are mainly concerned with the year-to-year magnitude of agricultural fluctuations within each of the periods taken as a whole. To the extent that instabilities between different periods are compared, the relatively higher average grain yields of the 1930s may imply a higher level of agricultural technology (e.g. better irrigation facilities), and hence a lesser degree of instability compared with, say, the 1950s. However, this is a matter subject to interpretation. And empirical analysis may or may not prove this to be the case. This point is discussed in greater detail in the text.

Another methodological point concerns the difference in the coverage of grain statistics between the pre-war and post-war periods. The difference as shown above is obviously too marginal as to distort the possible outcome of our national aggregate analysis.

The time-series used: This is the complete series generated by the familiar National Agricultural Research Bureau (NARB) for 1931–7. NARB was the only organization that released estimates of crop production on an annual basis before the war (cf. Liu and Yeh, *Economy of the Chinese Mainland*, 126). For a detailed discussion of the nature of the NARB estimates, see Thomas Wiens, 'Agricultural Statistics in the People's Republic of China', 275–306.

Sources: Tables AB.5 (sown area), AB.6 (output), and AB.7 (yield).

Appendix B

AB.2(*b*):   1952–1969

	Sown area		Output		Yield	
	Million ha. (1)	Index (2)	Million tonnes (3)	Index (4)	Kg./ha. (5)	Index (6)
1952	123.979	100.0	163.92	100.0	1,320	100.0
1953	126.637	102.1	166.83	101.8	1,320	100.0
1954	128.995	104.1	169.52	103.4	1,313	99.5
1955	129.839	104.7	183.94	112.2	1,418	107.4
1956	136.339	110.0	192.75	117.6	1,410	106.8
1957	133.633	107.8	195.05	119.0	1,463	110.8
1958	127.613	102.9	200.00	122.0	1,568	118.8
1959	116.023	93.6	170.00	103.7	1,463	110.8
1960	122.429	98.7	143.50	87.5	1,170	88.6
1961	121.443	98.0	147.50	90.0	1,215	92.0
1962	121.621	98.1	160.00	97.6	1,313	99.5
1963	120.741	97.4	170.00	103.7	1,410	106.8
1964	122.103	98.5	187.50	114.4	1,538	116.5
1965	119.627	96.5	194.53	118.7	1,628	123.3
1966	120.988	97.6	214.00	130.6	1,770	134.1
1967	119.230	96.2	217.85	132.9	1,830	138.6
1968	115.196	92.9	209.00	127.5	1,800	136.4
1969	117.604	94.9	210.97	128.7	1,793	135.8

Sources: *TJNJ 1983*, 154 (sown area) and 158 (output). Average yields per sown hectare are derived from output and sown area.

AB.2(*c*):   1970–1990

	Sown area		Output		Yield	
	Million ha. (1)	Index (2)	Million tonnes (3)	Index (4)	Kg./ha. (5)	Index (6)
1970	119.267	100.0	239.96	100.0	2,010	100.0
1971	120.846	101.3	250.14	104.2	2,070	103.0
1972	121.209	101.6	240.48	100.2	1,986	98.8
1973	121.156	101.6	264.94	110.4	2,190	109.0
1974	120.976	101.4	275.27	114.7	2,273	113.1
1975	121.062	101.5	284.52	118.6	2,348	116.8
1976	120.743	101.2	286.31	119.3	2,370	117.9
1977	120.400	100.9	282.73	117.8	2,348	116.8
1978	120.587	101.1	304.77	127.0	2,528	125.8
1979	119.263	100.3	332.12	138.4	2,783	138.5
1980	117.234	98.3	320.56	133.6	2,738	136.2
1981	114.958	96.4	325.02	135.4	2,828	140.7
1982	113.463	95.1	354.50	147.7	3,124	155.4
1983	114.047	95.6	387.28	161.4	3,396	169.0
1984	112.884	94.6	407.31	169.7	3,608	179.5
1985	108.845	91.3	379.11	158.0	3,480	173.1
1986	110.933	93.0	391.51	163.2	3,525	175.3
1987	111.268	93.3	402.98	167.9	3,615	179.9
1988	110.123	92.3	394.08	164.2	3,585	178.4
1989	112.205	94.1	407.55	169.8	3,630	180.6
1990	113.466	95.1	435.00	181.3	3,834	190.7

Sources: 1970–83: *TJNJ 1984*, 137 (sown area) and 141 (output); 1984–9: *TJNJ 1990*, 357 (sown area) and 363 (output); 1990: *TJZY 1991*, 56.

TABLE AB.3. Rice, wheat, and grain-sown area, output, and yield in China, 1931–1990, with computed instability indices
AB.3(a):1931–1937

	Sown area (million ha.)			Output (million tonnes)			Yield (kg./ha.)		
	Rice	Wheat	Grain	Rice	Wheat	Grain	Rice	Wheat	Grain
1931	16.221	19.485	61.391	40.87	21.48	95.46	2,520	1,102	1,555
1932	16.385	20.495	62.973	47.02	22.56	105.43	2,870	1,101	1,674
1933	16.674	19.364	69.198	43.95	22.39	108.16	2,636	1,156	1,563
1934	16.518	19.384	69.056	34.84	22.34	99.02	2,109	1,153	1,434
1935	16.710	20.500	71.016	43.53	20.97	110.43	2,605	1,023	1,555
1936	16.350	20.075	69.938	43.53	22.78	111.10	2,663	1,135	1,589
1937	15.262	17.699	67.060	40.83	15.56	97.95	2,676	879	1,461
Index	2.22	3.74	3.63	6.64	7.88	5.35	6.07	6.45	3.57

Notes: See Ch. 2 for the methodology of estimating the 'instability indices'.

Sources: Table AB.2(a) (grain-sown area, output, and yield); Tables AB.8 (rice yield) and AB.9 (wheat yield); and same sources as Tables AB.8 and AB.9 respectively for rice- and wheat-sown area and output.

AB.3(b): 1952–1966

	Sown area (million ha.)			Output (million tonnes)			Yield (kg./ha.)		
	Rice	Wheat	Grain	Rice	Wheat	Grain	Rice	Wheat	Grain
1952	28.382	24.780	123.979	68.43	18.13	163.92	2,415	735	1,320
1953	28.321	25.636	126.637	71.27	18.28	166.83	2,520	713	1,320
1954	28.722	26.967	128.995	70.85	23.34	169.52	2,468	863	1,313
1955	29.173	26.739	129.839	78.03	22.97	183.94	2,678	863	1,418
1956	33.312	27.272	136.339	82.48	24.80	192.75	2,475	908	1,410
1957	32.241	27.542	133.633	86.78	23.64	195.05	2,693	855	1,463
1958	31.915	25.775	127.613	80.85	22.59	200.00	2,535	878	1,568
Index 1952–8	2.82	2.67	1.81	2.93	7.08	1.45	3.01	5.03	2.02
1959	29.033	23.575	116.023	69.37	22.18	170.00	2,393	938	1,463
1960	29.607	27.294	122.429	59.73	22.17	143.50	2,018	810	1,170
1961	26.276	25.572	121.443	53.64	14.25	147.50	2,040	555	1,215
1962	26.935	24.075	121.621	62.99	16.67	160.00	2,340	690	1,313
1963	27.715	23.771	120.741	73.77	18.48	170.00	2,663	780	1,410
1964	29.607	25.408	122.103	83.00	20.84	187.50	2,805	818	1,538
1965	29.825	24.709	119.627	87.72	25.22	194.53	2,940	1,020	1,628
1966	30.529	23.919	120.988	95.39	25.28	214.00	3,128	1,058	1,770
Index 1952–66	5.18	3.83	2.50	11.58	13.54	8.92	8.06	11.93	7.19

Sources: Rice and wheat: *TJNJ 1984*, 138 (sown area) and 141 (output); and grain: Table AB.2(b).

AB.3(c): 1970–1990

	Sown area (million ha.)			Output (million tonnes)			Yield (kg./ha.)		
	Rice	Wheat	Grain	Rice	Wheat	Grain	Rice	Wheat	Grain
1970	32.358	25.458	119.267	109.99	29.19	239.96	3,398	1,148	2,010
1971	34.918	25.639	120.846	115.21	32.58	250.14	3,300	1,268	2,070
1972	35.143	26.302	121.209	113.36	35.99	240.48	3,225	1,365	1,986
1973	35.090	26.439	121.156	121.74	35.23	264.94	3,473	1,335	2,190
1974	35.512	27.061	120.976	123.91	40.87	275.27	3,488	1,508	2,273
1975	35.729	27.661	121.062	125.56	45.31	284.52	3,518	1,635	2,348
1976	36.217	28.417	120.743	125.81	50.39	286.31	3,473	1,770	2,370
1977	35.526	28.065	120.400	128.57	41.08	282.73	3,623	1,463	2,348
Index 1970–7	1.64	0.69	0.40	1.54	6.28	2.16	1.81	5.70	2.02
1978	34.421	29.183	120.587	136.93	53.84	304.77	3,975	1,845	2,528
1979	33.873	29.357	119.263	143.75	62.73	332.12	4,245	2,138	2,783
1980	33.879	29.228	117.234	139.91	55.21	320.56	4,133	1,890	2,738
1981	33.295	28.307	114.938	143.96	59.64	325.02	4,320	2,108	2,828
1982	33.071	27.941	113.463	161.60	68.47	354.50	4,886	2,449	3,124
1983	33.137	29.050	114.047	168.87	81.39	387.28	5,096	2,809	3,356
1984	33.179	29.577	112.884	178.26	87.82	407.31	5,373	2,969	3,608

Index								
1978–84	0.53	1.58	0.61	2.85	6.90	3.18	2.73	5.55	2.56
Index								
1970–84	2.13	1.99	1.23	2.97	6.45	3.27	4.60	6.19	3.66
1985	32.070	29.218	108.845	168.57	85.81	379.11	5,250	2,940	3,480
1986	32.266	29.616	110.933	172.22	90.04	391.51	5,340	3,045	3,525
1987	32.193	28.798	111.268	174.26	85.90	402.98	5,415	2,985	3,615
1988	31.987	28.785	110.123	169.11	85.43	394.08	5,280	2,970	3,585
1989	32.701	29.841	112.205	180.13	90.81	407.55	5,505	3,045	3,630
1990	—	—	113.466	—	—	435.00	—	—	—
Index								
1985–90	—	—	0.98	—	—	2.67	—	—	1.76
1978–90	—	—	1.47	—	—	3.24	—	—	3.99

Sources: Rice and wheat: *TJNJ 1984*, 138 (sown area) and 141 (output) for 1970–83; and *TJNJ 1990*, 357 (sown area) and 363 (output) for 1984–9 (statistics for 1990 not yet available as of July 1990).

Grain: Table AB.2(c).

TABLE AB.4. Maize, soya bean, and potato sown area, output, and yield in China, 1952–1984 (with computed instability indices for subperiods)

	Sown area (million ha.)			Output (million tonnes)			Yield (kg./ha.)		
	Maize	Soya bean	Potato	Maize	Soya bean	Potato	Maize	Soya bean	Potato
1952	12.566	11.679	8.488	16.85	9.52	16.33	1,343	815	1,880
1953	13.134	12.362	9.016	16.69	9.93	16.66	1,271	803	1,848
1954	13.171	12.654	9.781	17.14	9.08	16.98	1,301	718	1,736
1955	14.639	11.442	10.054	20.32	9.12	18.90	1,388	797	1,880
1956	17.662	12.647	10.992	23.05	10.24	21.85	1,305	850	1,988
1957	14.943	12.748	10.495	21.44	10.05	21.92	1,433	788	2,089
1958	—	9.551	15.382	—	8.67	32.73	—	908	2,128
Index 1952–8	4.39	6.73	6.15	5.33	5.18	8.26	3.10	5.05	3.02
1970	15.871	7.985	10.717	33.03	8.71	26.68	2,086	1,091	2,489
1971	16.726	7.791	10.405	35.85	8.61	25.07	2,143	1,105	2,410
1972	16.703	7.583	10.841	32.10	6.45	24.52	1,922	851	2,262
1973	16.571	7.408	11.306	38.63	8.37	31.56	2,331	1,130	2,791
1974	17.410	7.261	11.069	42.92	7.47	28.24	2,465	1,029	2,551
1975	18.598	6.999	10.969	47.22	7.24	28.57	2,539	1,034	2,605
1976	19.228	6.691	10.366	48.16	6.64	26.66	2,505	992	2,572
1977	19.658	6.845	11.229	49.35	7.26	29.67	2,512	1,061	2,642

	Index 1970–7								
	3.27	0.80	2.22	4.24	6.29	5.46	4.14	5.71	3.42
1978	19.961	7.144	4.867	55.95	7.57	31.74	2,805	1,058	2,653
1979	20.133	7.247	4.579	60.04	7.46	28.46	2,985	1,028	2,595
1980	20.395	7.227	4.920	62.60	7.94	28.73	3,075	1,095	2,828
1981	19.425	8.023	5.185	59.21	9.33	25.97	3,045	1,163	2,700
1982	18.543	8.419	5.829	60.56	9.03	27.05	3,266	1,073	2,887
1983	18.824	7.567	6.077	68.21	9.76	29.25	3,623	1,290	3,111
1984	18.537	7.286	6.923	73.41	9.70	28.48	3,960	1,331	3,169
Index 1978–84	0.97	4.64	4.02	2.95	3.43	4.54	2.88	3.73	2.89

Sources: *TJNJ 1984*, 138 (sown area) and 141 (output) for 1952–83; and *TJNJ 1989*, 242 (sown area) and 248 (output) for 1984.

TABLE AB.5. Provincial grain-sown area in China, 1931–1937 (1,000 ha.)

	1931	1932	1933	1934	1935	1936	1937	1931–7 Average
Hebei	6,178	6,589	6,354	6,341	6,572	5,915	5,153	6,158
Shanxi	3,347	3,399	3,839	3,790	3,599	3,605	3,585	3,598
Chahaer	—	—	—	—	—	—	—	—
Suiyuan	562	585	873	1,015	990	1,040	968	863
Liaoning	—	—	—	—	—	—	—	—
Jilin	—	—	—	—	—	—	—	—
Heilongjiang	—	—	—	—	—	—	—	—
Jiangsu	5,354	5,500	6,914	6,803	7,223	7,081	6,811	6,535
Zhejiang	2,647	2,810	3,133	3,160	3,245	3,160	3,163	3,046
Anhui	3,263	3,244	3,757	3,638	3,790	3,718	3,584	3,571
Fujian	1,168	1,197	1,333	1,382	1,433	1,463	1,546	1,360
Jiangxi	2,125	2,128	2,537	2,226	2,640	2,727	2,647	2,434
Shandong	7,422	7,460	7,171	7,305	7,341	7,104	6,467	7,182
Henan	7,681	8,039	8,210	8,159	8,350	8,324	7,535	8,042
Hubei	4,092	4,150	4,611	4,740	4,790	4,834	4,688	4,559
Hunan	2,277	2,350	2,727	2,731	2,811	2,834	2,880	2,659
Guangdong	3,846	3,904	3,957	4,080	3,830	3,518	3,476	3,802
Guangxi	—	—	—	—	—	—	—	—
Sichuan	5,797	5,823	7,125	7,130	7,304	7,347	7,243	6,825

Guizhou	1,166	1,207	1,293	1,281	1,364	1,343	1,287	1,278
Yunnan	1,635	1,651	2,017	1,941	1,828	1,996	2,019	1,870
Xizang	—	—	—	—	—	—	—	—
Shaanxi	1,749	1,801	2,001	1,948	1,969	1,984	2,017	1,923
Gansu	1,002	1,058	1,209	1,274	1,399	1,406	1,448	1,257
Qinghai	—	—	—	—	424	425	436	428
Ningxia	80	78	87	112	114	114	107	100
Xinjiang	—	—	—	—	—	—	—	—
China	61,391	62,973	69,198	69,056	71,016	69,938	67,060	67,233

Notes: For Qinghai province, grain-sown area statistics are available for 1935–7 only. These are included in the national total. Thus, the calculated totals are not exactly comparable between 1931–4 and 1935–7, although the differences between the two sets of national totals with and without Qinghai is very marginal. The figures for 1935–6 refer to harvested area rather than sown area. For Jiangxi province the figure for 1931–4 omits a total of 17 *xian*. This also applies to the grain output statistics (Table AB.6).

Sources: Xu Daofu, *Zhongguo Jindai Nongye Shengchan Yi Maoyi Tongji Ziliao* (Agricultural Production and Trade Statistical Materials of Modern China) (Shanghai: Renmin Chubanshe, 1983), 12–79, as compiled from the NARB, *Crop Reports*, iii–vii, various issues.

Appendix B

TABLE AB.6. Provincial grain output in China, 1931–1937 (thousand tonnes)

	1931	1932	1933	1934	1935	1936	1937	1931–7 Average
Hebei	7,071	7,381	2,090	6,711	6,712	6,967	4,774	6,655
Shanxi	2,516	3,059	3,220	3,507	2,933	3,157	2,832	3,023
Chahaer	—	—	—	—	—	—	—	—
Suiyuan	466	598	691	849	821	805	840	734
Liaoning	—	—	—	—	—	—	—	—
Jilin	—	—	—	—	—	—	—	—
Heilongjiang	—	—	—	—	—	—	—	—
Jiangsu	8,545	10,225	11,793	10,399	12,711	13,026	12,292	11,279
Zhejiang	5,212	6,488	5,819	4,104	6,363	6,399	6,283	5,834
Anhui	4,447	4,812	5,563	4,349	4,554	6,324	5,315	5,055
Fujian	2,753	2,723	2,658	3,015	3,171	3,195	3,702	3,031
Jiangxi	4,186	4,501	4,996	2,446	4,976	5,635	5,182	4,542
Shandong	10,075	10,065	9,445	9,264	9,034	9,650	7,872	9,335
Henan	7,895	8,722	10,043	9,584	9,520	10,068	6,584	8,886
Hubei	6,225	7,569	7,869	6,719	7,071	8,030	6,839	7,133
Hunan	5,196	6,916	6,026	4,478	6,590	7,299	6,876	6,223

Guangdong	9,188	9,561	9,210	9,481	9,039	7,106	8,704	8,893
Guangxi	—	—	—	—	—	—	—	—
Sichuan	13,636	15,378	15,388	15,415	16,066	13,877	10,902	14,536
Guizhou	2,313	2,208	2,204	2,016	2,200	2,204	2,154	2,195
Yunnan	—	—	—	—	—	—	—	—
Xizang	2,892	3,060	3,041	3,349	3,172	3,322	2,776	3,107
Shaanxi	1,726	1,297	1,814	2,352	2,405	2,088	1,597	1,892
Gansu	1,030	866	1,187	1,377	1,414	1,409	1,292	1,224
Qinghai					519	413	1,009	479
Ningxia	91	91	105	147	116	128	130	117
Xinjiang	—	—	—	—	—	—	—	—
China	95,463	105,431	108,162	99,022	110,431	111,102	97,954	103,942

Notes: Sweet potatoes have been converted to grain equivalent at the approximate calorie equivalent value of 5 tonnes potatoes to 1 tonne grain.

Source: As Table AB.5.

TABLE AB.7. Provincial grain yield per sown hectare in China, 1931–1937 (kg./ha.)

	1931	1932	1933	1934	1935	1936	1937	1931–7 Average
Hebei	1,145	1,120	1,116	1,058	1,021	1,178	926	1,081
Shanxi	752	900	839	925	815	876	790	842
Chahaer	—	—	—	—	—	—	—	—
Suiyuan	829	1,022	792	836	829	774	868	850
Liaoning	—	—	—	—	—	—	—	—
Jilin	—	—	—	—	—	—	—	—
Heilongjiang	—	—	—	—	—	—	—	—
Jiangsu	1,596	1,859	1,693	1,529	1,760	1,840	1,805	1,726
Zhejiang	1,969	2,309	1,857	1,299	1,961	2,025	1,986	1,915
Anhui	1,363	1,483	1,481	1,196	1,202	1,701	1,483	1,415
Fujian	2,358	2,274	1,993	1,281	2,213	2,183	2,396	2,228
Jiangxi	1,970	2,116	1,970	1,099	1,885	2,066	1,958	1,866
Shandong	1,357	1,349	1,314	1,268	1,231	1,359	1,217	1,300
Henan	1,028	1,085	1,223	1,175	1,140	1,210	874	1,105
Hubei	1,521	1,824	1,707	1,304	1,476	1,661	1,459	1,565
Hunan	2,282	2,944	2,210	1,640	2,345	2,575	2,388	2,341
Guangdong	2,389	2,449	2,328	2,324	2,360	2,020	2,504	2,339
Guangxi	—	—	—	—	—	—	—	—
Sichuan	2,352	2,641	2,160	2,162	2,200	1,889	1,505	2,130
Guizhou	1,984	1,830	1,704	1,575	1,614	1,642	1,674	1,717
Yunnan	1,769	1,853	1,508	1,726	1,735	1,665	1,375	1,661
Xizang	—	—	—	—	—	—	—	—
Shaanxi	987	720	906	1,208	1,221	1,052	792	984
Gansu	1,028	819	982	1,081	1,010	1,002	893	973
Qinghai	—	—	—	—	1,225	973	2,314	—
Ningxia	1,132	1,160	1,205	1,306	1,020	1,122	1,214	1,166
Xinjiang	—	—	—	—	—	—	—	—
China	1,555	1,674	1,563	1,434	1,555	1,589	1,461	1,546

Sources: Derived from sown area (Table AB.5) and output (Table AB.6).

TABLE AB.8. Provincial rice yield per sown hectare in China, 1931–1937 (kg./ha.)

	1931	1932	1933	1934	1935	1936	1937	1931–7 Average
Hebei	1,470	1,350	1,493	1,470	1,538	1,358	1,770	1,468
Shanxi	—	—	555	593	645	563	570	540
Chahaer	—	—	—	—	—	—	—	1,017
Suiyuan	—	—	—	—	—	—	—	—
Liaoning	—	—	—	—	—	—	—	—
Jilin	—	—	—	—	—	—	—	—
Heilongjiang	—	—	—	—	—	—	—	—
Jiangsu	2,153	3,150	2,918	2,153	2,880	3,083	2,993	2,784
Zhejiang	2,505	2,880	2,430	1,545	2,678	2,828	2,663	2,505
Anhui	2,153	2,498	2,460	1,170	1,538	2,415	2,678	2,175
Fujian	2,895	2,783	2,490	2,745	2,805	2,798	3,023	2,797
Jiangxi	2,595	2,738	2,633	1,313	2,633	2,753	2,543	2,485
Shandong	—	—	615	728	615	818	705	705
Henan	1,808	1,838	1,905	1,718	1,313	1,943	2,423	1,891
Hubei	1,988	2,528	2,595	1,620	2,123	2,520	2,610	2,277
Hunan	2,625	3,443	2,738	1,958	3,053	3,473	3,090	2,910
Guangdong	2,595	2,670	2,595	2,528	2,580	2,213	2,798	2,569
Guangxi	—	—	—	—	—	—	—	2,320
Sichuan	3,008	3,375	2,963	2,798	3,053	2,490	2,130	2,869
Guizhou	2,400	2,168	2,235	1,838	2,145	2,003	2,430	2,172
Yunnan	2,378	2,573	2,310	2,445	2,325	2,460	2,070	2,374
Xizang	—	—	—	—	—	—	—	—
Shaanxi	2,445	1,928	2,348	1,898	1,778	2,303	1,920	2,104
Gansu	—	—	375	375	210	255	158	264
Qinghai	—	—	—	—	—	—	—	—
Ningxia	—	—	—	675	848	825	915	798
Xinjiang	—	—	—	—	—	—	—	—
China	2,520	2,870	2,636	2,109	2,605	2,663	2,676	2,578

Notes: All figures exclude glutinous rice.

Source: As Table AB.5.

TABLE AB.9. Provincial wheat yield per sown hectare in China, 1931–1937 (kg./ha.)

	1931	1932	1933	1934	1935	1936	1937	1931–7 Average
Hebei	900	855	1,058	870	758	773	660	850
Shanxi	578	705	780	900	720	780	600	726
Chahaer	—	—	—	—	—	—	—	778
Suiyuan	510	728	758	863	698	653	840	728
Liaoning	—	—	—	—	—	—	—	—
Jilin	—	—	—	—	—	—	—	—
Heilongjiang	—	—	—	—	—	—	—	—
Jiangsu	1,448	1,395	1,335	1,410	1,260	1,283	1,245	1,340
Zhejiang	1,110	1,155	1,058	1,073	848	998	1,155	1,056
Anhui	1,020	1,043	1,110	1,283	998	1,275	810	1,077
Fujian	1,178	1,283	1,223	1,193	1,095	1,103	1,238	1,185
Jiangxi	1,058	1,110	1,110	1,110	690	923	938	968
Shandong	1,118	1,103	1,133	1,043	930	1,028	998	1,050
Henan	1,028	1,028	1,230	1,065	1,013	1,290	570	1,044
Hubei	1,358	1,418	1,433	1,200	1,208	1,320	1,050	1,283
Hunan	1,110	1,320	1,200	1,215	1,043	1,155	1,260	1,186
Guangdong	908	990	863	990	915	1,035	983	956
Guangxi	—	—	—	—	—	—	—	1,049
Sichuan	1,778	1,995	1,823	1,973	1,815	1,778	1,208	1,750
Guizhou	1,493	1,470	1,283	1,703	1,148	1,455	1,103	1,377
Yunnan	1,403	1,440	1,050	1,388	1,298	1,425	983	1,271
Xizang	—	—	—	—	—	—	—	—
Shaanxi	818	555	645	1,268	1,215	915	450	864
Gansu	735	623	698	1,103	923	795	758	805
Qinghai	—	—	—	—	1,313	1,035	1,170	1,171
Ningxia	998	1,080	983	1,125	743	1,073	1,125	1,025
Xinjiang	—	—	—	—	—	—	—	—
China	1,102	1,101	1,156	1,153	1,023	1,135	879	1,079

Source: As Table AB.5.

TABLE AB.10. Provincial output, sown area, and average yields of wheat and rice in a 'normal' year by the standard of the 1930s in China

	Wheat			Rice		
	Output (1,000 metric tonnes)	Sown area (1,000 ha.)	Yields kg./ha.	Output (1,000 metric tonnes)	Sown area (1,000 ha.)	Yields kg./ha.
Hebei	1,828	1,925	950	45	29	1,540
Shanxi	1,031	1,015	1,016	29	12	2,385
Chahaer	149	153	978	23	13	1,704
Suiyuan	137	165	836	—	—	—
Liaoning	229	169	1,355	240	96	2,502
Jilin	826	574	1,440	318	79	4,033
Heilongjiang	743	590	1,259	153	4	3,522
Jiangsu	3,313	2,588	1,280	4,294	1,592	2,698
Zhejiang	701	553	1,271	4,297	1,441	2,981
Anhui	1,585	1,308	1,212	3,512	1,274	2,757
Fujian	321	247	1,295	2,674	915	2,924
Jiangxi	297	270	1,102	4,945	1,761	2,808
Shangdong	3,641	3,053	1,193	26	10	2,529
Henan	3,710	3,657	1,014	377	212	1,775
Hubei	1,663	1,152	1,444	4,572	1,372	3,332
Hunan	306	212	1,445	6,067	1,522	3,988
Guangdong	156	74	2,108	8,450	3,029	2,790
Guangxi	—	—	—	—	—	—
Sichuan	1,579	1,133	1,394	791	2,551	3,234
Guizhou	273	163	1,679	1,886	561	3,362
Yunnan	368	273	1,347	1,900	693	2,741

Appendix B

TABLE AB.10. *cont.*

	Wheat			Rice		
	Output (1,000 metric tonnes)	Sown area (1,000 ha.)	Yields kg./ha.	Output (1,000 metric tonnes)	Sown area (1,000 ha.)	Yields kg./ha.
Xizang	—	—	—	—	—	—
Shaanxi	1,119	911	1,229	299	124	2,400
Gansu	745	532	1,400	528	33	1,601
Qinghai	—	—	—	—	—	—
Ningxia	63	31	2,031	532	18	3,003
Xinjiang	455	289	1,571	1,939	90	2,149
China	25,268	21,044	1,201	52,361	19,766	2,649

Notes: Definition and nature of 'normal yields'. Several concepts of 'yield' per sown hectare were given in the 1930s. They included, for example, the 'most frequent' yields (*tongchang chanliang*) developed by John L. Buck, *Land Utilization in China* (Nanjing: Nanjing University Press, 1937). Another concept is the 'complete year yield' (as translated from *shizunian chanliang* by Thomas Wiens, 'Agricultural Statistics in the PRC', in Alexander Eckstein (ed.), *Quantitative Measures of China's Economic Output* (Ann Arbor, Mich.: University of Michigan Press, 1980), 300). The latter measure was used by NARB from 1934–5 onwards as a standard for gauging the extent of yield losses in individual years.

However, the measure used in this table is the 'normal year yield' (*pingchangnian chanliang*). This series was generated by C. C. Chang (Zhang Xinyi), Head of Industrial (Chanye) Statistics Section under the Directorate of Statistics of the National Government in 1932. It was officially translated as 'yield in average years' and was used by NARB prior to 1935 for measuring yield losses before its being fully replaced by the 'complete'—or officially 'best' (*shizu*) year yield series in 1935 (cf. *NARBCR* 8 (20 Sept. 1934), 494–5, and 11 (10 Feb. 1935), 141.

The official translations of the terms are, however, not always consistent. NARB also used 'normal' for '*shizu*' in the English version of its *Crop Reports*, while in 1932 for example, the Directorate of Statistics applied 'average years' and 'normal years' interchangeably for '*pingchangnian*'.

It is difficult to distinguish exactly between the different yield measures. Buck tends to see the 'normal year yields' as something comparable to his 'most frequent yields', while Wiens sees it as implying 'normal yields', (i.e. 'yields in the absence of adverse conditions') (Wiens, 303).

Nevertheless, Wien interprets at the same time the 'complete (*shizu*) year yields' as 'normal yields' as well (being from 'fields unaffected by natural disasters') (Wiens, 300).

There is, however, a clear distinction between the 'complete year yields' and 'normal year yields', both as applied by NARB. The former refers obviously to the 'maximally available yields' (hence the term *shizu*) (see *SBNJ 1936*, 843). While it is not clear how this 'maximum yields' series was compiled in the 1930s, the relevant provincial figures are (with few exceptions) consistently very much higher than the 'normal year yields' series. For the country as a whole, it can be estimated that average yields of rice and wheat according to the former measure were respectively 36 and 40% higher than the latter.

The nature of the 'normal year yields' series is also not clear. Similar to the 'maximum yield series', it was presumably surveyed only once. Granted, however, that yields per sown hectare normally fluctuate between exceptionally good and bad years which are interspersed among the normal years, one would expect the value of the estimated 'normal year yields' to approach the long-run average. In fact at least for rice and wheat crops 'normal yields' are respectively only by 2.8 and 11% higher than the 1931–7 means (cf. Tables AB.8 and AB.9), although the discrepancy may vary more substantially from province to province.

For want of anything better, we make use of the 'normal yield' series as an alternative to the 1931–7 means for analysing, as is in the text, how deviations in rice and wheat yields from 'normal yields' in different provinces were related to the variations in precipitations in the respective growing periods.

Statistical coverage: All the rice yield figures exclude glutinous rice. The figures for Rehe province are included in those for Chahaer.

Sources: TIYB (Special issue on Agriculture), combined Jan. and Feb. 1932 issue, as reproduced in *SBNJ 1933*, M-246–247, and *MGJJNJ* i (Nanjing: Commercial Press, 1934), F-88–90.

Appendix B

TABLE AB.11. Provincial and regional instability indices of grain-sown area, output, and yield in China, selected periods, 1931–1984

	Sown area			Output				Yield		
	1931–7	1952–7	1979–84	1931–7	1952–7	1974–9	1979–84	1931–7	1952–7	1979–84
North-east										
Liaoning	—	4.26	1.03	—	3.37	3.98	6.03	—	3.78	5.20
Jilin	—	1.33	1.00	—	3.46	5.24	10.19	—	3.72	9.65
Heilongjiang	—	2.94	0.91	—	6.63	6.82	11.31	—	5.90	10.51
Regional	—	1.78	0.80	—	3.12	5.27	8.94	—	1.44	8.20
North-west										
Suiyuan	11.7	—	—	8.84	—	—	—	6.58	—	—
Neimenggu	—	2.00	0.81	—	11.42	—	5.81	—	10.64	5.39
Shaanxi	2.26	2.71	1.25	15.49	8.23	8.93	7.60	15.3_	6.47	7.61
Gansu	3.12	2.62	0.80	9.29	6.47	2.84	5.15	7.10	5.80	4.29
Qinghai	—	1.15	0.91	—	9.22	5.51	5.64	—	10.19	4.11
Ningxia	7.13	—	2.00	7.72	—	6.73	3.68	5.62	—	2.73
Xinjiang	—	3.17	0.91	—	1.36	3.45	3.65	—	4.44	2.69
Regional	4.48	1.14	0.91	10.83	6.26	5.56	3.99	8.77	9.39	3.26
North										
Beijing	—	—	0.81	—	15.13	—	2.56	—	—	2.02
Tianjin	—	—	2.91	—	8.81	—	7.99	—	—	7.39
Hebei	4.80	1.72	0.65	7.35	7.86	3.75	5.37	4.25	8.47	5.49
Shanxi	3.75	1.33	0.76	8.16	6.68	—	5.56	5.89	5.74	4.86
Shandong	2.50	1.48	1.65	3.25	6.10	3.53	5.25	4.25	6.79	3.96
Henan	3.12	1.31	1.26	12.35	3.50	2.97	4.68	9.33	2.84	3.83

	3.04	1.04	0.94	8.44	2.50	2.17	4.87	5.98	4.44	3.81
Regional										
Central										
Jiangxi	6.99	—	0.29	15.42	2.32	3.17	2.87	13.38	—	2.61
Hubei	3.24	4.16	1.23	7.86	8.42	2.96	6.95	8.56	6.60	6.14
Hunan	3.43	1.76	1.17	10.14	4.98	0.73	3.64	11.48	5.81	2.70
Regional	4.56	2.88	0.96	9.37	4.81	1.28	4.02	8.92	5.32	2.76
East										
Shanghai	—	—	4.03	—	9.54	6.09	11.49	—	—	10.05
Jiangsu	6.29	7.86	0.62	7.03	3.18	5.44	3.76	4.46	4.92	3.59
Zhejiang	3.90	4.58	0.56	10.05	1.10	11.78	6.83	10.90	5.17	6.41
Anhui	3.91	3.48	1.43	9.49	7.45	4.71	3.95	9.34	8.22	3.76
Regional	5.05	5.36	0.52	7.38	3.36	4.25	4.55	10.85	5.68	4.35
South-west										
Sichuan	4.85	1.12	1.12	8.90	1.09	2.90	1.53	7.58	1.77	1.34
Guizhou	2.74	2.62	2.00	2.39	1.32	—	4.86	4.23	1.67	1.60
Yunnan	4.54	2.97	0.66	5.09	1.06	3.14	1.94	6.79	3.03	2.50
Xizang	—	—	1.95	—	—	—	9.31	—	—	9.83
Regional	4.18	1.44	0.56	7.50	0.94	2.90	0.91	6.29	1.59	1.00
South										
Fujian	1.75	0.69	0.96	3.38	3.25	1.36	1.35	13.54	2.73	0.92
Guangdong	3.52	1.82	1.45	5.94	1.85	3.12	3.21	4.11	1.43	2.71
Guangxi	—	4.90	0.91	—	5.20	1.44	4.80	—	2.79	4.17
Regional	2.93	2.25	0.62	4.52	1.45	2.07	3.10	4.03	1.47	2.76

TABLE AB.11. *cont.*

Notes: (1) The regional totals are not strictly comparable between periods, due to differences in provincial coverage. This is especially so for the period 1974–9 (details below), and the inclusion in the North-west of Xinjiang and Qinghai in the post-war periods. But note that Suiyuan (1931–7) became part of Neimenggu in the later periods, and Ningxia was merged with Gansu in the 1950s. Thus, the discrepancy may not be too serious. Of course, in addition to Suiyuan, Neimenggu as established in the 1950s comprises Chahaer, and Rehe of the 1930s. But data for these two latter provinces are not available. Nevertheless, for the provincial instability indices, no effort has been made to adjust for possible discrepancies arising from the redemarcation of the provincial borders, especially Jiangsu in relation to Shanghai, and Hebei to Beijing and Tianjin. The three municipalities were part of the two related provinces in the 1930s, and their present boundaries are not exactly the same as in the 1950s. This may tend to undermine somewhat the intertemporal comparability for these individual provinces, but not the regional total.

(2) The provincial indices do not cover the whole period 1974–9 for Gansu and Ningxia (six years only 1976–81), Jiangxi (five years 1976–80), Yunnan (five years, 1977–81) and Fujian (five years, 1975–80). The regional totals do cover 1974–9, but for lack of data some provinces (as indicated in the note and the table) are excluded. The only exceptions are Ningxia and Jiangxi. Both lack only 1975 grain output data. I have used the average of 1974 and 1976 to fill the gap. The resultant distortion should be minimal in view of their small share within the region.

Sources: 1931–7: Tables AB.5 (sown area), AB.6 (output), and AB.7 (yield): 1952–7: Kenneth R. Walker, *Foodgrain Procurement and Consumption in China* (Cambridge: Cambridge University Press, 1984), 221 (sown area); 202 (output); and 239 (yield per sown hectare), supplemented by Chao Kang, *Agricultural Production in Communist China* (Madison, Wis.: University of Wisconsin Press, 1970), 296–7 (sown area), and 300–1 (yield); 1974–8: Walker, 319 (output); 1979–80: *NYNJ 1981*, 22; 1981: *NYNJ 1982*, 34; 1982: *NYNJ 1983*, 37; 1983: *NYNJ 1984*, 85; 1984: *NYNJ 1984*, 146.

TABLE AB.12. Gross value of agricultural output (GVAO) in China (in 1952 and 1980 yuan), 1952–1990 with computed instability indices (ISI) for subperiods AB.12(*a*): 1952–1984 (in 100 m.)

	Current prices	1952 yuan	Index		Current prices	1952 yuan	Index
1952	461	461	100.0	1970	1,058	767	100.0
1953	510	475	103.1	1971	1,107	790	103.1
1954	535	491	106.6	1972	1,123	789	102.9
1955	575	529	114.6	1973	1,226	855	111.5
1956	610	556	120.4	1974	1,277	891	116.2
1957	537	575	124.8	1975	1,343	932	121.5
1958	566	589	127.8	1976	1,378	955	124.5
				1977	1,400	971	126.6
ISI				ISI			
1952–8		1.14		1970–7		1.31	
1959	497	509	110.4	1978	1,567	1,058	138.1
1960	457	444	96.4	1979	1,896	1,150	150.0
1961	559	434	94.0	1980	2,180	1,194	155.8
1962	584	461	100.0	1981	2,460	1,273	166.1
1963	642	514	111.6	1982	2,785	1,414	184.5
1964	720	584	126.7	1983	3,123	1,549	202.2
1965	833	632	137.1	1984	3,790	1,823	237.8
1966	910	687	148.7				
1967	924						
1968	928						
ISI				ISI			
1952–66		9.62		1978–84		3.08	
				1970–84		4.48	

Notes: There are two different official series of GVAO, one including, and the other excluding industrial output from *cunban* (hamlet-run) enterprises. This is comparable to the output from the former brigade and production team industries under the commune system. It is however different from the *xiangban* (village-run) (formerly commune level) industries which form part of the gross value of industrial output (GVIO) proper.

From 1949 to 1970 the two GVAO series were identical (see *NCJJTJDQ* 46–7). But from 1971 onwards, *cunban* enterprises were reclassified from argiculture into industry (ibid. 50), thus giving rise to a new GVAO series which is net of *cunban* industrial output. A similar readjustment was made starting in 1984, as discussed in the notes to Table AB.12(*b*).

The figures in this table refer to GVAO gross of *cunban* industrial output. (This is different from the net series as given in e.g. *TJNJ 1990*, 49). The index is based on the official Chinese standard of *comparable prices*; and it is used to convert the GVAO at *current price* into the *1952 yuan* series.

Sources: *NCJJTJDQ* 46–50.

AB.12(b): (in 100 m. yuan)

	GVAO gross of *cunban* industries			GVAO net of *cunban* industries					GVAC from *cunban* industries		
	Current prices	Constant 1980 prices	Index	Current prices	Constant 1980 prices	Index	Comparable prices	Index	Current prices	Constant 1980 prices	Index
	(1)	(2)	(3)	(4)	(5)	(6)	(7)	(8)	(9)	(10)	(11)
1978	1,567	1,970	100.0	1,397	1,801	100.0	1,397	100.0	170	169	100.0
1979	1,896	2,150	109.1	1,698	1,937	107.6	1,501	107.5	198	213	126.0
1980	2,180	2,223	112.8	1,923	1,965	109.1	1,524	109.1	247	259	153.3
1981	2,460	2,369	120.3	2,182	2,091	116.1	1,612	115.4	278	278	164.5
1982	2,785	2,632	133.6	2,483	2,328	129.3	1,794	128.4	302	305	180.5
1983	3,123	2,884	146.4	2,750	2,508	139.3	1,933	138.4	376	376	222.5
1984	3,790	3,391	172.1	3,214	2,816	156.4	2,170	155.4	575	575	340.2
1985	4,580	3,873	196.6	3,620	2,912	161.7	2,245	160.7	961	961	568.6
1986	—	4,292	217.9	4,013	3,011	167.2	2,320	166.1	—	1,281	758.0
1987	—	4,601	233.6	4,676	3,185	176.8	2,455	175.7	—	1,416	837.9
1988	—	4,898	248.6	5,865	3,310	183.8	2,551	182.6	—	1,588	939.6
1989	—	5,102	259.0	6,535	3,414	189.6	2,631	188.3	—	1,688	998.8
1990	—	5,424	275.3	7,662	3,672	203.9	2,830	202.6	—	1,752	1,036.7
ISI 1978–84		3.22			2.53		2.58				
85–90		2.68			0.78		0.78				
78–90		4.36			2.66		2.58				

Notes: From 1984 onwards, industrial output from *cunban* enterprises (see notes to Table AB.12(a)) was reclassified into GVIO proper. Thus a distinction should be made between GVAO gross and net of *cunban* industries. (The net series now constitutes the standard GVAO statistics published in *TJNJ*, *NYNJ*, *JJNJ*, and the regular communiqués of the State Statistical Bureau.) Since a similar reclassification was made in 1971, the latest readjustment implies that after the early 1970s new *cunban* industrial enterprises continued to develop and their output (unlike that from the reclassified units) was treated as part of the GVAO until the renewed reclassification in 1984.

These changes made it increasingly difficult to obtain a consistent time-series for GVAO gross of *cunban* industrial output. Official figures for this gross series (which includes *cunban* industries reclassified in both 1971 and 1984) are available only up to 1985 and 1986 respectively in current and constant 1980 prices. An effort is made in the table to extend the latest series through 1990 by adding the comparable component of *cunban* industrial output to the present standard GVAO statistics. This 1978–90 series of GVAO gross of *cunban* output is therefore comparable to the series shown in the first part (AB.12(*a*)) of this table, save that different price bases are used to derive the deflated series.

The GVAO series net of *cunban* industries is less problematic, although at the time of this tabulation deflated figures (in constant 1980 prices) were not yet available for 1989 and 1990. The growth rates implied in the available index based on 'comparable prices' for these two years are used as a basis for deriving the necessary figures. Note also that the absolute *yuan* series based on 'comparable prices' is comparable, in terms of the deflator, to the GVAO series shown in the first part (AB.12(*a*)).

As with the gross GVAO series, absolute *yuan* figures are not available for its *cunban* industrial output component for the years 1987–90. To fill the gap we make use of the total output figures (in current prices) and growth rates given for the narrower measure of collective *cunban* industries as well as rural co-operative (*nongcun hezuo jingying*) and individual (*nongcun geti*) industries to derive the weighted average growth rates for these years (assuming that the latter two sub-branches of rural industries would have otherwise been part of the former *cunban* industries prior to the reclassification made in 1984). These estimated growth rates are then applied to the available absolute *yuan* figure for 1986 (in 1980 prices) to obtain the figures for the remaining years.

Sources: (1) *NCJJTJDQ* 47.
(2) 1978: *ibid.* 108; derived as follows: V1978 (in 1980 prices) = V1980 (in 1980 prices)/V1980 (in 1970 prices) V1978 (in 1970 prices); 1979: columns (5)+(10); 1980–6: *NCJJTJDQ* 108; 1987–90: columns (5)+(10).
(3) from column (2).
(4) *TJZY 1991*, 52.
(5) 1978–86: *NCJJTJDQ* 108; with figures for 1978 and 1979 being derived from the same formula as for 1978 in column (2); 1987–8: *NCTJNJ 1989*, 66; 1989–90: assumed same yearly growth rates as per column (7).
(6) from column (5).
(7) derived from columns (4) and (8).
(8) *TJZY 1991*, 53.
(9) columns (1) minus (4).
(10) 1978: columns (2) minus (5); 1979–86: *TJNJ 1987*, 157; 1987–9: *TJNJ 1988*, 310; *TJNJ 1989*, 263; and *TJNJ 1990*, 412; derived per method as explained in notes above; 1990: assumed same rate of increase from 1989 as for 'subsidiary output' (*fuye*) within the present standard GVAO statistics.
(11) from column (10).

TABLE AB.13. Chemical and natural fertilizer supplies in China, 1952–1981 (in nutrient weight kg./ha.)

	Chemical	Natural	Total		Chemical	Natural	Total
1952	0.55	53.33	53.88	1967	20.31	84.6	104.91
1953	0.75	53.78	54.53	1968	15.51	89.59	105.10
1954	0.94	55.8	56.74	1969	20.48	92.22	112.70
1955	1.54	56.4	57.94	1970	22.47	86.54	109.01
1956	1.86	56.52	58.38	1971	25.90	97.74	123.64
1957	2.37	60.23	62.60	1972	29.15	98.06	127.21
1958	3.69	65.47	69.16	1973	35.09	99.26	134.35
1959	3.67	65.76	69.43	1974	32.68	100.67	133.35
1960	4.31	56.73	61.04	1975	35.90	101.87	137.77
1961	3.19	58.01	61.20	1976	38.94	105.4	144.34
1962	4.49	61.88	66.37	1977	43.39	99.9	143.29
1963	6.68	68.22	74.90	1978	58.89	98.71	157.60
1964	8.03	73.36	81.39	1979	73.16	102.62	175.78
1965	13.55	81.36	94.91	1980	86.72	102.9	189.62
1966	18.68	81.9	100.58	1981	91.96	105.01	196.97

Sources: Y. Y. Kueh, 'Fertilizer supplies and foodgrain production in China, 1952–1982', in *Food Policy*, 9/3 (1984), 219–31.

TABLE AB.14. Provincial sown area covered by natural disasters in China 1931–1937 (thousand ha.)

AB.14(*a*): 1931

	Flood (1)	Drought (2)	Insects and diseases (3)	Other disasters (4)	Total (5)
Hebei	—	—	—	—	—
Shanxi	—	—	—	—	—
Chahaer	—	—	—	—	—
Suiyuan	—	—	—	—	—
Liaoning	—	—	—	—	—
Jilin	—	—	—	—	—
Heilongjiang	—	—	—	—	—
Jiangsu	3,775	—	—	—	3,775
Zhejiang	491	—	—	—	491
Anhui	1,913	—	—	—	1,913
Fujian	—	—	—	—	—
Jiangxi	575	—	—	—	575
Shandong	859	—	—	—	859

TABLE AB.14(a). *cont.*

	Flood	Drought	Insects and diseases	Other disasters	Total
	(1)	(2)	(3)	(4)	(5)
Henan	2,262	—	—	—	2,262
Hubei	1,450	—	—	—	1,450
Hunan	990	—	—	—	990
Guangdong	—	—	—	—	—
Guangxi	—	—	—	—	—
Sichuan	—	—	—	—	—
Guizhou	—	—	—	—	—
Yunnan	—	—	—	—	—
Xizang	—	—	—	—	—
Shaanxi	—	—	—	—	—
Gansu	—	—	—	—	—
Qinghai	—	—	—	—	—
Ningxia	—	—	—	—	—
Xinjiang	—	—	—	—	—
China	12,315	0	0	0	12,315

Notes: Coverage: The figures in this table refer almost exclusively to the farmland covered by the Yangzi and Huai Rivers floods in 1931. They are, however, not a complete record because some 16 provinces were reported to have been covered by the floods (*SBNJ 1933*, 70). The sample (8 provinces) shown here includes only those provinces which were most seriously affected, and accessible to the surveys (*ibid*. 70–1).

The figures cover total (i.e. both grain and non-grain crops) sown area. But in contrast to the natural disaster statistics available for 1949–90 (Table AB.15), the 1931–7 series do not distinguish between areas 'covered' (*shouzai*) in general terms and those areas 'affected' (*chengzai*), which sustained a crop output loss of above 30%.

Hardly any other provincial area statistics are available for the year (pertaining to the other types of natural disasters), although a number of provinces including Shaanxi, Gansu, Sichuan, and Rehe, are likely to have been seriously affected by drought (see Deng Yunte, *Chinese History of Famine Relief* (Beijing: Sanlian Sudian, 1958), 33).

Method of data collection: Several sources of statistics are available for the 1931 floods. They include the Directorate of Statistics (*Zujichu*) (National Government), National Land (Tudi) Commission, National Commission for Reconstruction Affairs (Zhenwu), the Nanjing University Survey (in co-operation with the National Flood Relief Commission), and various provincial relief committees. These sources differ in either the timing of the surveys or the extent of area surveyed. In general, the statistics from the relief agencies tend to have a smaller flood area per *xian* surveyed. This seems to imply a stricter definition of losses for the purpose of relief compensation, compared with the general surveys of the incidence of flood losses.

Against this background, the major criteria used in order to compile flood area data are to select the sources with the largest number of *xian* surveyed or the surveys which were conducted (as far as can be determined) at the latest dates. The figures chosen may, however, include those from the relief agencies or from sources which explicitly stated that the surveys concerned were not complete in terms of the number of *xian* affected. The latter

TABLE AB.14(*a*). *cont.*

includes, for example, the figures from the Directorate of Statistics. In either case, the cited statistics may not reflect the full extent of the flooded areas.

Where (as in the case of Henan, and especially Hunan province) relief agency figures show the largest number of *xian* covered by the floods with an exceedingly small size of hectares covered, the more complete area figures from the Directorate of Statistics are used, although they give a smaller number of *xian* flooded. However, these figures have been adjusted upwards by adding on a balance obtained from multiplying the provincial average of hectares covered per *xian* in 1931–5 (see note to Table AB.14(*b*) by the extra number of *xian* reported to have been affected by the relief agency. Since different sources are used, the provincial series shown (combining both relief and non-relief agency sources) should not be regarded as a consistent one, in that the figures are not exactly comparable between some of the provinces.

Sources: Jiangsu: *DZYJS* series, 87, *Jiangsu Volume*, Taipei, 1972, 45890 (figure originated from National Land Commission).

Zhejiang, Jiangxi, and Shandong: *TJYB*, Oct. 1931, Table 2 (also reproduced in *SBNJ 1933*, 71) (Directorate of Statistics figures).

Anhui and Hubei: *SBNJ 1933*, 70–1 (Reconstruction Affairs Commission figure).

Henan: *TJYB*, ibid., plus balance (as explained in notes above) for an extra 38 *xian* from *DJYJS* series, 88 (*Henan volume*, 46665). (The latter source gives a total of 2,065,000 ha. inundated by floods in 77 *xian*, and the figure seems to have originated in the provincial Relief Affairs Committee.)

Hunan: *TJYB*, ibid., plus balance for an extra 32 *xian* from *SBNJ 1933*, 71. The latter source gives a total of only 466,000 ha. covered by the floods in 66 *xian* (Reconstruction Affairs figures).

AB.14(*b*): 1932

	Flood (1)	Drought (2)	Insects and diseases (3)	Other disasters (4)	Total (5)
Hebei	649	—	8	—	657
Shanxi	106	—	1	—	107
Chahaer	—	—	—	—	—
Suiyuan	—	—	—	—	—
Liaoning	—	—	—	—	—
Jilin	3,946	—	—	—	3,946
Heilongjiang	3,093	—	—	—	3,093
Jiangsu	—	—	1	—	1
Zhejiang	—	—	—	—	—
Anhui	129	924	10	—	1,063
Fujian	—	—	—	—	—
Jiangxi	146	—	—	3	149
Shandong	134	41	110	2	288
Henan	463	916	13	264	1,657

Table AB.14(*b*). *cont.*

	Flood (1)	Drought (2)	Insects and diseases (3)	Other disasters (4)	Total (5)
Hubei	—	—	—	—	—
Hunan	165	—	—	—	165
Guangdong	6	—	—	—	6
Guangxi	—	—	—	—	—
Sichuan	—	—	—	—	—
Guizhou	—	—	—	—	—
Yunnan	—	—	—	—	—
Xizang	—	—	—	—	—
Shaanxi	—	244	—	16	260
Gansu	—	20	—	21	41
Qinghai	—	6	—	2	8
Ningxia	—	—	—	—	—
Xinjiang	—	—	—	—	—
China	8,837	2,151	144	308	11,441

Notes: Estimation procedure: Unless otherwise indicated in the sources below, all figures for 1932, as for 1933 and 1934, are derived as follows:

$$A_{it} = \sum_{t=1931}^{1935} \frac{a_i}{x_i} \cdot Y_{it}$$

where A_{it} stands for the estimated provincial areas (ha.) covered by a particular type of natural disasters (i) in year (t); Σ_{ai} and Σ_{xi} respectively the sum total of areas covered in each *xian* in 1931–5 and the number of the corresponding *xian*; and Y_{it}, the number of *xian* reported to have been affected by the same natural disaster, with or without area figures being given in the year concerned. Where different Y_{it} figures are given by different sources, the largest number is chosen. Where individual *xian* are named (as is normally the case), extra care has been taken to avoid double counting in obtaining the complete Y_{it} figures. However, the values of *ai/xi* are estimated by pooling data from different sources. This may involve identical *xian* with different *ai* estimates. Likewise, summary figures from various sources for *ai* and *xi* (without naming the individual *xian*) have been added together, disregarding possible overlaps of a large number of *xian*. This procedure helps to balance possible underestimates, or overestimates by relief or non-relief agencies of the extent of the farm area covered by natural disasters (see notes to Table AB.14(*a*)).

With minor exceptions, the provincial samples for *xi* for the years 1931–5 combined seem to be large enough to be considered representative. This is particularly so with floods and droughts, comprising a national average of 91 and 75 *xian* respectively for a total 21 and 14 major provinces. For a number of minor provinces, for which information limited to the number of *xian* covered (but no *ai* and *xi* samples) are available, the averaged *ai/xi* figures from the neighbouring provinces are used for estimating their A_i values.

Nature and coverage of the data: The data are collected from a great variety of sources as compiled notably by the Ministry of Industry in the 1930s. Apart from the national-level sources mentioned in notes to Table AB.14(*a*), many of the *xian* data were generated by

TABLE AB.14(*b*).   *cont.*

special correspondents dispatched by the Ministry to the localities concerned, and from provincial newspapers. The majority of the data seem, however, to have come from various local *xian* or sub-*xian* governments. To the extent that most of the data represent consolidated accounts by the various provincial relief authorities, they tend to be biased downwards. For all the years during 1932–4, *xian* reported to have been affected by natural disasters for which no area figures are available, greatly outnumber those which do provide hectare statistics. The reports are full of descriptive accounts of losses in terms of property, life, acreage, rural infrastructure, etc., although these are less useful for the analytical purposes of this study.

Hardly any hectare figures are available for 1932, and there are also few descriptive accounts of affected *xian*. This may be partly due to the fact that the Ministry of Industry did not begin to compile such reports until 1933–4. However, 1932 also happened to enjoy the best weather in China during all the 1930s (see e.g. Zhongguo Wenhua Zianshe Xiehui (China Cultural Reconstruction Association), *Shinianlai de Zhongguo* (China in the Past Ten Years), i (Nanjing: Commercial Press, 1937), 191–2), apart from the disastrous floods in Jilin and Heilongjiang noted below.

Sources: *ai*, *xi*, and Y_{it}: Ministry of Industry, *MGJJNJ 1934*, P1–72; *1935*, S1–74; and *1936*, S1–81. *SBNJ 1933*, P74–5; *1934*, Q3–5 and Q14–16; *1935*, Q20–33; *1936*, Q15–17. *NARBCR* ii (1934) nos. 9–11. (NARB made available from 1935 onwards a fairly complete set of statistics showing the number of *xian* affected by natural disasters, but these data are not used here, because NARB used the estimated 'maximum', instead of 'normal yields' as its standard to define yield losses—see notes to Table AB.10 for the distinction and Table AB.14(*e*) for further explanation.) *MGTJTY 1935*, 530–3. DZYJS series, no. 87 (*Jiangsu* vol.), 45890; no. 88 (*Henan* vol.), 46665–6; no. 55 (*Hunan* vol.), 28178–96; no. 48 (*Anhui* vol.), 24529–30; no. 47 (*Hubei* vol.), 24239–47. *TJYB*, Oct. 1932, 16–17; University of Nanking (Agricultural Economics Dept., School of Agricultural Science), in co-operation with the National Flood Relief Commission, 'The 1931 Flood in China: An Economic Survey', in *JLXB*, 2/1 (May 1932), 223. Chen Hui, '1933 Niande Zhongguo Nongye Zaihuang' (China's Agricultural calamities and famine in 1933), in Qian Jiaju, *Zhongguo Nongcun Jingji Lunwenji* (Collected Essays on China's Rural Economy) (Nanjing: Zhonghua Shuju, 1936), 204–13. Wang Hungfa, *Guomin Jingji Jianshe Zhi Jichu* (Foundation of National Economic Construction) (Shanghai: Commercial Press, 1937), 149; Li Shutian, *Zhongguo Shuili Wenti* (China's Water Conservancy Problems) (Shanghai: Commercial Press, 1937), 397.

Jilin and Heilongjiang: *SBNJ 1933*, P74 reported that 80% of the cultivated area in both provinces were seriously flooded. The figures are derived by applying this percentage to the total area as given in Dwight Perkins, *Agricultural Development in China 1938–1968*, 236. The rough estimate for Heilongjiang is consistent with a report in *HLJRB* (16 Aug. 1957) that the 'affected' (*chengzai*) area in 1932 was more than three times the size of that of 1956. The absolute size affected in 1956 was given in *HLJRB* (13 Aug. 1957) to be 930,000 ha.

AB.14(*c*): 1933

	Flood (1)	Drought (2)	Insects and diseases (3)	Other disasters (4)	Total (5)
Hebei	401	36	127	6	570
Shanxi	81	—	—	22	103
Chahaer	—	—	—	—	—
Suiyuan	108	—	—	—	108
Liaoning	—	—	—	—	—
Jilin	—	—	—	—	—
Heilongjiang	—	—	—	—	—
Jiangsu	635	270	6	5	916
Zhejiang	342	123	7	27	499
Anhui	810	693	28	30	1,560
Fujian	4	321	—	—	325
Jiangxi	214	609	—	—	823
Shandong	388	—	127	—	515
Henan	863	952	50	528	2,393
Hubei	673	42	—	1	715
Hunan	347	1,047	—	1	1,395
Guangdong	2	325	14	15	356
Guangxi	—	99	64	—	163
Sichuan	12	38	—	4	54
Guizhou	1	23	—	2	26
Yunnan	2	8	—	—	10
Xizang	—	—	—	—	—
Shaanxi	147	964	1	18	1,129
Gansu	26	13	2	27	69
Qinghai	1	1	—	1	3
Ningxia	3	—	—	—	3
Xinjiang	—	—	—	—	—
China	5,060	5,564	425	686	11,736

Notes and sources: The procedure and sources used for the estimation are same as those adopted for Table AB.14(*b*). However, exception is made, as noted below, where an area figure cited by a single source is the largest among all sources and is also accompanied by the largest number of *xian* reported to be covered by the respective type of natural disasters. In this case, the area figure concerned is used, instead of the one obtained by applying the *ai/xi* value to the number of *xian*. The exceptions are as follows:

Floods: Shanxi: Chen Hui, 'China's agricultural calamities and famines in 1993', 212.

Insects and diseases: Hebei: *MGTJTY 1935*, 532–3; Shandong: *MGTJTY 1935*, 531; Guangxi: ibid. 530.

Other disasters: Shanxi: *MGJJNJ 1934*, P48–9 (the figure refers to wind and hailstorms).

AB.14(*d*): 1934

	Flood (1)	Drought (2)	Insects and diseases (3)	Other disasters (4)	Total (5)
Hebei	401	3,851	75	17	4,344
Shanxi	86	974	1	37	1,097
Chahaer	15	—	—	3	1,780
Suiyuan	322	—	—	4	325
Liaoning	—	—	—	—	—
Jilin	—	—	—	—	—
Heilongjiang	—	—	—	—	—
Jiangsu	404	3,297	11	19	3,731
Zhejiang	—	1,453	5	—	1,458
Anhui	65	2,468	22	9	2,563
Fujian	128	198	—	—	326
Jiangxi	203	1,619	1	—	1,823
Shandong	553	2,571	55	3	3,182
Henan	1,011	2,313	9	360	3,693
Hubei	416	1,199	4	—	1,618
Hunan	248	1,720	14	—	1,983
Guangdong	13	87	—	35	135
Guangxi	2	100	—	—	101
Sichuan	23	13	3	2	42
Guizhou	2	5	3	1	11
Yunnan	—	—	—	—	—
Xizang	—	—	—	—	—
Shaanxi	115	922	3	19	1,059
Gansu	32	9	—	24	65
Qinghai	1	—	—	1	2
Ningxia	5	7	—	1	13
Xinjiang	—	—	—	—	—
China	4,042	22,807	205	537	27,590

Notes and sources: The procedure and sources used for the estimation are same as those adopted for Table AB.14(*b*), with similar exceptions as made in Table AB.14(*c*). The exceptions are as follows:

Floods: Chahaer, Suiyuan, Shanxi and Sichuan: Wang Hungfa, *Foundation of National Economic Construction*, 149.

Droughts: Shaanxi, Shanxi, Hebei, Shandong, Jiangsu, Anhui, Henan, Hubei and Zhejiang: all from *NARBCR* 2 (1934) no. 8, 494; no. 10, 640, and no. 11, 757 (also reproduced in *SBNJ 1935*, Q23). The same sources give a figure of 1,493,000 and 1,220,000 ha. respectively for Hunan and Jiangxi; but the higher figures shown in the table for the two provinces have been obtained by applying the described procedure and sources. Since the drought area statistics are dominated by the NARB samples (virtually all from the great Yangzi and Huai River basins droughts occurring in the summer), it should also be noted that they cover only the summer (i.e. autumn-harvested) crops. Comparable data for the winter (i.e. summer-harvested) crops are not available.

AB.14(*e*):   1935

	Flood (1)	Drought (2)	Insects and diseases (3)	Other disasters (4)	Total (5)
Hebei	30	2,107	1	6	2,207
Shanxi	23	1,371	13	49	1,456
Chahaer	—	1	1	1	2
Suiyuan	—	102	—	8	110
Liaoning	—	—	—	—	—
Jilin	—	—	—	—	—
Heilongjiang	—	—	—	—	—
Jiangsu	2	90	7	298	397
Zhejiang	103	71	404	537	1,114
Anhui	17	618	11	21	666
Fujian	13	—	—	93	105
Jiangxi	89	21	23	83	215
Shandong	305	389	281	46	1,021
Henan	230	7	0	—	237
Hubei	309	269	188	782	1,549
Hunan	124	28	37	438	626
Guangdong	97	116	33	110	356
Guangxi	—	—	—	—	—
Sichuan	—	127	24	234	384
Guizhou	—	55	—	19	74
Yunnan	—	—	—	9	9
Xizang	—	—	—	—	—
Shaanxi	24	135	—	35	194
Gansu	—	—	90	231	321
Qinghai	—	—	—	—	—
Ningxia	—	37	2	2	42
Xinjiang	—	—	—	—	—
China	1,365	5,607	1,114	3,001	11,087

Notes: Nature of the new statistics: From 1935 onwards, the NARB generated a fairly complete set of provincial statistics about the impact of various natural disasters on crop production in China. These data have been used for our estimates of the size of farm area covered to the exclusion of other circumstantial evidence. However, the new data are not directly comparable with the earlier ones. In contrast to the 1934 summer drought statistics, for example, NARB made use of 'maximum', rather than 'normal yields' (see Table AB.10 for the distinction) as a basis for measuring the percentages of output and yield losses. This is explicitly stated in the case of the 1935 summer (i.e. autumn-harvested) crops, but the rule seems to apply to the winter (1934–5) crops as well. The new practice implies that farmland which, under the less stringent standard of 'normal yields' might not have been classified as being covered by natural disasters, was now included in the statistics. An adjustment of the new statistics should therefore be undertaken.

AB.14(*e*). *cont.*

Moreover, for the winter (1934–5) crops, NARB provides only information about the number of *xian* covered by natural disasters, the absolute amount of crop output losses (in kg.), and loss percentages for the *xian* concerned. But no relevant area data from the stricken *xian* are given. Thus, an attempt must also be made to derive the size of farm area covered by natural disasters from other available data.

Method and problems of estimation: (i) The winter-crops area: The procedure for the estimation involves two steps. The first step is to derive the absolute size of farm area covered by natural disasters according to the standard (i.e. crop loss measured against the 'maximum' yields), adopted for the summer crops. The second step is to adjust the estimated area size to make it comparable to the pre-1935 series which are based on 'normal' rather than 'maximum' yields. This second step applies also to the summer crops, and is explained in (ii) below.

The formula used in the first step of estimation is as follows:

$$A_{ij} = \frac{SA_i}{NX_i} \times CX_{ij} \times \frac{XL_{ij}}{CA_{ij}}$$

where A_{ij} stands for the estimated provincial areas (ha.) of the crop (*i*) covered by a particular type of natural disasters (*j*); SA_i the total provincial sown area for the crop (*i*); NX_i the total number of *xian* which reported the respective crop acreage; CX_{ij} the number of *xian* reported to have the crop (*i*) covered by natural disaster (*j*); XL_{ij} the mean percentage of crop losses sustained by the *xian* concerned, taking the total *xian* output of the crops (*i*) as a basis (ignoring that part of the sown area for the crop (*i*) might not have been covered by the natural disaster (*j*); and CA_{ij} the mean percentage of crop losses sustained within that part of the *xian* sown area which was host to the natural disaster.

The NARB statistics for 1935 give figures for SA_i, NX_i, CX_{ij}, and XL_{ij}, but not CA_{ij}. The latter is taken as the average of the respective figures for 1936 and 1937 in the case of drought. For the other types of natural calamities comparable figures are not available for 1937; and only the respective figures for 1936 are used. All these CA_{ij} figures are based on 'maximum' rather than 'average' yields. What is the important point in this context is that the formula represents only a very rough estimate for A_{ij}, with the assumption that every *xian* covered by natural disasters had an identical sown-area size for the crop concerned (which is equal to the provincial average per *xian*). Moreover, by applying XL_{ij} and CA_{ij}, the size of the *xian* area, presumed covered by natural disasters, is, in fact, projected from the relative magnitude of crop losses. Viewed this way, the approach adopted for the estimates is of course not entirely satisfactory, especially in relation to our main purpose of analysing the impact of the weather (as captured by the sown area covered by natural disasters) on crop yields. Fortunately, this is the only case in 1931–7 where we have to resort to such a method to fill the gap in the time-series required for our aggregate analysis of the weather and yield relationship for the pre-war period.

(ii) The summer-crops area: The NARB gives statistics on the absolute size of farm area covered by natural disasters in addition to both absolute and relative magnitudes of crop losses based on 'maximum' yields. The method used to convert the reported area, or the estimated area as in the case of winter crops into area covered by natural disasters (*shouzai*) according to the 'normal' yields, is simply to determine whether the reported/estimated areas meet the criterion,

$$\frac{L_{ij}}{A_{ij}} > (Y_{mi} - Y_{ni})$$

AB.14(*e*). *cont.*

where L_{ij} stands for the reported provincial total of losses (tonnes) in crop (i) output due to a particular type of natural calamities (j); A_{ij} provincial sown area (ha.) of the respective crop reported/estimated to be covered by the natural calamities; and Y_{mi} and Y_{ni} respectively the provincial average of 'maximum' and 'normal' yields per sown ha. for the crop concerned.

In other words, if the reported average loss (kg.) per sown ha. covered by natural disasters (i.e. L_{ij}/A_{ij}) is smaller than (or equal to) the discrepancy between 'maximum' and 'normal' yields (i.e. $Y_{mi} - Y_{ni}$) the implied effective yield per sown ha. (Y_{ei}) is actually even higher than (or equal to) the 'normal' yield. This can be shown in the diagram below.

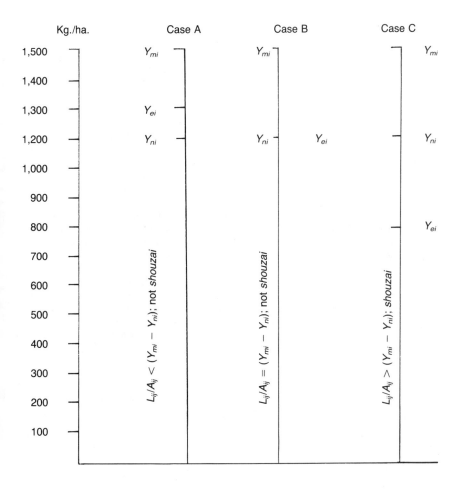

In this case, the reported area is not considered as covered by any natural disasters. If, however, the above criteria are fulfilled, the implied effective yield per sown hectare is lower than the 'normal' yield by an amount equal to the positive balance from substracting $(Y_{mi} - Y_{ni})$ from L_{ij}/A_{ij}. The area concerned should then be regarded as covered by natural disasters.

The criteria shown above again reflect a simplified formula. Subject to possible sampling bias for Y_{mi} and Y_{ni} (see notes to Table AB.10), the indiscriminate application of the formula may result in scores of provincial samples being unduly ruled out of (or left in) the estimated category of sown area covered (*shouzai*) by adverse weather on the basis of 'normal' yields. However, the biases in both directions may cancel out each other, so that the estimated national aggregative *shouzai* area may be less problematic than the individual provincial samples.

Nevertheless, it is interesting to note that in our collection of NARB data, the number of provincial samples, which are 'ruled out' of the estimated category, tends to be consistently higher for the northern crops than the southern crops. This applies to 1935, 1936, and 1937 as well; the implication is that the discrepancy between Y_{mi} and Y_{ni} is considerably larger in the North (compared with the South), to allow for a greater number of A_{ij} not to be considered affected by adverse weather. This finding is not surprising. It precisely reflects the fact that crop yields in the North are much more vulnerable to weather conditions than in the South. An exceptionally good spring weather in the North China plain can greatly help boost wheat yields per ha. relative to what can be achieved in normal years when there is insufficient rainfall. By contrast, in the more stable weather conditions of the South, the gap between Y_{mi} and Y_{ni} (e.g. rice yields) can be expected to be smaller.

Coverage of the estimation: The NARB data used to estimatic winter-crop areas cover, in terms of the notation (i), only such winter crops as wheat, barley, broad beans, field peas, and oats; and in terms of (j) five categories of natural calamities, namely drought, wind storms, diseases, insects and the residual category for snow, frost, flood, hailstorms, etc. The winter crops therefore embrace exclusively the grain crops as defined by NARB (see Table AB.2(*a*)), and exclude such minor non-grain crops as oilseeds.

As for the summer crops, rice, gaoliang, millet, proso-millet, and maize are included. Area data for soya bean and cotton reported as affected by adverse weather are also provided in the *NARBCR*, but they are excluded from our estimates, in order that the computed area series can be made comparable to that for winter crops. More importantly, the exclusion of non-grain crops (by NARB definitions), for 1935 and 1936–7 makes it possible consistently to relate estimated *shouzai* areas to the total provincial grain-sown area series, in the attempt to derive the 'weather' (*shouzai*) index (see Appendix A).

However, it should be noted that the 1935 summer crops area series is not exactly comparable to that of the 1934 series which excludes proso-millet, but includes both soya bean and cotton. Unlike the 1935 series, no separate area figures are given in the 1934 series for individual crops (including these two non-grain crops), and so they cannot be excluded from the general provincial *shouzai* area statistics provided by NARB. The same applies to the other 1931–4 series which all refer to *shouzai* area in general, rather than a summation of the grain-crop areas only. Note also that the 1935 summer crops area covers floods and droughts only. Data for other kinds of natural calamities are not available.

Sources:

The winter (1934–5) crops

SA_i and NX_i: NARBCR (English version) 3/6 (1936), 123.

CX_{ij}, L_{ij}, and XL_{ij}: NARBCR 3/9 (1936), 296–7 and 575–9.

CA_{ij}: NARBCR 4/9 (1937), 229 and 5/11 (1938), 314 (for droughts); and 4/9 (1937), 230–3 (all other types of natural calamities).

Y_{mi}: NARBCR 3/2 (1935), 359–60. (These are the 1934 figures); all derived from yields per sown hectare and the harvested percentages in relation to the 'maximum' (or 'best' year) yields.

Y_{ni}: *TJYB*, combined Jan./Feb. issue, 1932 (as reproduced in *SBNJ 1935*, 674–83). For Qinghai province, Y_{ni} is not available; the 1931–7 mean yields are used instead as derived from Xu Daofu, *Zhongguo Jindai Nongye Shengchan Yi Maoyi Tongji Ziliao*.

The summer (1935) crops

L_{ij} and A_{ij}: *NARBCR* 3/11 (1936), 935–6 and 975–6.

Y_{mi}: *NARBCR* 4/1 (1936, English version), 2–3; derived from yields per sown hectare and the harvested percentage in relation to the 'best year' ('maximum') yields. Note that the official NARB translation for *shizunian* is 'normal year'.

Y_{ni}: ibid.

AB.14(f): 1936

	Flood (1)	Drought (2)	Insects and diseases (3)	Other disasters (4)	Total (5)
Hebei	—	500	0	20	518
Shanxi	43	486	48	316	893
Chahaer	—	71	—	—	71
Suiyuan	—	105	3	7	115
Liaoning	—	—	—	—	—
Jilin	—	—	—	—	—
Heilongjiang	—	—	—	—	—
Jiangsu	735	50	118	153	1,057
Zhejiang	393	198	67	48	706
Anhui	117	172	2	53	344
Fujian	28	113	10	8	159
Jiangxi	214	141	3	10	368
Shandong	69	210	1,828	30	2,137
Henan	—	1,069	—	—	1,069
Hubei	236	261	35	259	791
Hunan	114	50	73	45	311
Guangdong	13	384	10	3	410
Guangxi	—	—	—	—	—
Sichuan	79	1,604	6	172	1,860
Guizhou	6	65	3	32	106
Yunnan	1	128	46	2	177
Xizang	—	—	—	—	—
Shaanxi	—	141	7	298	446
Gansu	19	59	24	90	193
Qinghai	5	14	0	5	24
Ningxia	—	0	—	—	0
Xinjiang	—	—	—	—	—
TOTAL	2,102	5,818	2,284	1,551	11,756

Notes: The procedure adopted in order to estimate the grain crops areas affected (*shouzai*) by various types of natural calamities is the same as that applied for the summer (1935) crops (Table AB.14(*e*)). The coverage is, however, slightly different. While the winter (1935–6) crops involved are the same as in 1934–5, there is now a separate account for the impact of floods. The other kinds of weather adversities surveyed remain the same as in 1935. As for the summer (1936) crops, only drought data are available and proso-millet is replaced by sweet potatoes among the grain crops investigated.

Sources: By using the same notations as in Table AB.14(*e*), for the summer (1935) crops:

L_{ij} and A_{ij}: NARBCR 4/9 (1936, English version), 229–33 (for winter crops); and 5/1 (1937), 2–3 (summer crops).

Y_{mi}: NARBCR 4/9 (1936, English version), 224–5 (winter crops); and 4/12 (1936), 324–5 (summer crops); all derived from yields per sown hectare and the harvested percentages in relation to the 'best year' (*shizunian*) yields (or 'maximum' yields). Note the official NARB's translation of *shizunian* as 'normal year'.

Y_{ni}: as in Table AB.14(*e*).

AB.14(*g*):　1937

	Flood	Drought	Insects and diseases	Other disasters	Total
	(1)	(2)	(3)	(4)	(5)
Hebei	734	848	—	—	1,582
Shanxi	114	536	—	—	650
Chahaer	—	—	—	—	—
Suiyuan	—	28	—	—	28
Liaoning	—	—	—	—	—
Jilin	—	—	—	—	—
Heilongjiang	—	—	—	—	—
Jiangsu	21	509	—	—	530
Zhejiang	97	55	—	—	152
Anhui	61	312	—	—	373
Fujian	—	2	—	—	2
Jiangxi	30	4	—	—	34
Shandong	195	815	—	—	1,010
Henan	802	2,710	—	—	3,073
Hubei	156	598	—	—	754
Hunan	140	24	—	—	164
Guangdong	98	62	—	—	160
Guangxi	69	157	—	—	226
Sichuan	119	2,267	—	—	2,368
Guizhou	9	139	—	—	148
Yunnan	2	345	—	—	347
Xizang	—	—	—	—	—
Shaanxi	62	638	—	—	700
Gansu	16	128	—	—	144
Qinghai	—	12	—	—	12
Ningxia	—	—	—	—	—
Xinjiang	—	—	—	—	—
TOTAL	2,726	10,189	—	—	12,915

Notes: The estimation procedure is the same as that adopted for Tables AB.14(*e*) and AB.14(*f*). Both the winter (1936–7) and summer (1937) crops under study include the same grain crops as in Table AB.14(*f*). However, only drought data are available for winter crops; for summer crops both droughts and floods are surveyed.

Sources: By using the same notations as in Table AB.14(*e*): L_{ij} and A_{ij}: *NARBCR* 5/11 (1937), 312–13 (for winter crops); and 6/3 (1938), 28 (Guangxi province) and 30–1 (summer crops under drought), and 33–4 (summer crops under floods).

Y_{mi}: *NARBCR* 5/8 (1937), 256–7 (for winter crops); and 6/2 (1937), 14 (Guangxi province) and 16–17 (summer crops).

Y_{ni}: as in Table AB.14(*e*).

TABLE AB.15. Official national total of sown area covered (*shouzai*) and affected (*chengzai*) by natural disasters in China, 1949–1990 (million ha.)

	Floods		Droughts		Other disasters		Total	
	shouzai	*chengzai*	*shouzai*	*chengzai*	*shouzai*	*chengzai*	*shouzai*	*chengzai*
1949	—	8.53	—	—	—	—	—	—
1950	6.56	4.71	2.40	0.41	1.05	—	10.01	5.12
1951	4.17	1.48	7.83	2.30	0.56	—	12.56	3.78
1952	2.79	1.84	4.24	2.59	1.16	—	8.19	4.43
1953	7.41	3.20	8.62	0.68	7.39	3.20	23.42	7.87
1954	16.13	11.31	2.99	0.26	2.33	1.02	21.45	12.59
1955	5.25	3.07	13.43	4.14	1.31	0.66	19.99	7.07
1956	14.38	10.99	3.13	2.06	4.68	2.18	22.19	15.23
1957	8.08	6.03	17.21	7.40	3.86	1.55	29.15	14.98
1958	4.28	1.44	22.36	5.03	4.32	1.35	30.96	7.82
1959	4.81	1.82	33.81	11.17	6.01	0.74	44.63	13.73
1960	10.16	4.98	38.13	16.18	17.17	3.82	65.46	24.98
1961	8.87	5.40	37.85	18.65	15.03	4.78	61.75	28.83
1962	9.81	6.32	20.81	8.69	6.56	1.66	37.18	16.67
1963	14.07	10.48	16.87	9.02	1.24	0.52	32.18	20.02
1964	14.93	10.04	4.22	1.42	2.49	1.18	21.64	12.64
1965	5.59	2.81	13.63	8.11	1.58	0.30	20.80	11.22
1966	2.51	0.95	20.02	8.11	1.68	0.70	24.21	9.76
1970	3.13	1.23	5.72	1.93	1.12	0.14	9.97	3.30
1971	3.99	1.48	25.05	5.32	2.01	0.65	31.05	7.45
1972	4.08	1.26	30.70	13.61	5.68	2.31	40.46	17.18
1973	6.24	2.58	27.20	3.93	3.05	1.11	36.49	7.62
1974	6.40	2.30	25.55	2.74	6.70	1.49	38.65	6.53
1975	6.82	3.47	24.83	5.32	3.73	1.45	35.38	10.24
1976	4.20	1.33	27.49	7.85	10.81	2.26	42.50	11.44
1977	9.10	4.99	29.85	7.01	13.07	3.16	52.02	15.16
1978	2.85	0.92	40.17	17.97	7.77	2.91	50.79	21.80
1979	6.76	2.87	24.65	9.32	7.96	2.93	39.37	15.12
1980	9.15	5.03	26.11	12.49	9.27	4.80	44.53	22.32
1981	8.62	3.97	25.69	12.13	5.48	2.64	39.79	18.74
1982	8.36	4.46	20.70	9.97	3.70	1.56	32.76	15.99
1983	12.13	5.73	16.07	7.60	6.67	2.87	34.87	16.20
1984	10.63	5.40	15.82	7.02	5.44	2.89	31.89	15.26
1985	14.20	8.95	22.99	10.06	7.18	3.70	44.37	22.71
1986	9.16	5.58	31.04	14.76	6.94	3.32	47.14	23.66
1987	8.69	4.10	24.92	13.03	8.48	3.26	42.03	20.39
1988	11.95	6.13	32.90	15.30	6.02	2.51	50.87	23.94
1989	11.33	5.92	29.36	15.26	6.30	3.77	46.99	24.45
1990	11.80	5.60	18.17	7.81	8.50	4.41	38.47	20.65

Notes: *Shouzai* area is defined as farmland covered generally by natural disasters, whereas *chengzai* area refers to farmland within the general category of *shouzai* which sustained a crop output loss of over 30%. Natural disasters other than floods and droughts include frosts, winds, hailstorms, etc.

Sources: 1949–82: *TJNJ 1983*, 212; 1983–9: *TJNJ 1990*, 389; 1990: *TJZY 1991*, 64.

TABLE AB.16. Scattered provincial statistics of total sown area covered (*shouzai*) affected (*chengzai*) by floods and droughts 1952–1961 (thousand ha.)

AB.16(*a*): Floods

	1952	1953	1954	1955	1956	1957	1958	1959	1960	1961
Beijing	—	—	—	—	—	31	—	—	—	—
Tianjin	—	—	—	—	—	—	—	—	—	—
Hebei	72	1,986	2,644	1,180	3,032	—	n.a.	801	—	201
Shanxi	—	—	299	—	600	—	—	400	—	—
Neimonggu	—	—	—	—	—	—	—	67	—	—
Liaoning	—	—	—	—	97	650	55	606	69	—
Jilin	63	521	371	177	970	708	23	86	605	15
Heilongjiang	—	—	n.a.	—	930	n.a.	—	41	n.a.	49
Shanghai	—	—	—	—	—	—	—	—	—	—
Jiangsu	322	n.a.	2,103	n.a.	2,160	867	—	48	509	395
Zhejiang	200	661	507	—	223	133	7	175	26	413
Anhui	1,000	229	2,351	305	1,773	412	129	39	370	284
Fujian	6	—	n.a.	267	49	—	10	137	193	233
Jiangxi	—	267	507	79	—	—	280	47	—	133
Shandong	680	2,856	1,633	1,253	1,523	2,327	67	67	183	600
Henan	417	79	2,667	—	3,333	2,067	447	—	—	—
Hubei	—	n.a.	1,343	n.a.	—	—	n.a.	13	—	127
Hunan	100	20	156	23	—	—	67	—	—	—
Guangdong	4	200	n.a.	n.a.	27	553	19	472	1,000	641
Guangxi	63	—	200	—	n.a.	63	52	107	21	37
Sichuan	5	—	n.a.	—	211	—	—	7	—	13
Guizhou	45	31	96	32	25	72	40	69	54	13
Yunnan	n.a.	—	n.a.	—	—	37	—	43	—	—
Xizang	—	—	—	—	—	—	—	—	—	—
Shaanxi	—	—	—	—	—	—	—	7	1	—
Gansu	—	—	—	—	—	—	—	—	—	—
Qinghai	—	—	—	—	—	—	—	—	—	—
Ningxia	—	—	—	—	—	—	—	—	—	—
Xinjiang	—	—	—	—	—	—	—	—	—	—
TOTAL	2,977	6,850	14,877	3,316	14,953	7,920	1,196	3,232	3,031	3,154
Official total										
Shouzai	2,790	7,410	16,130	5,250	14,380	8,080	4,280	4,810	10,160	8,870
Chengzai	1,840	3,200	11,310	3,070	10,990	6,030	1,440	1,820	4,980	5,400

Notes: These scattered data of provincial farm areas covered by floods (as well as those by droughts shown in Table AB.16(*b*)) were collected from a wide variety of sources from the 1950s and early 1960s, including various Chinese provincial newspapers held at the School of Oriental and African Studies, University of London, and the newspaper clippings prepared by the former Union Research Institute (now housed at the Hong Kong Baptist College). The collection was completed in 1984–5, before more complete recently-available sets of similar figures began to emerge from China. However, no attempt is made to incorporate all the new statistics. Time and resources available did not permit a continuous updating, and the data used in the study do not in any case occupy such a critical position as to warrant renewed systematic search. A quick check on some of the new data reveals no gross inconsistencies in our collection, although it is not always easy to make a direct comparison between some of the old and new data.

TABLE AB.16(*a*)   *cont.*

A number of points need to be kept in mind in interpreting the data.

(1) The figures refer to 'sown area', rather than 'arable' or 'cultivated area'. That is to say, the same cultivated area may be covered by drought (or floods) during both the wheat-growing season in the spring and coarse-grain (say gaoliang or maize) growing period which follows after the wheat crops are harvested. In this case, the cultivated area covered by the drought (which prevailed either continuously from the spring through the summer, or in two separate incidents) is counted twice in obtaining the total sown area under drought (or floods). A personal communication from the State Statistical Bureau confirms that this is the standard procedure for their statistical compilation.

(2) Drought may prevail with intervals during the same cropping seasons. A 'cumulative' total of drought areas can thus be obtained. Chinese sources occasionally give such statistics, but in the final counting, the overlaps seem to have been discounted. However, one and the same cropping may be successively affected by floods and drought. This is a real possibility, although it is unlikely to happen frequently. In this case, the drought and flood areas should obviously be counted separately. A further complication is that the same farm area covered by drought may simultaneously suffer from plant diseases and locust damage. Such natural disasters also form part of official Chinese statistics and, taken together, they all tend to 'inflate' the size of the sown area reported to be subject to weather adversities.

(3) Occasionally, the yearly total of farm area covered by natural disasters is given in terms of 'cultivated area' as a summary statistic. A clear case is the combined total cited for Hebei, Shandong, Henan, and Shanxi in 1960. As the drought continued through the various growing seasons in the year, the affected areas in terms of sown-area accounting were understated. It is difficult to adjust for such biases by simply applying the relevant multiple-cropping index to the given cultivated area base, because the spring and autumn fields may have been affected to a different extent.

(4) In contrast to new official practice, the original sources of the 1950s and early 1960s often do not distinguish between *shouzai* and *chengzai* areas, i.e. areas covered by natural disasters in general, and those areas which sustained a crop loss of over 30% respectively. Where such distinction is made (as noted where possible in the sources), it is not clear whether the criterion applied to the *chengzai* area is exactly comparable to that adopted in recent years. In one source (*HNRB*, 25 Aug. 1957), e.g. a 20% loss is used as the yardstick. Our collection of data thus represents a combination of *shougzai* and *chengzai* area data.

(5) Many of the figures shown in Tables AB.16(*a*) and AB.16(*b*) are preliminary data reported when droughts or floods were still continuing. They may or may not be followed up with conclusive surveys. Where conflicting figures are given, the ones dated the latest are used. Moreover, many data are collected from sporadic locality samples or derived from rough percentage figures or such incomplete information as to the extent of farm areas 'rescued' from serious drought or flood impact. These should not be regarded as a complete accounting.

(6) Almost all the original area statistics are given in the usual Chinese unit of *wan mu* (10,000 *mu*), and many of them are given in such highly rounded-up terms as 1,000 or 2,000 *wan mu*. This is clearly discernible in our collection of data, as the Chinese *mu* figures are converted into hectares at the ratio of one *mu* to 0.067 ha.

(7) The abbreviation n.a. stands as usual for 'not available' where drought or floods were reported but no hectare figures given. Only for the more significant cases of such reports are the original literature sources given below. Nevertheless, the sign—in the two tables does not necessarily imply a total absence of natural disasters for the province concerned.

(8) The nature of the data illustrated above naturally implies that some of the collected provincial samples may eventually prove to deviate substantially from the consolidated official figures. This has yet to be verified. However, our reconstructed national totals are

TABLE AB.16(*a*)   *cont.*

still quite close to official figures published in recent years. They generally fall within the 10% (plus or minus) margin of the official totals. Two different categories of exception should be noted, however. The first concerns the years 1952 (total of provincial samples for drought by 23% less than official total) and 1955 (by 27 and 37% less than official totals for drought and floods respectively). The explanation for the relatively large discrepancy may be that weather hazards in most of the provinces (note the large number of them indicated by n/a) were too minor to warrant press reports.

(9) Another category of exception relates to the years 1958–61. With the notable exception of the 1959 drought area, the reconstructed national total for all the other years (for both drought and flood areas) falls short of the consolidated official totals by a wide margin. A number of explanations can be given. The first is that 1959 (for which the national total for drought area from the provincial samples is virtually identical to the official total) was characterized by disastrous droughts (rather than floods) on a nation-wide scale; hence the appearance of widespread and sensational press reports much the same as in 1934 in connection with the catastrophic Yangzi and Huai River basin droughts. The second explanation is that in 1958, in terms of *chengzai* areas, the scale of flood menace was actually the smallest since 1951, and that of drought was in no way to be compared with 1957 (see the official *chengzai* figures). The comparative lack of press reports in 1958 about floods and droughts may reflect the euphoria of the Great Leap Forward; but we cannot be sure. As for 1960 and 1961, it is clearly increased censorship which led to reduced statistical reports in this and other respects.

Finally, some may regard the compilation of the provincial statistics as unnecessary and unworthy of the effort, because summary official statistics are readily available. However, our independent accounting not only provides a useful source of verification, but originating from the provincial newspapers, it also helps to suggest (as the notes on sources show) the complexity of how natural disasters evolved at the local level and how peasants coped with the adversities.

Sources: *Beijing*: **1957**: *BJRB*, 24 Apr. 1957.

Hebei: **1952–6**: *RMRB*, 17 Mar. 1957, as reproduced in *XHBYK* 9 (1957), 104. The figures seem to be based on areas before the changes in the provincial boundary were made in 1957. Breakdowns are available, in the original sources, for each year into areas flooded by river overflows (Heyan) and inundated (Lilao). **1959**: This is a rough estimate based on a report in *LNRB*, 6 Sept. 1959, that a total of 1,333,000 ha. (20 million *mu*) were flooded in Beijing, Hebei, Liaoning, and Neimenggu combined due to disastrous rainstorms in July 1959. The area flooded in Neimenggu was given in *NMGRB*, 30 Dec. 1959, as 67,000 ha. The balance of 1,266,000 ha. is apportioned to Hebei (including Beijing) (i.e. 801,000 ha.), and Liaoning (466,000 ha.) according to their respective total grain-sown area size as given in Walker, *Foodgrain Procurement and Consumption in China*, 221. This may overstate the affected proportion for Hebei to the extent that it enjoyed a higher cropping index, but the areas affected in Liaoning were nevertheless mainly in the east and south of the province bordering Hebei. **1961**: *RMRB*, 4 Aug. 1961, Congzhou *diqu* only.

Shanxi: **1954**: *SXRB*, 17 July 1957. **1956**: *RMRB*, 22 Dec. 1956. **1959**: *SXRB*, 31 Dec. 1959.

Neimenggu: **1959**: *NMGRB*, 30 Dec. 1959.

Liaoning: **1956**: *LNRB*, 30 Sept. 1957 (*Chengzai* figure). **1957**: *LNRB*, 5 May 1959 and *LNGQTYB*, 16 July 1957. **1958**: *LNRB*, 3 July and 8 Aug. 1958 (scattered figures from Jinzhou *qu*, Andong *xian*, and Fengcheng *xian*). **1959**: 466,000 ha. from the combined share with Hebei (see Hebei 1959), plus 140,000 ha. resulting from separate rainstorms in mid- and late July 1959 as reported in *LNRB*, 23 Dec. 1959. See also *LNRB*, 21 Dec. 1959 for the location of floodings. **1960**: total of scattered data as reported in *LNRB*, 9, 12, 13, 18, 23, and 28 Aug. 1960, covering nine *xian* in the prefectures (*diqu*) of Wushu, Shenyang, Yingkou, and Dandong.

Jilin: **1952–61**: Yu Derun, 'Jilin Sheng laozai de xingcheng yu zhili' (The formation of waterlogging calamities and their control in Jilin province), in *ZGNB* 7 (1963), 4–10. The 1956 figure is from *JLRB*, 7 Dec. 1956. Yu gives a lower figure of 930,000 ha. His 1952–61 series may thus represent *chengzai* (rather than the greater *shouzai*) figures.

Heilongjiang: **1956**: *HLJRB*, 13 Aug. 1957. **1959**: *HLJRB*, 18 and 23 July 1959—combined figures from three *xian* (Jixiang, Gaojin, and Gannan). **1961**: *HLJRB*, 23 Aug. 1961.

Jiangsu: **1952**: 247,300 ha. inundated in the summer as given in Huai River Commission Engineering Dept.'s, 'Guanyu ruhe jinyibu jiejue Huaihe liuyu neilao wenti de chubu yijian' (Preliminary opinions on how to further solve the problems of inundation in the Huai River basin), in *ZGSL* 5 (1957), 91; plus 74,700 ha. of spring flood in South Jiangsu as cited in *JFRB*, 28 Apr. 1952. **1954**: derived from *DGBBJ*, 10 June 1957 which stated that the areas covered in 1956 in both summer and autumn amounted to 3,133,000 ha. (47 million *mu*), which is presumably 49% higher than the *chengzai* area in 1954, hence 3,133/1.49 = 2,103,000 ha. (*chengzai* figures). **1955**: *XHRBNJ*, 13 Sept. 1955, reported that Xuzhou *diqu* was flooded in July; but no area figure is given. **1956**: *DGBBJ*, 10 June 1957 and *XHRBNJ*, 23 July 1957, both cited a total of 3,133,000 ha. covered (*shouzai*) in spring and summer. The figure of 2,160,000 ha. given here is from the latter source which stated that out of the total, 360,000 ha. sustained crop losses of 50 to 100%, and 1,800,000 ha. suffered from various degrees of loss; hence 360,000 + 1,800,000 = 2,160,000 ha. The figures may have overstated the extent of flooding, as damage due to related windstorms (which did not necessarily bring about flooding) seems to have been included as well. **1957**: *XHRB*, 16 Aug. 1957. **1959**: combined figures for 6 *xian* in the prefectures (*diqu*) of Yancheng and Yangzhou as a result of rainstorms on 12–13 July 1959, as reported in *XHRB*, 15 July 1959. **1960**: *XHRBNJ*, 10 Aug. 1960. **1961**: *MGNJ 1962*, 644 reported that rainstorms occurring on 3 to 6 July 1961 affected the prefectures (*diqu*) Xuzhou, Yancheng, and Huaiyin. A total of 12 *xian* in Xuzhou and Huaiyin were seriously flooded, with five of them having two-thirds, and the remaining seven *xian* about one-third of their farmland covered by the floods. The figure of 395,000 ha. cited is a rough estimate obtained as follows: $(5,224/75 \times 12) \times [(5/12 \times 2/3) + (7/12 \times 1/3)]$, where the figure of 5,224 refers to total arable area, as given in *XHRB*, 23 Jan. 1960, and 75 to the total number of *xian* in the province.

Zhejiang: **1952**: *NFRB*, 22 Aug. 1952. **1953**: *ZJRB*, 22 Dec. 1956 reported that farm area covered by both floods and drought amounted to 717,000 ha. in 1953. *RMRB*, 27 Sept. 1953 cited a total of 56,000 ha. covered by drought in 14 *xian* in the same year; hence 717 − 56 = 616,000 ha. **1954**: *ZJRB*, 22 Dec. 1956. The figure may cover drought as well, but floods were clearly the major weather hazard in 1954. **1956**: *JFRB*, 15 Aug. 1957. **1957**: *ZJRB*, 7 July 1957; Jiaxing *diqu* only. **1958**: incomplete local figure from *ZJRB*, 6 May 1958. **1959**: *ZJRB*, 7–9 Sept. 1959, incomplete figures (in thousand ha.) from the prefectures of Wenzhou (100), Ningbo (35.3), Jinhua (14.9), and Taizhou (25 with Huangyan *xian* 22 and Linhai *xian* 3). **1960**: Scattered local data from *ZJRB*, 14 Apr. 1960, and 12 and 15 Aug. 1960, covering Jiaxing *diqu*, Jinhuaxian, Hangzhou *shi*, Yiwu *xian*, Linhai *xian*, and Laiqing *xian*. **1961**: Provincial radio broadcast, 28 May 1961 quoted in *JRDL* (Materials Office) 'Jinnian shangbannian dalu ziran jaihai shilu' (A true record of natural calamities on the mainland in the first half of the current year), 139, 10 July 1961, 10, and *MGNJ 1962*, 665. The two sources give respectively a total of 280,000 and 133,000 ha. inundated separately due to Pacific Typhoon nos. 3 and 4.

Table AB.16(*a*)   *cont.*

Anhui: **1952**: *RMRB*, 27 Sept. 1954 and Huai River Commission Engineering Dept. (as quoted for Jiangsu 1952), 90. The latter source states that 570,000 and 430,000 ha. were respectively seriously (*zhongzai*) and lightly (*qingzai*) affected. The combined total should therefore be interpreted as a *shouzai* (rather than *chengzai*) figure. The original source also indicates that two prefectures (Liuan and Xuxian) and two municipalities (Huainan and Bangfu) are omitted from the figure. Note also that a more recent source, *AHSQ 1949–83*, 73 gives a complete provincial *chengzai* figure of 723,000 ha. **1953–61**: *AHSQ 1949–83*, 73–4 (*chengzai* figure). Except for 1957, the area figures which can be reconstructed from original source of the 1950s for 1954 (2,000,000 ha.) and 1956 (1,633,000 ha.), are lower than the complete provincial *chengzai* figures given in the table. The reconstructed figure for 1954 is from *CJRB*, 4 Sept. 1954 and 29 Nov. 1954. The latter source reported that a total of 1,200,000 ha. of late autumn crops covered by floods were drained and replanted, and the former stated that up to late Aug. farm areas so drained made up 60% of the total inundated; hence 1,200 ÷ 0.60 = 2,000,000 ha. For the reconstructed 1956 figure, see *RMRB*, 15 Sept. 1956. However, *RMRB*, 5 July 1957 also reported that the same autumn crops were covered consecutively by nine disastrous rainstorms, resulting in a *cumulative* total of 5,333,000 ha. being so covered. As for 1957, *AHRB*, 17 Aug. 1957 reported that by mid-August, 667,000 ha. were drained of flood waters, with 93,000 ha. still inundated; hence a total of 760,000 ha. may be given as affected (*shouzai*), in comparison with the complete provincial *chengzai* figure of 412,000 ha. shown in the table.

Fujian: **1952**: *NFRB*, 22 Aug. 1952 (incomplete figure). **1955**: *FJRB*, 15 Sept. 1955. **1956**: *FJRB*, 30 Jan. 1957 (figures for the prefectures of Jinjiang and Longxi only). **1958**: *FJRB*, 24 May 1958 and 18 June 1958 (incomplete local figures). **1959**: summation of various prefecture and *xian* data reported in *FJRB* various days June through Sept. 1959. **1960**: summation of prefecture data from *DGBHK*, 14 June 1960; *ZGXWS*, 2 and 22 June 1960 and *FJRB*, 21 May 1960. **1961**: *JRDL* 139, 10 July 1961, 10.

Jiangxi: **1953**: *JXRB*, 11 Dec. 1954. **1954**: *RMRB*, 27 Jan. 1956. **1955**: *JXRB*, 14 Nov. 1956 (incomplete figure, referring to Xiu River floods only). **1958**: *JXRB*, 30 Nov. and 8 Dec. 1958. **1959**: *JXRB*, 2 Jan. and 18 Mar. 1960. **1961**: *MGNJ 1962*, 665 citing *XHS*, 28 Apr., 18 June, 3 July, and 6 Aug. 1961 (incomplete figure referring to the middle reach of Ganjiang only).

Shandong: **1952 and 1954**: derived by subtracting the given drought *shouzai* area (*DZRB*, 23 Jan. 1957) from the overall *shouzai* area (*DZRB*, 19 Jan. 1957). This may overstate the flood areas to the extent that the overall figure also includes such minor natural calamities as hails, windstorms, and plant diseases. **1953 and 1956**: *DZRB*, 19 Jan. 1957 reported that farm areas flooded (*shouzai*) in 1956 were more than 1,333,000 ha. less than the area covered in 1953. The 1956 area is taken to be 1,523,000 ha. (*DZRB*, ibid.), including wind and hailstorms, but excluding insect and plant diseases. **1955**: *DZRB*, 19 Jan. 1957 (*shouzai* figure) including wind, hailstorms, and frost damage. **1957**: *DZRB*, 11 Feb. 1958 (*shouzai* figure). **1958**: *DZRB*, 20 Sept. 1958. The figure which refers to the 'effectively' affected area could be a gross understatement, as the same source also reported that according to the amount and pattern of rainfall, a total of 1,333,000 ha. was predicted. **1959**: *DZRB*, 30 Dec. 1959 (incomplete figure). **1960**: summation of various prefecture and *xian* data as given in *DZRB*, 7, 8, and 30 July; and 2, 3, and 4 Aug. 1960. **1961**: *MGNJ 1962* citing *XHS*, 12, 27, and 31 July, and 24 Oct. 1961 (referring to the rainstorms occurring in North-west Shandong during 11–13 July 1961).

Henan: **1952**: *ZGSL* 5 (1957), 90. **1953**: *RMRB*, 23 Aug. 1953. **1954**: *HNRB*, 29 Nov. 1956 reported that farm areas covered by floods during June to August 1956 amounted to a total of more than 3,333,000 ha. which is 667,000 ha. more than the areas similarly covered

TABLE AB.16(*a*)   *cont.*

in 1954; hence 3,333,000 − 667,000 = 2,667,000 ha. for 1954 (*shouzai* figure). **1956**: *HNRB*, ibid. **1957**: *HNRB*, 23 Aug. 1957. **1958**: *RMRB*, 14 July 1958 (an estimate based on precipitation pattern).

Hubei: **1954**: *DGBTJ*, 18 Apr. 1955 and *JFRB*, 31 July 1955 reported that a total of 806,000 ha. were drained and replanted with late autumn and cross-winter crops in 1954. Assuming roughly that as in Anhui, the areas drained represented 60% of the total flooded, then the size affected would be 806,000:0.60 = 1,343,000 ha. This may understate the extent of areas flooded given the fact that Hubei stood in the centre of the disastrous Yangzi River floods in 1954. **1959**: *HUBRB*, 10 June 1959 (incomplete figure). **1961**: *JRDL* 139 (10 July 1961), 10, citing *XHS*, Wuhan dispatch 13 Mar. 1961.

Hunan: **1952**: *CJRB*, 7 Sept. 1952. **1953**: *ZGZK* 3/7 (8 Mar. 1953), 11, referring to areas surrounding the Dongting Lake. **1954**: combined figure from *XHNB*, 29 May 1954, *XHS* Beijing dispatch 16 Aug. 1954, and *RMRB*, 22 July 1954. **1955**: *XHNB*, 28 Aug. and 5 Sept. 1955. **1958**: *XHNB*, 12 and 19 May 1958.

Guangdong: **1952**: combined figure from *NFRB*, 7 July 1952 and *GSRB*, 25 May 1952. **1953**: *RMRB*, 20 June 1953 and *GSRB*, 24 May 1953. **1956**: *NFRB*, 17 Aug. 1956. **1957**: *RMRB*, 2 Nov. 1957. **1958**: *NFRB*, 10 June and 18 Nov. 1958. **1959**: *NFRB*, 17 Mar. 1960 (400,000 ha. of paddy-field flooded), and *RMRB*, 16, 19, and 24 June 1959 (72,000 ha. of potato inundated). **1960**: *NFRB*, 4 Dec. 1960. **1961**: *MGNJ 1962*, 666, citing *XHS*, 22 and 26 Apr. 1961 (for 267,000 ha.), *XHS*, 18 June 1961 (7,000 ha.), and *XHS*, 18 Sept. 1961 (67,000 ha.); and *ZGZK*, 9 Jan. 1962, 10 (301,000 ha.).

Guangxi: **1952**: *NFRB*, 6 Aug. 1952. **1954**: *HKSB*, 12 Aug. 1954 citing the provincial Flood Control Bulletin. **1957**: *GXRB*, 29 June 1957. **1958**: *GXRB*, 17 July, and 18, 21, and 23 Sept. 1958 (summation of various prefecture and *xian* data). **1959**: *GXRB*, 21 June 1959. **1960**: *GXRB*, 23 May 1960 (Guilin *diqu* only). **1961**: *JRDL* 139 (10 July 1961), 10, citing provincial radio broadcast, 14 June 1961.

Sichnan: **1952**: *XHRBCQ*, 20 July 1952 (Daxian figure only). **1956**: *SCRB*, 3 Sept. 1957. **1959**: *RMRB*, 17 July and 13 Aug. 1959 (scattered *xian* figures). **1961**: *ZGZK* (Hong Kong), 9 Jan. 1962 citing *XHS*, 4 July 1961 (Luoshan *diqu* only).

Guizhou: **1952–61**: *GZJJSC*, 152.

Yunnan: **1957**: *YNRB*, 6 Aug. 1957. **1959**: *YNRB*, 10 Aug. 1959.

Shaanxi: **1959**: *SAXRB*, 24 July 1959. **1960**: *SAXRB*, 4 Sept. 1959 (both years scattered *xian* data).

Official totals: Table AB.15.

AB.16(*b*): Droughts

	1952	1953	1954	1955	1956	1957	1958	1959	1960	1961
Beijing	—	—	—	—	—	—	—	—	—	—
Tianjin	—	—	—	—	—	—	—	—	—	—
Hebei	—	1,333	—	n.a.	—	n.a.	1,600	1,067	5,655	3,570
Shanxi	—	267	267	1,333	—	2,000	1,067	829	3,049	1,832
Neimenggu	n.a.	—	—	n.a.	n.a.	n.a.	1,467	—	n.a.	—
Liaoning	n.a.	—	—	n.a.	n.a.	716	197	67	n.a.	1,000
Jilin	—	—	—	n.a.	—	—	1,500	170	—	—
Heilongjiang	—	—	—	—	—	—	505	93	—	—
Shanghai	—	—	—	16	—	—	—	—	—	—
Jiangsu	37	n.a.	n.a.	n.a.	588	—	n.a.	2,000	n.a.	1,038
Zhejiang	n.a.	56	243	—	61	285	333	105	407	n.a.
Anhui	337	1,201	n.a.	147	325	222	4,589	3,601	807	1,214
Fujian	n.a.	400	540	201	n.a.	280	—	27	175	8
Jiangxi	n.a.	586	—	n.a.	—	n.a.	533	780	42	—
Shandong	887	3,000	1,037	4,533	—	5,467	2,117	4,867	4,667	2,647
Henan	800	—	—	n.a.	n.a.	4,733	800	7,430	5,742	2,000
Hubei	667	667	—	n.a.	667	1,600	341	3,600	—	1,267
Hunan	n.a.	n.a.	—	n.a.	501	1,067	294	1,333	1,074	1,658
Guangdong	9	1,200	793	2,507	1,133	200	367	467	1,333	93
Guangxi	133	—	27	837	n.a.	247	1,067	300	333	1,734
Sichuan	n.a.	406	—	n.a.	n.a.	233	897	4,000	2,997	2,543
Guizhou	392	314	97	217	276	223	265	676	658	375
Yunnan	n.a.	—	25	n.a.	—	207	17	227	83	—
Xizang	—	—	—	—	—	—	—	—	—	—
Shaanxi	—	—	—	n.a.	—	n.a.	—	1,349	255	723

TABLE AB.16(b) *cont.*

	1952	1953	1954	1955	1956	1957	1958	1959	1960	1961
Gansu	—	—	—	n.a.	—	—	n.a.	—	n.a.	n.a.
Qinghai	—	200	—	—	67	38	—	—	—	180
Ningxia	—	—	—	—	—	—	—	—	—	—
Xinjiang	—	n.a.	—	—	—	16	—	—	—	—
TOTAL	3,262	9,630	3,029	9,791	3,617	17,534	17,956	32,988	30,308	21,882
Official total										
Shouzai	4,240	8,620	2,990	13,430	3,130	17,210	22,360	33,810	38,130	37,850
Chengzai	2,590	680	260	4,140	2,060	7,400	5,030	11,170	16,180	18,650

Notes: see Table AB.16(*a*).

Sources:

Hebei: **1953**: *ZNRB*, 29 Sept. 1953; *RMRB*, 28 Apr. 1953 also reported that farmland which could not be sown due to water shortfall already amounted to 1,000,000 ha. **1955**: *RMRB*, 9 Aug. 1955. **1957**: *RMRB*, 17 June 1957. **1958**: *HBRB*, 4 May 1958 (667,000 ha. of wheat crops), plus *HBRB*, 21 June 1958 (933,000 ha. of autumn crops in Xingtai *shi*). The actual drought area is likely to have been much more substantial, as *HBRB*, 1 Aug. 1958 reported that there was a continuing shortfall of rain from Sept. last through July. According to *HBRB*, 10 May 1958, this made spring sowing extremely difficult. See also *HBRB*, 5 and 9 July 1958 on large-scale labour mobilization to fight drought in Baoding and Tianjin *diqu*. **1959**: *HBRB*, 18 May 1959. **1960**: a rough estimate based on information in *RMRB*, 29 Dec. 1960 that the combined total of farmland under drought in Hebei, Henan, Shandong, and Shanxi was 60% of the total arable area of the four provinces. The average arable area for 1955–7 (given in Kenneth Walker, *Foodgrain Procurement and Consumption in China*, 302) is used as base to derive the absolute drought area; after subtracting the known drought-affected area of Shandong and Shanxi (see relevant sources), the balance is apportioned to Hebei and Henan according to their arable area size. This procedure tends to understate the drought area in terms of sown area (rather than arable area) given the fact that the drought prevailed over both the summer-harvested and autumn crops in 1960. **1961**: *XHS* (Beijing), 11 June 1961 stated that 30 to 50% of farmland in Hebei and Shanxi was affected by drought. The same procedure (as for 1960) is applied to derive the provincial drought areas by assuming that 40% of the combined arable area of the two provinces was affected.

Shanxi: **1953:** *SXRB*, 4 Sept. 1957. **1954:** *ZGZK* 9/2 (10 Jan. 1955), 40. **1955:** *SXRB*, 4 Sept. 1957 (*chengzai* figure). **1957:** *SXRB*, 4 Sept. and 9 Oct. 1957 (*shouzai* figure). **1958:** *SXRB*, 25 Apr. 1958. **1959:** *SXRB*, 10 Jan. 1960 (for 600,000 ha. in south Shanxi) and *SXRB*, 10 Aug. 1959 (229,000 ha. in south-east Shanxi). **1960:** *SXRB*, 30 Dec. 1960. **1961:** same estimate as for Hebei, 1961.

Neimenggu: **1958:** *NMGRB*, 7 July 1958.

Liaoning: **1957:** *LNRB*, 30 Sept. 1957. **1958:** *LNRB*, 9 Nov. 1958 (3,500 ha. for rice crops), *LNRB*, 19 Aug. 1958 (39,200 ha. wheat crops, plus 154,100 ha. for potatoes). **1959:** *LNRB*, 23 Aug. 1959. **1961:** *RMRB*, 30 May 1961.

Jilin: **1958:** *JLRB*, 6 Aug. 1958. **1959:** *JLRB*, 30 May and 5 July 1959.

Heilongjiang: **1958:** summation of various prefecture and *xian* data given in *HLJRB*, 6, 11, 23, 27, and 28 June 1958; cf. also *RMRB*, 4 July 1958 for similar summary report. **1959:** *HLJRB*, 24 Aug. 1959, wheat crops only.

Shanghai: **1955:** *JFRB*, 16 Nov. 1955.

Jiangsu: **1952:** *JFRB*, 2 July 1952 and *RMRB*, 6 July 1952. **1955:** *XHRBNJ*, 30 Sept. 1955, noted serious droughts, but no figures given. **1959:** *XHRBNJ*, 23 Jan. 1960, for rice crops only. **1961:** rough estimate based on Xiang Dakun and Zhou Jiadong, 'Jinnian shangbannian dalu jaihuang fazhan qingkuang zhi fengxi' (An analysis of the developments of calamities on the mainland in the first half of the year), in *FQYJ*, July–Sept. 1961, 10, citing Jiangsu provincial broadcast 4 June 1961, and *XHS* (Nanjing), 9 July 1961 (cf. also *MGNJ* 1962, 664). One-third of arable land in Jiangsu north of the Yangzi River, notably in the prefectures of Huaiyin, Yancheng, Xuzhou, and Yangzhou was reported to be affected by drought. Assuming roughly that the prefectures concerned accounted for one-half of the provincial farm area, then by applying 33% to half of the 1955–7 average provincial arable area-base (given in Kenneth Walker, *Foodgrain Procurement and Consumption in China*, 302), the combined area covered by drought can be given as 1,038,000 ha.

Zhejiang: **1953:** *RMRB*, 27 Sept. and 4 Oct. 1953. **1956:** *JZRB*, 22 Dec. 1956 gives a total arable area of 811,000 ha. covered by drought and floods in 1956. Subtracting the flooded area of 223,000 ha. from the total, gives 588,000 ha. **1957:** *WHBHK*, 30 Sept. 1957. **1958:** *ZJRB*, 3 Oct. 1960. **1959:** *RMRB*, 2 Sept. 1959. **1960:** *ZJRB*, 3 Oct. 1960 reported that a total of 533,000 ha. were affected by drought, floods, and other natural calamities. Drought seemed to be the major weather hazard. Subtracting 100,000 ha. for insects (*ZJRB*, 11 Mar. 1960), and 26,000 ha. for floods (see Table AB.16(*a*)), the balance of 407,000 ha. may largely be attributed to drought, although it is probably an overestimate, given that the insect, and especially flood, areas as cited are not complete figures.

Anhui: **1952–7 and 1961:** *AHSQ 1949–83*, complete provincial *chengzai* figures. Similar figures (in thousand ha.) are also available for 1958 (1,222), 1959 (2,285), and 1960 (551), but the larger contemporary *shouzai* figures as noted below are used here. **1958:** *AHRB*, 1 Oct. 1958. **1959:** *AHRB*, 21 Dec. 1959 (autumn crops only). **1960:** *AHRB*, 4 Sept. 1960. Note also that the *AHSQ* figure for 1956 is the smallest of all the years. This is consistent with the conclusion of *DLXB* 32/1 (Mar. 1966), 38 that there was no drought damage in 1956.

Fujian: **1953:** *DGBHK*, 14 Dec. 1953. **1955:** *FJRB*, 12 Oct. 1957. **1956:** *FJRB*, 23 Dec. 1956, 30 Jan. 1957, and *RMRB*, 1 Jan. 1957. **1957:** *FJRB*, 6 Oct. 1957. **1959:** *FJRB*, 4 Aug. and 3 Nov. 1959, incomplete data from Jinjiang and Minhou. **1960:** *FJRB*, 9 Mar. 1960 (wheat and barley 87,000 ha.), and *ZGXWS*, 2 Feb. 1961 (80,000 ha. of early rice in Minnan area and 8,000 ha. of late rice in Lunghai plain). **1961:** *JRDL* 139 (10 July 1961), 9, citing provincial broadcasts of 19, 20, 23, 24, and 26 June 1961; more than 20 *xian* affected in north Fujian; incomplete local data.

Appendix B

Table AB.16(b) *cont.*

Jiangxi: **1953**: *JXRB*, 11 Dec. 1954. **1954**: *RMRB*, 27 Jan. 1956; the figure represents areas replanted with late autumn crops as a result of prolonged drought from Aug. to Oct. **1956**: *JXRB*, 8 Nov. 1956, serious drought in south Jiangxi. **1958**: *JXRB*, 1 Sept. 1958. **1959**: *JXRB*, 2 Jan. and 18 Mar. 1960. **1960**: *JXRB*, 16, 27, and 29 July; 6 and 9 Aug.; 1, 25, 26, and 27 Sept.; and 7 Oct. **1960**: summation of various *xian* data.

Shandong: **1952–5**: *DZRB*, 23 Jan. 1957 (*shouzai* figures). **1957**: *DZRB*, 11 Feb. 1958 (*shouzai* figure; the *chengzai* figure is given as 2,667,000 ha.). **1958**: rough estimates based on *DZRB*, 6 Mar. 1958 that a total of 3,947,000 ha. cross-winter crops were planted under drought conditions, and *RMRB*, 24 Nov. 1957 that generally 30% of the wheat areas failed to sprout. This gives 1,184,000 ha. (= 3,947,000 × 0.3). The drought persisted through June 1958 (*RMRB*, 24 June 1958 and *DZRB*, 20 Sept. 1958), and seriously affected spring sowing (*DZRB*, 29 and 31 May 1958). A total of 4,667,000 ha. were sown by mid-June under drought conditions (*DZRB*, 21 June 1958); earlier (*DZRB*, 29 May 1959), 20 to 50% of cotton areas and 50 to 65% of potato areas failed to sprout. The 20% failure rate is applied to the total area sown (i.e. 4,667,000 × 0.2) to give another 933,400 ha. **1959**: *DZRB*, 30 Dec. 1959. **1960**: *DZRB*, 20 and 27 Nov. 1960 (the year-end figure seems to refer to cultivated (rather than sown) area, and may cover other natural calamities too). *DZRB*, 11 June 1960 suggests that a total of 3,333,000 and 2,667,000 ha. respectively of wheat crops and spring-sown crops were covered by drought, as well as (*DZRB*, 24 Sept. 1960) 1,333,000 ha. of autumn crops sown after the wheat harvest. If correct, the drought area figures as estimated for Shandong, Henan, Hebei, and Shanxi on the basis of cultivated areas (see note on Hebei, 1960) greatly understate the real magnitude of the disaster in 1960. **1961**: *JRDL* 139 (10 July 1961), 8 citing *XHS*, 14 Apr. 1961; see also *RMRB*, 15 Apr. 1961. The figure refers to spring-sown areas in three prefectures. *MGNG 1962*, 668 also reported that vast areas of cross-winter crops (especially wheat) were seriously affected, but no area figures are available.

Henan: **1952**: *CJRB*, 29 July 1952 reported that a combined total of 1,467,000 ha. for Hubei and Henan were seriously affected by drought. The Henan figure is obtained by subtracting the given Hubei figure of 667,000 ha. **1957**: *HNRB*, 14 and 25 Aug. 1957 give 1,400,000 ha. (*chengzai* figure) of summer-harvested crops, and *HNRB*, 14 Aug. 1957 mentioned that 3,333,000 ha. of autumn crops were 'threatened' by drought. *RMRB*, 13 Nov. 1957 reported that the drought persisted through October rendering the sowing of the cross-winter wheat crops difficult. **1958**: *RMRB*, 24 Nov. 1959 reported that 4,000,000 ha. of winter (1957–8) wheat were sown, but 20% failed to sprout, hence 4,000,000 × 0.2 = 800,000 ha. **1959**: *HNRB*, 23 Oct. 1959 and 1 Mar. 1960; more than 80% of the arable areas affecting 91 *xian* were affected by drought continuously in July through September. Applying this percentage to the average 1955–7 total arable (given in Walker, *Foodgrain Procurement and Consumption in China*, 302) gives 7,430,000 ha. **1960**: derived from the combined total for Hebei, Henan, Shandong, and Shanxi (see Hebei 1960). **1961**: *JRDL* 139, 10 July 1961 (2,000,000 ha.) of wheat crops affected by drought, citing *XHS* (Zhengzhou), 16 Feb., 14 Apr., and 3 May 1961; cf. also *ZGZK*, 9 Jan. 1962. *MGNJ 1962*, 667 also reported prolonged drought from late May through mid-July 1961 seriously inhibiting summer sowing; many farm areas were forced to replant with non-scheduled crops once again. No exact drought-area figures were given, however.

Hubei: **1952**: *CJRB*, 29 Dec. 1952 reported that output losses on 667,000 ha. were mitigated after various rescue measures were implemented. Earlier on (*CJZB*, 1 Oct. 1952), 533,000 ha. (*chengzai* figure) were given as being hit by the serious drought occurring during May through July

1952; cf. also *JFRB*, 28 June 1952 and *RMRB*, 29 June 1952. **1953**: *RMRB*, 1 May 1953 and *ZNRB*, 29 Sept. 1953 (citing 21 May 1953) reported that rice seedlings for 667,000 ha. could not be transplanted due to shortfall of rain. **1957**: *DGBBJ*, 21 Nov. 1957, autumn drought; cf. also *HUBRB*, 14 Dec. 1957 for similar report. **1958**: *HUBRB*, 20 Oct. 1957 reported that out of the total of 487,000 ha. in Xiaogan *diqu* scheduled for winter (1957–8) sowing, only 30% were not short of moisture, hence 487,000 × 0.70 = 341,000 ha. **1959**: *HUBRB*, 24 Nov. 1959. **1961**: *RMRB*, 28 Sept. 1961.

Hunan: **1956**: *RMRB*, 19 Aug. and 1 Sept. 1956 and *JFRB*, 20 Aug. 1956; cf. also *XHNB*, 21 Dec. 1957 which states that the scale of the drought in 1956 was unprecedented. **1957**: *RMRB*, 29 Sept. 1957. **1958**: *XHNB*, 7 Sept. 1958 (40,500 ha. in the east), and 1 Oct. 1958 (for 253,300 ha. in the centre and south of the province). **1959**: *XHNB*, 31 Aug. 1959 (for 933,000 ha. of early and mid-season rice crops), and *XHNB*, 17 Nov. 1959 (400,000 ha. of late rice crops); *XHNB*, 5 Aug. 1959 also reported that other non-rice crops were affected by drought. **1960**: rough estimate based on *XHNB*, 28 July 1960 that 25 out of a total of 89 *xian* were seriously affected by drought. This gives 1,067,000 ha. by applying the ratio 25/89 to the 1955–7 average provincial arable area as given in Kenneth Walker's, *Foodgrain Procurement and Consumption in China*, 302. **1961**: rough estimate based on *RMRB*, 28 July 1961 that 25% of the sown area were affected by drought. The total sown area-base as given in Walker (p. 306) is used to derive the absolute *shouzai* figure.

Guangdong: **1952**: *NFRB*, 27 Apr. 1952, incomplete figure from Hainan Island only, to the effect that a total of 8,900 ha. were 'rescued' from serious drought impact. **1953**: *NFRB*, 5 Dec. 1953 and *DGBHK*, 1 Jan. 1954. The drought covered 87 *xian* and 2 municipalities. **1954**: *NFRB*, 15 July 1955. **1955**: *NFRB*, 2 Apr. 1955 and *WHBHK*, 1 Jan. 1956, citing provincial Agriculture Bureau for 933,000 ha. cross-winter (1954–5) crops (*shouzai* figure). (*NFRB*, 28 Dec. 1955 gives a *chengzai* total of 520,000 ha. for the early crops). *NFRB*, 14 July 1955 citing Irrigation Bureau chief that in the spring rice seedlings could be transplanted on only 800,000 ha. without difficulty. Transplanting on 602,000 ha. was completed thanks to rigorous anti-drought measures; and on another 667,000 ha. a 'crash-transplanting' was made possible only thanks to delayed rainfall on 5 July. This still left 38,000 ha. on which transplanting could not take place due to water shortages. We add up the three figures, i.e. 602 + 667 + 38, to give a subtotal of 1,307,000 ha. for the spring and summer droughts. Autumn drought occurring in early Oct. 1955 as reported in *DGBHK* 1 Jan. 1956 accounted for another 267,000 ha. Hence the grand total of 2,507,000 ha. **1956**: *NFRB*, 10 Aug. 1956 (*shouzai*, the *chengzai* figure is given as 354,000 ha.). **1957**: *NFRB*, 15 Aug. 1957. **1958**: *NFRB*, 7 May 1958 (incomplete figure). **1959**: *RMRB*, 8 Nov. 1959. **1960**: *NFRB*, 21 Mar. 1960. **1961**: *RMRB*, 4 Aug. 1961 (Hainan Island only).

Guangxi: **1952**: *CJRB*, 13 July 1952, covering 52 *xian* in 8 prefectures. This seems to be a *shouzai* figure because *NFRB*, 14 July 1952 cites a lower total of 67,000 ha. for the same number of *xian*, emphasizing that it involved *severe* drought. **1954**: rough estimate (*chengzai* figure) based on *GXRB*, 30 Sept. and 7 Oct. 1954. A total of 35 *xian* were hit by drought, and the *chengzai* area per *xian* can be assumed to be an average of 0.7749 thousand ha. for 10 of the *xian*. This gives a total of 0.7749 × 35 = 27,000 ha. **1955**: *RMRB*, 18 June 1957, *shouzai* figure (*chengzai* area 419,000 ha.), including floods; but the major hazard was clearly drought (as in Guangdong); cf. *GXRB*, 19 June (and 20 July 1955) which reported that a major part of the 1,300,000 ha. of early rice (767,000) and early maize (533,000) could not be transplanted/sown until after the

Appendix B

TABLE AB.16(b) *cont.*

date of *Lixia* (establishment of summer). **1956**: *GXRB*, 27 Dec. 1956 and 13 Jan. 1957; *RMRB* reported that in 1956 550 persons starved to death and 14,700 persons were forced to flee their homesteads. **1957**: *GXRB*, 21 Oct. 1957. **1958**: *GXRB*, 3 Oct. 1959. **1959**: *GXRB*, 27 Nov. 1959. **1960**: *GXRB*, 29 Dec. 1960. **1961**: *JRDL* 139 (10 July 1961), 9 reported that 70% of the provincial arable was covered by drought. The absolute figure is derived by applying the percentage to the total arable given in Walker, *Foodgrain Procurement and Consumption in China*, 302.

Sichuan: **1953**: rough estimate based on *RMRB*, 1 July 1953 (half of the *xian* were affected by drought), and *XHRBCQ*, 25 February 1954 (16 *xian* in south Sichuan with a total of 66,670 ha. covered by drought; hence 66.67/16 × 195/2 = 406,000 ha., where the figure 195 refers to the total number of *xian* in the province. **1956**: *SCRB*, 1 October 1956 reported that north Sichuan was affected by drought, but no figures are given. However, it is noteworthy that a starting total of 3,600,000 ha. were host to disastrous locusts in 1956 (*SCRB*, 1 Nov. 1957). **1957**: rough estimate based on *CQRB*, 9 August 1957 (a total of 66,670 ha. of rice crops in two prefectures were rescued from drought damage), and *RMRB*, 29 Sept. 1957 (part of late rice crops in 7 prefectures were hit by drought); hence 66.67/2 × 7 = 233,000 ha. This is clearly an underestimate. Locust damage on a scale similar to that in 1956 was reported, however, in *SCRB*, 22 Aug. 1958. **1958**: rough and incomplete estimate based on *RMRB*, 14 Apr. 1958 (out of the 1,333,000 ha. of winter-wheat crops and 2,000,000 ha. of other non-rice crops, 20% were affected by drought, i.e. 3,333 × 0.2 = 667,000 ha.), and *SCRB*, 4, 19, and 21 Apr. and 2 May 1958 (scattered prefecture and *xian* data of rice crops covered by drought totalling 231,000 ha.). **1959**: *RMRB*, 10 Aug. 1959 and *SCRB*, 27 May 1960, including 85 *xian* in 10 prefectures, and 2 municipalities, and amounting to 56% of the 'major spring' (Dachun) crops. **1960**: as elsewhere, there was a glaring lack of drought-area figures reported in the Chinese press; not even locality figures. This seems to signal the beginning of the statistical black-out initiated by the central authorities. Nevertheless, there were ample sporadic reports (*SCRB*, 28 Jan., 1, 2, 8, 10, 15, and 20 June, 10 July, 13 and 20 Aug., 3 Sept., 3 Nov., and 31 Dec. 1960) which showed that 9 prefectures and 2 municipalities were affected by drought in 1960. This is comparable with 1959, as well as 1961. The only clues to the likely drought-area size are *SCRB*, 8 June 1960 (on Daxian prefecture to the effect that only half of the area under paddy and dry crops could be planted, leaving a total of 333,000 ha. unplanted); and *SCRB*, 10 June 1960 (planting of a total of 4,000,000 ha. of major crops pending the arrival of rainfall; a substantial area earmarked for mid rice crops had swiftly to be switched to maize and potatoes). If the figure of 333,000 ha. for Daxian prefecture is a representative one, then the 9 prefectures and 2 municipalities may imply a total of at least 2,997,000 (333,000 × 9) ha. covered by drought in 1960. **1961**: *RMRB*, 15 Aug. 1961 reported that drought began to spread in late April, and covered 10 prefectures by May and June. In June, *chengzai* areas amounted to one-third of the provincial arable area, i.e. a total of 2,543,000 ha. based on the 1955–7 provincial average of arable area as per Walker, *Foodgrain Procurement and Consumption in China*, 302.

Guizhou: **1952–61**: *GZJJSC* 152. Note that the figures for 1959 and 1960 are very different from the rounded figures of 533,000 ha. (i.e. 8 million *mu*) and 667,000 ha. (i.e. 10 million *mu*), reported in *GZRB*, 30 Sept. 1959 and 12 Dec. 1960.

Yunnan: **1954**: *GMRB*, 6 Apr. 1954 (summation of scattered local data, incomplete). **1957**: *YNRB*, 27 Nov. 1957. **1958**: *YNRB*, 13, 24, and 26 Apr., 4 and 18 May, and 4 June 1958 (summation of various prefecture and *xian* data; incomplete). **1959**: *YNRB*, 1 July, 6, 10, and 26 Aug. 1959 (summation of scattered prefecture data). **1960**: *YNRB*, 26 Apr., 4 and 18 May, 1, 4, 7, and 10 July, and 18, 24, 25, 27, and 28 Aug. 1960 (summation of scattered local data).

Shaanxi: **1955**: *JFRB*, 13 Nov. 1955; *RMRB*, 19 Aug. 1955; and *XHBYK* 15 (1956), 93 all reported serious droughts; but no area figures are available. **1959**: *SAXRB*, 31 Aug. and 13 Sept. 1959; autumn drought affecting both grain and cotton. **1960**: *RMRB*, 10 Feb. 1960 (spring drought covering 240,000 ha.) and *SAXRB*, 3 July 1960 (summer drought in Yulin of 15,000 ha.). **1961**: *ZGZK*, 9 Jan. 1962 citing *RMRB*, 28 Feb. and 3 Mar. 1960, affecting wheat crops only.

Gansu: **1955**: *GSURB*, 15 Mar. 1956; no area figures are available.

Qinghai: **1953**: *QHRB*, 20 Feb. 1955. **1956**: *QHRB*, 30 Aug. 1957, covering hailstorms, floods, and frost damage; but the major weather hazard seems to have been drought (see also *QHRB*, 20 Nov. 1956. **1957**: *QHRB*, 3 July 1957. **1961**: *JRDL* 139 (10 July 1961), citing provincial broadcasts on 5, 6, 7, and 9 June 1961.

Xinjiang: **1953**: *XJRB*, 19 Sept. 1954, negligible *xin* data. **1957**: *XJRB*, 6 June 1957.

Official total: Table AB.19.

TABLE AB.17.　Monthly provincial average of precipitation in China, 1931–1935 (mm.)

AB.17(*a*):　1930–1

		Sept.	Oct.	Nov.	Dec.	Jan.	Feb.	Mar.	Apr.	May	June	July	Aug.
Hebei	(5)	42.3	29.3	3.0	0.6	1.7	1.5	7.2	11.3	61.0	75.6	113.0	190.6
Shandong	(3)	53.6	36.9	58.4	1.3	14.3	8.7	3.6	22.0	59.1	83.4	47.4	127.1
Henan	(0)	—	—	—	—	—	—	—	—	—	—	—	—
Shanxi	(1)	5.6	35.3	0	5.1	2.0	0.9	0	0.6	42.8	53.6	102.7	91.6
Shaanxi	(0)	—	—	—	—	—	—	—	—	—	—	—	—
Gansu	(0)	—	—	—	—	—	—	—	—	—	—	—	—
Sichuan	(1)	131.6	66.2	33.9	11.9	5.6	10.2	34.1	98.9	190.6	55.1	112.6	125.3
Hubei	(2)	85.6	64.8	32.7	21.6	22.3	97.3	63.2	145.3	218.3	184.8	425.4	161.3
Anhui	(1)	86.2	44.9	51.4	25.4	60.9	154.9	43.1	132.8	192.6	188.1	348.8	13.0
Jiangsu	(10)	80.4	85.6	33.4	35.5	33.3	96.9	28.7	102.5	142.7	142.7	466.5	92.4
Zhejiang	(4)	160.6	144.3	64.3	77.9	59.2	151.6	101.6	108.1	276.4	128.4	166.5	172.9
Fujian	(2)	31.2	25.6	9.4	23.7	66.8	55.2	98.4	229.7	161.1	165.0	79.1	510.1
Jiangxi	(1)	78.4	75.6	62.0	34.7	59.4	159.4	97.7	184.0	246.6	83.4	417.3	31.6
Hunan	(2)	—	126.9	57.3	29.9	39.4	68.9	145.7	223.0	228.6	200.6	453.9	110.0
Guizhou	(0)	—	—	—	—	—	—	—	—	—	—	—	—
Yunnan	(2)	186.2	103.0	100.5	1.5	0.5	0	10.3	8.2	29.6	210.1	177.5	228.5
Guangdong	(4)	326.8	27.5	8.4	55.1	20.6	20.9	117.4	173.8	218.2	207.8	182.3	243.7
Guangxi	(2)	51.5	112.3	8.4	38.4	17.4	17.4	84.4	157.0	91.3	168.2	211.7	278.9

AB.17(b): 1931–2

		Sept.	Oct.	Nov.	Dec.	Jan.	Feb.	Mar.	Apr.	May	June	July	Aug.
Hebei	(5)	36.3	0.1	11.3	7.0	0.2	3.8	0	26.0	62.5	43.9	263.0	75.6
Shandong	(5)	119.1	2.8	33.7	40.5	3.0	3.6	11.6	18.8	50.6	69.3	239.9	150.6
Henan	(2)	—	—	—	—	31.9	28.0	24.0	47.5	108.3	75.5	—	—
Shanxi	(2)	3.2	7.1	0	0	0	1.3	0	7.6	29.8	38.6	198.9	239.8
Shaanxi	(0)	—	—	—	—	—	—	—	—	—	—	—	—
Gansu	(0)	—	—	—	—	—	—	—	—	—	—	—	—
Sichuan	(2)	122.2	95.4	46.9	21.5	17.3	20.1	58.5	53.0	153.3	152.0	149.5	149.0
Hubei	(2)	91.1	5.7	24.8	28.4	73.3	53.8	35.0	87.8	298.1	181.2	72.9	262.5
Anhui	(3)	63.8	0	149.5	35.9	9.4	17.6	54.1	100.8	236.2	154.3	60.8	131.2
Jiangsu	(11)	178.9	4.9	77.4	37.4	17.9	21.3	16.9	117.5	138.4	118.5	43.6	112.6
Zhejiang	(5)	231.7	18.0	118.4	78.6	4.1	78.5	92.3	105.1	242.2	197.6	70.4	234.0
Fujian	(2)	351.9	8.7	24.0	94.4	0.3	53.9	84.2	233.3	109.8	237.3	67.0	154.6
Jiangxi	(1)	217.6	0.3	72.1	37.2	11.0	57.8	20.4	43.6	265.3	232.4	91.4	68.1
Hunan	(2)	24.4	3.1	22.6	49.3	13.9	82.0	128.1	106.8	324.1	195.6	20.1	112.7
Guizhou	(0)	—	—	—	—	—	—	—	—	—	—	—	—
Yunnan	(2)	205.1	39.8	22.4	50.4	11.3	60.4	31.5	46.3	87.7	330.2	331.6	247.2
Guangdong	(4)	229.8	125.6	33.8	55.5	1.0	61.2	57.6	75.3	73.9	273.1	313.9	216.5
Guangxi	(4)	126.1	13.6	15.7	56.0	3.7	88.9	40.6	63.6	93.3	223.9	209.6	201.8

AB.17(c): 1932–3

	Sept.	Oct.	Nov.	Dec.	Jan.	Feb.	Mar.	Apr.	May	June	July	Aug.
Hebei (5)	53.2	4.3	0.3	1.9	0.8	0.5	20.4	19.1	38.6	200.7	117.7	142.1
Shandong (3)	42.9	9.6	0.8	22.4	9.7	6.8	14.9	33.8	87.7	78.2	80.1	111.4
Henan (3)	—	—	—	—	—	—	—	—	—	—	—	—
Shanxi (2)	139.8	5.3	0.7	1.5	—	—	16.0	19.2	32.1	97.0	—	—
Shaanxi (2)	—	—	—	—	1.8	0.8	34.5	51.5	56.6	57.7	123.0	102.6
Gansu (1)	—	—	—	—	0.6	5.3	13.5	17.0	24.3	23.1	124.0	77.6
Sichuan (2)	245.1	87.7	29.6	9.4	11.2	9.5	21.3	24.6	173.5	154.6	71.2	134.8
Hubei (2)	104.8	19.7	3.7	41.6	22.8	24.9	33.2	145.7	245.1	196.3	256.0	205.6
Anhui (3)	59.4	9.4	29.0	58.2	25.4	42.3	72.2	139.8	153.9	160.7	46.2	102.8
Jiangsu (11)	79.1	7.1	7.9	50.1	59.4	48.5	65.1	111.9	113.9	85.5	62.7	134.6
Zhejiang (4)	166.0	68.9	51.2	30.2	78.9	108.5	111.2	130.8	158.6	249.8	153.5	215.4
Fujian (2)	199.7	9.0	29.2	54.5	42.5	25.1	65.8	40.2	88.8	340.1	150.1	78.0
Jiangxi (2)	54.4	27.8	27.6	61.4	73.2	110.9	132.3	217.9	189.7	267.4	100.6	182.9
Hunan (4)	141.5	44.2	35.8	71.2	43.0	86.3	55.5	186.6	209.5	337.0	129.6	126.0
Guizhou (1)	—	—	—	—	—	—	—	—	—	—	171.9	75.0
Yunnan (2)	161.5	67.2	44.2	4.6	4.3	3.8	3.8	41.2	224.0	225.2	309.4	133.8
Guangdong (4)	139.8	20.8	5.6	33.0	13.5	20.5	28.4	48.2	95.3	134.3	148.3	112.8
Guangxi (4)	211.1	108.8	27.8	34.3	9.5	43.9	62.9	107.8	170.8	224.9	176.4	217.8

AB.17(*d*): 1933–4

		Sept.	Oct.	Nov.	Dec.	Jan.	Feb.	Mar.	Apr.	May	June	July	Aug.
Hebei	(7)	21.5	7.6	14.8	12.3	0.5	8.9	4.6	12.5	45.3	108.0	166.8	178.6
Shandong	(5)	40.4	7.1	11.0	26.0	7.1	2.4	13.6	21.5	16.1	146.2	189.2	255.5
Henan	(3)	34.0	38.5	7.0	1.8	10.0	6.6	82.1	28.7	110.4	77.4	135.6	45.2
Shanxi	(3)	28.0	11.8	0.1	4.1	6.6	5.8	8.6	12.6	53.7	75.7	117.1	115.5
Shaanxi	(2)	84.0	61.1	24.0	9.8	3.3	22.8	24.6	58.4	74.1	56.0	87.4	52.8
Gansu	(1)	39.5	27.9	0	3.7	2.0	9.0	0	0.8	24.8	5.0	54.0	218.4
Sichuan	(2)	103.0	153.5	27.6	27.1	28.4	18.0	38.8	51.0	122.4	247.7	199.3	263.4
Hubei	(2)	60.5	139.4	62.5	47.7	20.8	54.1	61.4	161.5	193.5	111.3	96.7	121.8
Anhui	(4)	83.0	126.1	38.9	27.4	12.0	37.4	77.1	145.6	118.5	85.3	30.8	110.6
Jiangsu	(12)	111.8	72.3	11.5	25.9	19.4	16.9	60.3	65.1	61.7	83.5	97.1	107.5
Zhejiang	(5)	284.1	60.2	39.7	19.0	57.0	29.3	145.6	138.1	93.4	135.9	120.8	48.9
Fujian	(2)	115.5	1.3	42.2	75.7	14.9	35.2	136.3	57.8	151.0	348.4	197.1	104.2
Jiangxi	(2)	95.7	91.4	22.7	31.6	29.9	59.1	135.0	224.0	103.9	88.1	5.2	64.9
Hunan	(5)	65.1	91.0	58.8	44.5	28.6	85.2	93.3	247.8	141.3	212.6	55.8	68.3
Guizhou	(1)	184.9	144.3	52.7	58.6	24.7	70.5	15.3	94.3	232.2	341.3	247.3	89.3
Yunnan	(3)	113.7	94.2	23.8	3.1	6.4	55.1	7.9	19.5	75.8	253.6	454.7	235.9
Guangdong	(5)	142.6	50.1	72.7	48.6	6.8	29.4	50.6	85.7	191.3	222.3	254.0	159.6
Guangxi	(5)	139.2	94.1	29.0	7.9	22.0	55.1	20.1	82.5	263.3	325.8	310.9	234.4

Appendix B

AB.17(*e*): 1934–5

		Sept.	Oct.	Nov.	Dec.	Jan.	Feb.	Mar.	Apr.	May	June	July	Aug.
Hebei	(6)	89.9	13.8	4.4	7.0	0.5	(4.0)	1.5	(5.5)	9.8	59.1	226.3	66.2
Shandong	(3)	88.3	38.1	14.3	33.6	17.3	(4.0)	2.6	6.0	53.5	54.1	190.0	93.4
Henan	(4)	82.5	37.4	11.6	17.6	4.2	5.2	8.6	12.5	51.8	31.9	382.9	86.2
Shanxi	(3)	100.1	24.3	1.9	11.0	5.1	(6.1)	4.6	7.0	8.8	22.5	78.0	46.5
Shaanxi	(2)	175.3	105.8	34.5	12.7	7.3	5.8	19.2	34.7	60.0	18.2	180.3	148.7
Gansu	(1)	38.4	19.9	0.8	9.1	7.3	4.7	6.7	37.4	17.0	—	—	—
Sichuan	(2)	153.4	109.9	48.6	10.7	11.8	36.3	54.5	40.3	157.1	205.5	53.0	117.3
Hubei	(2)	239.3	96.0	25.9	43.5	26.3	82.6	76.8	51.9	110.7	320.1	571.0	75.7
Anhui	(4)	86.4	69.6	34.6	50.3	27.0	60.9	61.3	65.5	83.0	179.0	113.5	35.5
Jiangsu	(12)	185.5	40.3	50.2	45.4	30.0	61.4	56.7	41.9	31.4	178.7	55.5	127.7
Zhejiang	(6)	127.5	34.7	65.8	31.7	72.6	111.2	129.6	151.0	131.5	204.7	160.7	284.8
Fujian	(2)	105.1	29.7	37.0	58.5	73.8	111.0	66.2	226.4	118.1	142.7	298.4	190.3
Jiangxi	(2)	137.6	106.8	36.5	56.8	46.5	16.2	175.2	94.3	246.5	192.0	74.8	83.6
Hunan	(5)	95.2	109.8	117.6	49.9	58.7	177.2	176.8	207.3	230.7	387.3	76.8	40.1
Guizhou	(1)	111.8	166.3	87.4	0.5	14.7	81.6	11.1	51.0	321.3	327.3	120.3	61.4
Yunnan	(2)	115.3	102.6	108.2	7.2	—	36.4	3.1	46.0	43.2	233.8	233.5	398.1
Guangdong	(4)	159.6	113.4	16.4	29.3	54.1	28.8	51.9	173.3	108.8	253.0	327.2	184.3
Guangxi	(4)	85.7	207.0	194.9	39.8	74.6	66.2	55.8	94.8	255.0	298.5	243.6	175.1

AB.17(f): Long-term mean

		Sept.	Oct.	Nov.	Dec.	Jan.	Feb.	Mar.	Apr.	May	June	July	Aug.
Hebei	(4)	54.2	16.2	14.0	4.2	3.2	3.2	8.8	18.8	48.0	67.4	170.2	142.8
Shandong	(4)	83.5	21.8	29.0	24.3	14.0	10.5	15.5	26.8	38.8	65.5	163.0	142.0
Henan	(3)	59.0	17.7	11.0	18.3	9.7	11.7	13.3	38.0	54.0	60.3	166.0	108.3
Shanxi	(2)	43.0	15.0	5.5	4.0	3.0	3.0	8.5	14.0	27.0	45.0	102.0	90.0
Shaanxi	(3)	85.3	42.3	17.7	5.3	4.7	6.7	14.7	34.3	47.7	58.0	108.3	106.0
Gansu	(1)	22.5	16.5	3.3	1.8	1.8	2.3	7.5	15.0	29.0	35.0	57.3	55.8
Sichuan	(5)	169.8	94.6	38.6	14.2	9.8	18.6	32.4	66.2	109.0	166.0	249.6	236.0
Hubei	(3)	93.0	85.7	50.0	22.3	36.7	40.0	75.3	127.0	143.3	186.7	208.0	148.3
Anhui	(3)	109.3	61.0	54.3	36.7	37.0	63.0	104.0	134.0	139.0	192.0	153.7	104.3
Jiangsu	(4)	94.5	48.3	40.3	34.3	36.5	42.8	61.8	83.3	77.3	148.0	167.3	126.8
Zhejiang	(3)	151.7	74.3	57.3	46.3	51.3	64.7	109.0	122.3	138.3	211.0	141.0	167.0
Fujian	(2)	153.0	42.5	36.5	40.5	40.5	82.0	106.0	128.5	162.5	190.0	149.0	182.0
Jiangxi	(3)	76.0	66.7	72.0	48.3	55.7	98.0	156.7	213.7	211.0	274.3	125.0	126.7
Hunan	(4)	68.5	89.0	89.5	44.8	48.5	89.0	112.5	164.3	190.0	221.0	108.3	117.8
Guizhou	(3)	110.3	90.0	46.3	25.0	15.3	29.3	49.0	93.0	173.3	202.0	200.0	158.7
Yunnan	(4)	148.3	101.0	41.3	15.0	14.0	32.0	43.5	45.5	116.3	198.0	256.0	245.0
Guangdong	(4)	170.5	84.0	51.3	35.5	26.8	48.0	67.5	124.3	200.8	236.8	238.8	240.8
Guangxi	(5)	131.6	76.0	39.0	34.2	29.2	47.8	73.8	106.0	190.2	234.8	270.4	265.2

Notes: Figures in parenthesis shown after the provinces indicate the number of reporting weather-stations. However, not all the stations give a complete series of monthly precipitation figures. For most of the coastal provinces, precipitation figures from the offshore islands are omitted in calculating the provincial averages. For a number of provinces, figures from neighbouring provinces are borrowed to fill some minor gaps. These include Hebei and Shandong for Feb. 1935, for which the precipitation figures from Kaifeng (Henan) are used. Similarly, for Shanxi, Feb. 1935, and Gansu, Jan. 1935, the respective monthly figures from Xian (Shaanxi) are used.

Sources: 1931–5: *QXYB* various issues for 1931, 1932, 1933, and 1934; supplemented by *SBNJ 1933*, 21–3 (for July–Dec. 1931), C26–31 (Jan.–June 1932); *1934*, B74–8 (July 1932–June 1933); *1935*, B67–83 (July 1933–Aug. 1934); *1936*, B66–78 (Sept. 1934–Aug. 1935). Long-term mean: Hatakeyama Hisanao (ed.), *Ajia No Kiko-Seikai Kikoshi* (Climate of Asia-Record of World Climate), i (Tokyo: Kokin Shoin Press, 1964), 186–260.

TABLE AB.18. Monthly and seasonal average rainfall in the nine major agricultural areas in China, 1958–1961

	Soya bean, Gaoliang	Spring wheat	Winter wheat, Gaoliang	Winter wheat, millet	Sichuan rice	South-west rice	Yangzi rice, wheat	Rice, tea	Double cropping
1958–9									
Winter	11.6	7.0	23.0	21.5	33.1	61.3	122.4	261.2	114.3
December	3.3	0.3	8.8	5.0	12.9	5.0	19.1	22.2	3.9
January	5.7	0.0	3.0	2.5	5.0	16.2	33.4	38.7	8.4
February	2.6	6.7	11.2	14.0	15.2	40.1	69.9	200.3	102.0
Spring	51.6	12.8	44.1	39.8	132.6	174.2	304.6	312.8	390.5
March	16.7	3.7	10.1	9.7	27.7	21.4	48.6	51.1	46.6
April	23.1	3.6	11.4	5.0	33.5	50.8	105.6	94.2	143.3
May	11.8	5.6	22.6	25.1	71.5	102.1	150.4	167.6	200.6
Summer	336.8	117.0	146.4	181.9	171.6	309.1	206.9	402.8	651.0
June	43.8	25.7	26.4	47.1	51.5	134.9	100.9	225.1	279.8
July	160.0	31.1	55.7	60.0	53.6	82.9	53.0	109.1	144.5
August	133.0	60.2	64.4	74.7	66.6	91.4	53.0	68.7	226.7
Autumn	87.3	22.3	59.4	48.5	109.5	120.4	134.3	191.5	216.2
September	51.9	16.4	33.4	32.9	57.5	68.9	63.1	129.8	156.1
October	16.1	4.1	10.5	10.4	38.2	46.6	13.0	19.4	20.8
November	19.3	1.8	15.6	5.2	13.8	4.9	58.1	42.3	39.3
1959–60									
Winter	7.5	1.0	11.4	4.3	32.8	36.6	95.3	121.5	51.6
December	4.9	0.8	7.7	1.4	9.9	14.9	56.8	45.4	28.8
January	2.6	0.2	2.1	0.4	1.7	10.7	14.9	48.2	19.0
February	0.5	0.0	1.6	2.4	21.2	10.9	23.6	27.8	3.9

Spring	61.6	20.7	28.4	52.8	127.9	108.4	225.4	591.5	429.8
March	4.4	9.9	9.0	12.7	40.1	24.2	98.8	159.2	92.1
April	8.0	5.9	8.3	14.4	20.9	9.4	43.5	186.3	81.4
May	49.2	4.9	11.1	25.7	67.0	74.8	83.1	246.0	256.3
Summer	248.8	22.9	152.8	142.9	266.4	344.6	366.4	380.3	566.1
June	64.0	3.0	38.1	14.2	82.4	72.3	178.8	178.4	177.6
July	111.7	13.4	71.8	69.9	124.6	225.0	117.2	114.4	193.5
August	73.1	6.6	42.9	58.8	59.4	47.3	70.4	87.5	195.0
Autumn	62.1	23.2	67.6	55.9	139.4	92.7	184.4	178.5	298.6
September	47.0	17.8	42.6	31.3	80.7	55.5	91.1	87.1	147.4
October	7.9	4.7	16.4	16.2	43.8	23.5	27.6	22.6	84.6
November	7.2	0.7	8.7	8.3	14.9	13.7	65.7	68.9	66.6
1960–1									
Winter	3.9	3.1	11.8	3.7	22.8	22.7	37.8	189.5	105.2
December	3.1	1.0	7.7	1.2	6.7	10.1	7.8	24.1	22.9
January	0.8	1.0	2.4	1.4	2.7	2.9	18.3	44.1	17.2
February	0.0	1.1	1.7	1.0	13.5	9.7	11.7	121.3	65.1
Spring	52.4	20.4	78.2	56.7	132.3	136.5	125.8	463.5	497.0
March	3.5	4.5	9.9	9.1	40.8	24.0	48.8	66.0	111.5
April	6.7	11.6	10.3	17.4	39.3	42.7	26.7	163.2	239.0
May	42.3	4.4	58.1	30.2	52.2	69.8	50.3	234.3	146.6
Summer	218.2	140.3	226.2	179.7	271.2	290.4	324.0	432.5	690.0
June	52.1	14.7	61.1	63.0	75.1	85.3	125.4	246.0	216.2
July	90.7	48.5	98.8	47.0	96.1	82.0	74.9	66.6	165.8
August	75.4	77.1	66.4	69.7	100.1	123.1	123.7	120.0	308.0
Autumn	43.2	54.2	110.8	114.1	166.3	114.3	313.8	256.2	387.7
September	32.4	28.9	69.7	25.1	72.5	41.7	202.1	135.4	285.6
October	4.0	20.8	23.8	76.2	68.8	51.7	70.0	56.0	41.4
November	6.8	4.6	17.4	12.8	25.0	21.0	41.7	64.9	60.7

Appendix B

T ABLE AB.18. *cont.*

	Soya bean, Gaoliang	Spring wheat	Winter wheat, Gaoliang	Winter wheat, millet	Sichuan rice	South-west rice	Yangzi rice, wheat	Rice, tea	Double cropping
Long-term means									
Winter	15.4	5.5	34.9	15.2	36.2	64.7	136.5	177.8	119.3
December	6.0	1.5	16.5	5.2	11.4	19.3	36.1	43.6	35.1
January	4.7	1.6	9.5	5.2	8.3	14.6	41.8	47.4	32.9
February	4.7	2.4	8.9	4.8	16.4	30.9	58.6	86.8	51.4
Spring	74.7	33.5	86.9	100.0	197.3	252.4	349.8	438.5	380.1
March	10.9	4.9	13.7	13.4	29.9	45.9	93.9	118.1	73.2
April	19.6	11.5	30.2	31.8	62.6	65.9	124.6	150.8	111.9
May	44.2	17.1	43.1	54.8	104.9	140.7	131.3	169.6	195.0
Summer	340.0	146.8	378.0	260.0	568.3	640.1	468.7	463.4	704.2
June	78.4	34.0	66.9	59.0	156.4	200.1	185.3	217.5	251.3
July	141.5	56.3	169.4	103.8	213.1	232.0	160.2	112.6	221.3
August	120.1	56.5	138.7	97.2	198.7	208.0	123.2	133.3	231.6
Autumn	110.5	47.5	110.1	101.4	274.0	271.7	229.5	240.3	278.5
September	66.4	35.6	67.7	58.4	155.1	132.0	102.7	109.6	160.4
October	29.0	8.4	22.8	33.6	86.0	96.3	70.2	67.4	75.5
November	15.1	3.5	19.7	9.4	32.9	43.4	56.5	63.3	42.6

Notes: All the precipitation figures are compiled from the daily rainfall charts kept by the Royal Hong Kong Observatory which receives telegraphed daily dispatches from the various meteorological stations in China. The daily figures are added to make the monthly and seasonal totals and divided by the total number of stations to derive the respective periodic averages for the individual agricultural areas.

The numbers of stations available at the beginning of the year (1 Jan.) and mid-year (1 June) are applied to derive the average rainfall respectively for the two periods (Oct., Nov., Dec., Jan., Feb., and Mar.), *and* (Apr., May, June, Aug., and Sept.). The numbers of stations are distributed as follows:

	SG	SW	WWG	WWM	SR	SWR	YRW	RT	DCR
1959									
1 Jan.	7	9	14	6	10	16	7	19	22
1 Jun.	8	9	14	7	11	17	7	19	18
1960									
1 Jan.	8	9	14	9	11	18	8	22	21
1 Jun.	9	11	15	9	13	19	10	20	21
1961									
1 Jan.	10	10	17*	9	12	23	12	21	21
1 Jun.	12	11	20	10	13	24	11	20	21
1962									
1 Jan.	13	12	20*	11	13	27	14	22	21

* Assumed figures.

Sources: 1958–61: Royal Hong Kong Observatory; long-term average: Hatakeyama Hisanao (ed.), *Ajia No Kiko-Seikai Kikoshi* (Climate of Asia-Record of World Climate), i (Tokyo: Kokin Shoin Press, 1964), 186–260. I am grateful to Mr Hok-chi Kong of the Royal Hong Kong Observatory for drawing my attention to this important source.

TABLE AB.19. Precipitation-based flood and drought indices in comparison with the size of *shouzai* area as a proportion of total sown area in China, 1921–1984

	Flood		Drought		Flood and drought combined	
	Index	Area	Index	Area	Index	Area
1921	1.05	—	0.50	—	1.55	—
1922	0.91	—	0.58	—	1.49	—
1923	0.95	—	0.50	—	1.45	—
1924	0.90	—	0.67	—	1.57	—
1925	0.53	—	0.92	—	1.45	—
1926	0.67	—	0.86	—	1.53	—
1927	0.32	—	0.83	—	1.15	—
1928	0.57	—	1.16	—	1.73	—
1929	0.44	—	1.11	—	1.55	—
1930	0.65	—	0.70	—	1.35	—
1931	1.17	20.1	—	—	—	—
1932	0.88	14.0	0.52	3.4	1.40	17.4
1933	0.70	7.3	0.68	8.0	1.38	15.3
1934	0.76	5.9	0.92	33.0	1.68	38.9
1935	0.96	1.9	0.82	7.9	1.78	9.8
1936	0.42	3.0	1.23	8.3	1.65	11.3
1937	1.00	4.1	0.52	15.2	1.52	19.3
1938	0.99	—	0.53	—	1.52	—
1939	0.60	—	0.83	—	1.43	—
1940	0.60	—	0.75	—	1.35	—
1941	0.39	—	1.04	—	1.43	—
1942	0.44	—	1.10	—	1.55	—
1943	0.71	—	0.83	—	1.54	—
1944	0.51	—	0.92	—	1.43	—
1945	0.60	—	0.86	—	1.46	—
1946	0.66	—	0.56	—	1.22	—
1947	0.86	—	0.61	—	1.47	—
1948	0.92	—	0.47	—	1.39	—
1949	1.20	7.3	—	—	—	—
1950	0.83	5.4	0.73	2.0	1.56	7.4
1951	0.48	3.2	0.89	6.1	1.37	9.3
1952	0.77	2.0	0.56	3.0	1.33	5.0
1953	0.66	5.1	0.74	6.0	1.40	11.1
1954	1.24	10.9	0.41	2.0	1.65	12.9
1955	0.64	3.5	0.80	8.9	1.44	12.4
1956	1.11	9.0	0.41	2.0	1.52	11.0
1957	0.59	5.1	0.90	10.9	1.49	16.0
1958	0.73	2.8	0.67	14.7	1.40	17.5

TABLE AB.19. *cont.*

	Flood		Drought		Flood and drought combined	
	Index	Area	Index	Area	Index	Area
1959	0.73	3.4	0.69	23.7	1.42	27.1
1960	0.57	6.7	0.90	25.3	1.47	32.0
1961	0.80	6.2	0.73	26.4	1.53	32.6
1962	0.88	7.0	0.47	14.8	1.35	21.8
1963	0.82	10.0	0.84	12.0	1.66	22.0
1964	1.04	10.4	0.57	2.9	1.61	13.3
1965	0.29	3.9	1.20	9.5	1.49	13.4
1966	0.53	1.7	1.06	13.6	1.59	15.3
1967	0.59	—	0.69	—	1.28	—
1968	0.51	—	0.82	—	1.33	—
1969	0.80	—	0.59	—	1.39	—
1970	0.71	2.2	0.43	4.0	1.14	6.2
1971	0.47	2.7	0.69	17.2	1.16	19.9
1972	0.33	2.8	1.29	20.6	1.62	23.4
1973	1.22	4.2	0.20	18.3	1.42	22.5
1974	0.47	4.3	0.82	17.2	1.29	21.5
1975	0.73	4.6	0.59	16.6	1.32	21.2
1976	0.57	2.8	0.67	18.4	1.24	21.2
1977	0.63	6.1	0.59	20.0	1.22	26.1
1978	0.37	1.9	1.04	26.8	1.41	28.7
1979	0.45	4.6	0.84	16.6	1.29	21.2
1980	—	6.3	—	17.8	—	24.1
1981	—	5.9	—	17.7	—	23.6
1982	—	5.8	—	14.3	—	20.1
1983	—	8.4	—	11.2	—	19.6
1984	—	7.4	—	11.0	—	18.4

Notes: The methodology for deriving the flood and drought indices is explained in detail in Appendix A (sect. 'Record of the SMB'). In contrast to the weather (*shouzai*) area index (as derived in Appendix A (sect. '*Shouzai* area as a weather proxy')), the *shouzai* areas used in this table were not weighted by the relative degree of severity of the natural disasters. The total sown-area base used for estimating the *shouzai* area proportion for 1949–51 is not a complete record. It omits some of the minor non-grain crops.

Sources: Flood and drought indices: State Meteorological Bureau, *Zhongguo Jin Wubainian Hanlao Fenbu Tuji* (Yearly Charts of Dryness/Wetness in China for the last 500-Year Period) (Beijing: Ditu Chubanshe, 1981), 331–2. *Shouzai* area: 1931–7: Table AB.14(*a–g*), 1949–84: Table AB.15. Total sown area: 1931–7: Table AB.2(*a*) (Grain only); 1949–51: *NYNJ 1980*, 34–5; 1952–79: *TJNJ 1983*, 154; 1980–4: *TJNJ 1990*, 357.

Index